Exploring Research Frontiers in Contemporary Statistics and Econometrics

Léopold Simar became Professor of Statistics at the Université catholique de Louvain (UCL) in 1992, after moving from the Faculté des Sciences Economiques Sociales et Politiques (FSESP) at Facultés Universitaires Saint-Louis in Brussels, where he had served as Professor of Statistics since 1974 and as Dean of the FSESP from 1978 to 1990. He founded the Institute of Statistics at the UCL in 1992, and chaired the Institute from its creation until 2004. During this period, the Institute became recognized as a leading center for research in mathematical statistics. Over his long and successful career, Léopold Simar has worked on a variety of topics in statistics, including count-data models, Bayesian estimation and inference, and frontier estimation. He is regarded as one of the world's leading experts on frontier estimation; his work in this area has found application in a broad variety of fields, including efficiency studies in industry, air traffic control, the research output of universities, insurance companies, etc.

Ingrid Van Keilegom • Paul W. Wilson
Editors

Exploring Research Frontiers in Contemporary Statistics and Econometrics

A Festschrift for Léopold Simar

Physica-Verlag

Editors

Prof. Ingrid Van Keilegom
Institut de statistique,
biostatistique et sciences actuarielles
Université catholique de Louvain
Voie du Roman Pays 20
1348 Louvain-la-Neuve
Belgium
ingrid.vankeilegom@uclouvain.be

Prof. Paul W. Wilson
Department of Economics
222 Sirrine Hall
Clemson University
Clemson
South Carolina 29634-1309
USA
pww@clemson.edu

ISBN 978-3-7908-2927-3 ISBN 978-3-7908-2349-3 (eBook)
DOI 10.1007/978-3-7908-2349-3
Springer Heidelberg Dordrecht London New York

Cover design: WMXDesign GmbH, Heidelberg, Germany

Printed on acid-free paper

Springer is part of Springer Science+Business Media (www.springer.com)

Preface

Léopold Simar became Professor of Statistics at the Université catholique de Louvain (UCL) in 1992, after moving from the Faculté des Sciences Economiques Sociales et Politiques (FSESP) at Facultés Universitaires Saint-Louis in Brussels where he had served as Professor of Statistics since 1974 and as Dean of the FSESP from 1978 to 1990. He founded the Institute of Statistics at UCL in 1992, and chaired the Institute from its creation until 2004. During this period, the Institute became recognized as a leading center for research in mathematical statistics. Over his long and successful career, Léopold Simar has worked on a variety of topics in statistics, including count-data models, Bayesian estimation and inference, and frontier estimation. He is regarded as one of the world's leading experts on frontier estimation; his work in this area has found applications in a broad variety of fields, including efficiency studies in industry, air traffic control, research output of universities, insurance companies, etc. He has published over 100 refereed works in academic journals and books, and has given over 100 invited talks at conferences and universities around the world. Léopold Simar is also a gifted and inspirational teacher. He is very well appreciated by his students at all levels, especially by his Ph.D. students.

In May 2009, 60–70 statisticians, econometricians, students, and others gathered in Louvain-la-Neuve to honor our friend Léopold Simar at his retirement. The group included Léopold's colleagues, coauthors, and students, as well as university administrators and other academics who have over the years worked with Léopold, served on committees, organized conferences, or otherwise participated in academic life with Léopold. The sets of colleagues, coauthors, and students included both current and former members, and each of the possible intersections of these three sets are nonempty.

The celebration in Louvain-la-Neuve was organized around a small, 2-day research conference where researchers presented results from some of their current work. This volume contains the papers that followed from those presentations. Between presentations, there was much discussion and learning over coffee and other refreshments; in the evening, there was still more discussion (and still more learning, and no less fun) over dinner. For many at the conference, there was the chance to see old friends; possibly for everyone except Léopold (who already knew everyone attending), there was the chance to meet new friends. As is common at academic gatherings such as this, the transition between organized presentations, coffee breaks, more presentations, lunch, still more presentations, and dinner was almost seamless. Conversations flowed smoothly, and the boundary between work and fun was not discrete but rather blurred. In fact, for us, there remain questions about the existence of such a boundary. We have not merely worked with our friend Léopold, but we have also had much fun together over the years. And, we have not finished.

To the outsider, statistics and econometrics might seem to be the same. Often in our professions, however, there is little discussion between statisticians and econometricians, and sometimes where there is discussion between the two, it is not polite. This is not the tradition in Louvain-la-Neuve. The Institute of Statistics founded by Léopold in 1992 at the Université catholique de Louvain in Louvain-la-Neuve has always been a place for the two groups to meet and learn from each other. Over the years in the Institute, both members and visitors, including both econometricians and statisticians, have benefited from each other with econometricians adopting the mathematical rigor that is more common among statisticians, and with statisticians appreciating the subtlety of estimation problems arising in economics.

The 11 papers in this volume are at the frontier of current research. Not coincidentally, ten of the papers are related (at least to some extent) to estimation of frontiers, a topic that Léopold has contributed much to over the past two decades. The eleventh paper deals with estimation of single index models, an area that Léopold has also contributed to. Both frontier estimation and single index models are important topics. In particular, frontier estimation is important in economics, management, management science, finance, and other fields for purposes of *benchmarking* and making relative comparisons. Some of the ideas in the papers collected here may also be useful for image resolution as well as other purposes. The chapters in this volume reflect the varied interests of their authors; some chapters are mostly theoretical, while others involve varied applications to real data to answer important questions; we expect that everyone in a very wide audience will find *something* useful in this collection.

Abdelaati Daouia, Laurent Gardes, and Stéphane Girard use extreme value theory to extend asymptotic results for (Nadaraya 1964) kernel estimators to estimation of extreme quantiles of univariate distributions. The results are then applied to derive the asymptotic distribution of smooth nonparametric estimators of monotone support curves. While estimation of efficiency in production has traditionally relied on linear programming techniques, the research in this chapter is a nice example of the new approaches that are currently being developed.

Alice Shiu and Valentin Zelenyuk examine the effect of ownership type on the efficiency of heavy and light industry in China. In recent years, a number of countries have attempted to reorganize their industry after turning away from direct central planning and state-owned enterprises. China is currently in the middle of this process, and for the moment presents an opportunity to examine differences in ownership structure as reflected in operating efficiency. The results potentially provide guidance for future transitions in other countries.

Maik Schwarz, Sébastien Van Bellegem, and Jean-Pierre Florens consider an additive measurement error model in which the error on measured inputs is assumed to be normal, but with unknown variance. They study a new estimator of the production frontier under this model and establish conditions required for identification and the consistency of their estimator. Their estimator is a modified version of the order-m frontier considered earlier by Cazals et al. (2002), and as such involves estimation of a survivor function for inputs conditioned on output variables.

Irène Gijbels and Abdelaati Daouia propose a new class of robust estimators of the production frontier in a deterministic frontier model. The class is based on so-called extremiles, which include the family of expected minimum-input frontiers and parallels the class of quantile-type frontiers. The class is motivated from several viewpoints, revealing its specific merits and strengths.

Alois Kneip and Robin Sickles revisit the (Solow 1957) residual and discuss how it has been interpreted by both the neoclassical production literature and the literature on productive efficiency. Kneip and Sickles argue that panel data are needed to estimate productive efficiency and innovation, and in doing so attempt to link the two strands of literature.

Paul Wilson extends theoretical results obtained by Kneip et al. (1998) for input- and output-oriented distance function estimators to hyperbolic distance function estimators. Asymptotic properties of two different hyperbolic data envelopment analysis distance function estimators are derived, and computationally efficient algorithms for computing the estimators are given. In addition, asymptotic results obtained by Cazals et al. (2002) for the input- and output-oriented order-m estimator are extended to the corresponding estimators in the hyperbolic orientation. The hyperbolic orientation is particularly useful for making cross-period comparisons, as in the case of estimation of changes in productivity, etc. using Malmquist or other indices, where infeasibilities often arise when working in the input or output directions due to shifts in the frontier over time.

Luiza Bădin and Cinzia Daraio consider the problem of estimating productive efficiency conditionally on environmental variables, some of which may be discrete. Bădin et al. (2010) proposed a data-driven, cross-validation technique for choosing bandwidths when the environmental variables are continuous. In this chapter, Bădin and Daraio describe how conditional estimation can be carried out when some environmental variables are continuous while others are discrete. The work builds on the work of Li and Racine (2008) which considers a similar problem in a regression context. Here, Bădin and Daraio also propose an heterogeneous bootstrap that allows one to make inference about the influence of environmental variables.

Cédric Heuchenne and Ingrid Van Keilegom cast the problem of regressing a random variable Y on a covariate X in terms of a location functional (which might be a conditional mean function, but which can also be a quantile or some other feature of interest) by focusing on L-functionals. Censoring is a common problem in such problems, and Heuchenne and Van Keilegom propose a new method for estimation that involves nonparametric imputation for the censored observations. The approach should be useful in a variety of applications, but particularly so in survival analysis.

Seok-Oh Jeong and Byeong Park consider the general problem of estimating a convex set as well as the upper boundary of that convex set. Estimation of a production set and its frontier function with multivariate covariates are examples of such estimation tasks. The developed methodology is applied to conical hull estimators and data envelopment analysis. In addition, Jeong and Park discuss practical considerations for bias-correction and interval estimation in the frontier context.

Pavlos Almanidis and Robin Sickles consider parametric frontier estimation. Simar and Wilson (2010) examined problems for inference in the (Aigner et al. 1977) model in finite samples situations where residuals are sometimes skewed in an unexpected direction, even if the (Aigner et al. 1977) model is the correct specification. Almanidis and Sickles extend the (Aigner et al. 1977) model to allow for skewness in either direction using a doubly truncated distribution for inefficiency.

Yingcun Xia, Wolfgang Härdle, and Oliver Linton consider the problem of estimating a single index regression model. The authors propose a refinement of the minimum average conditional variance estimation (MAVE) method involving a practical and easy-to-implement iterative algorithm that allows the MAVE estimator to be readily derived. The authors also discuss the difficult problem of how to select the bandwidth for estimating the index. In addition, they show that under appropriate assumptions, their estimator is asymptotically normal and most efficient in a semi-parametric sense.

We (i.e., the editors of this volume) have been partially supported by an Inter-university Attraction Pole (IAP) research network grant from the Belgian government and by a grant from the European Research Council. We also thank the editor at Springer, Niels Peter Thomas, and the staff at Springer for their efficient handling of the manuscript.

Louvain-la-Neuve
Clemson
January 2011

Ingrid Van Keilegom
Paul W. Wilson

References

Aigner, D., Lovell, C.A.K., & Schmidt, P. (1977). Formulation and estimation of stochastic frontier production function models. *Journal of Econometrics, 6*, 21–37.

Bădin, L., Daraio, C., & Simar, L. (2010). Optimal bandwidth selection for conditional efficiency measures: A data-driven approach. *European Journal of Operational Research, 201*, 633–664.

Cazals, C., Florens, J.P., & Simar, L. (2002). Nonparametric frontier estimation: A robust approach. *Journal of Econometrics, 106*, 1–25.

Kneip, A., Park, B., & Simar, L. (1998). A note on the convergence of nonparametric DEA efficiency measures. *Econometric Theory, 14*, 783–793.

Li, Q., & Racine, J. (2008). Nonparametric estimation of conditional CDF and quantile funcitons with mixed categorical and continuous data. *Journal of Business and Econonomic Statistics, 26*, 423–434.

Nadaraya, E.A. (1964). On estimating regression. *Theory of Probability and its Applications, 10*, 186–190.

Simar, L., & Wilson, P.W. (2010). Estimation and inference in cross-sectional, stochastic frontier models. *Econometric Reviews, 29*, 62–98.

Solow, R.M. (1957). Technical change and the aggregate production function. *Review of Economics and Statistics, 39*, 312–320.

Contents

Contributors

Pavlos Almanidis Department of Economics - MS 22, Rice University, 6100 S. Main Street, Houston, Texas 77005-1892, USA, pa1@rice.edu

Luiza Bădin Department of Applied Mathematics, Bucharest Academy of Economic Studies, Piata Romana nr. 6, 010374 Bucharest, Romania

Department of Statistical Inference, Gh. Mihoc - C. Iacob Institute of Mathematical Statistics and Applied Mathematics, Calea 13 Septembrie nr. 13, Bucharest, Romania, luiza.badin@csie.ase.ro

Abdelaati Daouia Toulouse School of Economics (GREMAQ), University of Toulouse, Manufacture des Tabacs, Aile J.J. Laffont, 21 Allée de Brienne, 31000 Toulouse, France, daouia@cict.fr

Cinzia Daraio Department of Management, CIEG - Centro Studi di Ingegneria Economico-Gestionale, University of Bologna, Via Umberto Terracini 28, 40131 Bologna, Italy, cinzia.daraio@unibo.it

Jean-Pierre Florens Toulouse School of Economics (GREMAQ), University of Toulouse, Manufacture des Tabacs, Aile J.J. Laffont, 21 Allée de Brienne, 31000 Toulouse, France, orens@cict.fr

Laurent Gardes Team Mistis, INRIA Rhône-Alpes and Laboratoire Jean Kuntzmann, 655 avenue de l'Europe, Montbonnot, 38334 Saint-Ismier cedex, France, laurent.gardes@inrialpes.fr

Irène Gijbels Department of Mathematics and Leuven Statistics Research Center, Katholieke Universiteit Leuven, Celestijnenlaan 200B, Box 2400, 3001 Leuven (Heverlee), Belgium, Irene.Gijbels@wis.kuleuven.be

Stéphane Girard Team Mistis, INRIA Rhône-Alpes and Laboratoire Jean Kuntzmann, 655 avenue de l'Europe, Montbonnot, 38334 Saint-Ismier cedex, France, stephane.girard@inrialpes.fr

Wolfgang Karl Härdle C.A.S.E. Centre for Applied Statistics and Economics, School of Business and Economics, Humboldt-Universität zu Berlin, Unter den Linden 6, 10099 Berlin, Germany, haerdle@wiwi.hu-berlin.de

Cédric Heuchenne QuantOM (Centre for Quantitative Methods and Operations Management), HEC-Management School of University of Liège, Rue Louvrex 14, 4000 Liège, Belgium

Institut de statistique, biostatistique et sciences actuarielles, Université catholique de Louvain, Voie du Roman Pays 20, 1348 Louvain-la-Neuve, Belgium, cedric.heuchenne@uclouvain.be

Seok-Oh Jeong Department of Statistics, Hankuk University of Foreign Studies, Mo-Hyeon Yong-In, Gyeong-Gi, 449-791, South Korea, seokohj@hufs.ac.kr

Alois Kneip Department of Economics, University of Bonn, Adenauerallee 24-26, 53113 Bonn, Germany, akneip@uni-bonn.de

Oliver Linton Faculty of Economics, Austin Robinson Building, Sidgwick Avenue, Cambridge, CB3 9DD, UK, obl20@cam.ac.uk

Byeong U. Park Department of Statistics, Seoul National University, Seoul 151-747, South Korea, bupark@stats.snu.ac.kr

Maik Schwarz Institut de statistique, biostatistique et sciences actuarielles, Université catholique de Louvain, Voie du Roman Pays 20, 1348 Louvain-la-Neuve, Belgium, maik.schwarz@uclouvain.be

Alice Shiu School of Accounting and Finance, The Hong Kong Polytechnic University, 1 Yuk Choi Road, Hung Hom, Kowloon, Hong Kong, afshiu@inet.polyu.edu.hk

Robin C. Sickles Department of Economics - MS 22, Rice University, 6100 S. Main Street, Houston, Texas 77005-1892, USA, rsickles@rice.edu

Sébastien Van Bellegem Toulouse School of Economics (GREMAQ), University of Toulouse, Manufacture des Tabacs, Aile J.J. Laffont, 21 Allée de Brienne, 31000 Toulouse, France

Center for Operations Research and Econometrics, Université catholique de Louvain, Voie du Roman Pays 34, 1348 Louvain-la-Neuve, Belgium, sebastien.vanbellegem@uclouvain.be

Ingrid Van Keilegom Institut de statistique, biostatistique et sciences actuarielles, Université catholique de Louvain, Voie du Roman Pays 20, 1348 Louvain-la-Neuve, Belgium, ingrid.vankeilegom@uclouvain.be

Paul W. Wilson Department of Economics, 222 Sirrine Hall, Clemson University, Clemson, South Carolina 29634-1309, USA, pww@clemson.edu

Yingcun Xia Department of Statistics and Applied Probability and Risk Management Institute, National University of Singapore, Singapore, staxyc@nus.edu.sg

Valentin Zelenyuk School of Economics and Centre for Efficiency and Productivity Analysis, The University of Queensland, 530, Colin Clark Building (39), StLucia, Brisbane, QLD4072, Australia, v.zelenyuk@uq.edu.au

Chapter 1
Nadaraya's Estimates for Large Quantiles and Free Disposal Support Curves

Abdelaati Daouia, Laurent Gardes, and Stéphane Girard

Abstract A new characterization of partial boundaries of a free disposal multivariate support, lying near the true support curve, is introduced by making use of large quantiles of a simple transformation of the underlying multivariate distribution. Pointwise empirical and smoothed estimators of the full and partial support curves are built as extreme sample and smoothed quantiles. The extreme-value theory holds then automatically for the empirical frontiers and we show that some fundamental properties of extreme order statistics carry over to Nadaraya's estimates of upper quantile-based frontiers. The benefits of the new class of partial boundaries are illustrated through simulated examples and a real data set, and both empirical and smoothed estimates are compared via Monte Carlo experiments. When the transformed distribution is attracted to the Weibull extreme-value type distribution, the smoothed estimator of the full frontier outperforms frankly the sample estimator in terms of both bias and Mean-Squared Error, under optimal bandwidth. In this domain of attraction, Nadaraya's estimates of extreme quantiles might be superior to the sample versions in terms of MSE although they have a higher bias. However, smoothing seems to be useless in the heavy tailed case.

A. Daouia (✉)
Toulouse School of Economics (GREMAQ), University of Toulouse, Manufacture des Tabacs, Aile J.J. Laffont, 21 Allée de Brienne, 31000 Toulouse, France
e-mail: daouia@cict.fr

L. Gardes · S. Girard
Team Mistis, INRIA Rhône-Alpes and Laboratoire Jean Kuntzmann, 655 avenue de l'Europe, Montbonnot, 38334 Saint-Ismier cedex, France
e-mail: laurent.gardes@inrialpes.fr; stephane.girard@inrialpes.fr

I. van Keilegom and P.W. Wilson (eds.), *Exploring Research Frontiers in Contemporary Statistics and Econometrics*, DOI 10.1007/978-3-7908-2349-3_1,

1.1 Main Results

Let $(X,Y),(X_1,Y_1),(X_2,Y_2)\dots$ be independent random vectors from a common probability distribution on $\mathbb{R}_+^p \times \mathbb{R}_+$ whose support boundary is assumed to be nondecreasing. For $x \in \mathbb{R}_+^p$ such that $\mathbb{P}(X \le x) > 0$, the graph of the frontier function[1]

$$\xi_1(x) = \inf\{y \ge 0 : \mathbb{P}(Y \le y|X \le x) = 1\}$$

coincides with the monotone surface of the joint support Ψ of (X, Y) (Cazals et al. 2002). As a matter of fact, the graph of $\xi_1(\cdot)$ is the lowest nondecreasing curve larger than or equal to the upper frontier of Ψ. In applied econometrics for instance, the support Ψ is interpreted as the set of all feasible production units, i.e., $(x, y) \in \Psi$ in a certain sector of technology if and only if it is possible for a given firm to produce a quantity y of goods by making use of a quantity x of resources. The production set Ψ is by construction free disposal. This means that its optimal frontier which represents the set of the most efficient firms is nondecreasing. The free disposal hull (FDH) estimator of $\xi_1(\cdot)$ is given by

$$\xi_{1,n}(x) = \max\{Y_i|i : X_i \le x\}$$

(Deprins et al. 1984). The FDH frontier is clearly the lowest step and monotone curve which envelopes all the data points (X_i, Y_i) and so it is very non-robust to extreme observations. To reduce this vexing defect, instead of estimating the frontier of the support Ψ, Aragon et al. (2005) have suggested to estimate a partial boundary of Ψ of order $\alpha \in (0,1)$ lying near its true full boundary. The frontier function $\xi_1(x)$ being the quantile function of order one of the distribution of Y given $X \le x$, they rather proposed to estimate the αth quantile function of this non-standard conditional distribution

$$q_\alpha(x) = \inf\{y \ge 0 : \mathbb{P}(Y \le y|X \le x) \ge \alpha\}.$$

The resulting sample quantile function, obtained by plugging the empirical version of the conditional distribution function may suffer from a lack of efficiency due to the large variation of the extreme observations involved in its construction. A smoothed variant $\hat{q}_\alpha(x)$ of this sample estimator may be then preferable as shown in Martins-Filho and Yao (2008), where $\mathbb{P}(Y \le y|X \le x)$ is estimated by

$$\hat{F}_{Y^x}(y) = \sum_{i=1}^{n} \mathbb{1}(X_i \le x)H\left((y - Y_i)/h\right) / \sum_{i=1}^{n} \mathbb{1}(X_i \le x),$$

[1]For two vectors x_1 and x_2 with $x_1 \le x_2$ componentwise, $\xi_1(\cdot)$ satisfies $\xi_1(x_1) \le \xi_1(x_2)$.

with $h = h_n \to 0$, $H(y) = \int_{-\infty}^{y} K(u)du$ and $K(\cdot)$ being a density kernel. However, no attention was devoted to the limit distribution of $\hat{q}_\alpha(x) := \hat{F}_{Y^x}^{-1}(\alpha) = \inf\{y \geq 0 : \hat{F}_{Y^x}(y) \geq \alpha\}$ when it estimates the optimal boundary itself. Daouia et al. (2009) have addressed this problem by specifying the asymptotic distribution of the smoothed α-frontier for fixed orders $\alpha \in (0,1]$ as well as for sequences $\alpha = \alpha_n$ tending to one as $n \to \infty$.

It is important to note that the distribution of Y being conditioned by $X \leq x$, the estimation of the corresponding quantiles does not require a smoothing procedure in x which would be the case if the distribution was conditioned by $X = x$ (see e.g. Girard and Jacob 2004). It should be also clear that although the simple nature of the conditioning $X \leq x$, it requires more powerful techniques of proof than the unconditional quantile setting. Our main contribution in this note is to get rid of this conditioning by exploiting the fact that

$$\xi_1(x) = \inf\{y \geq 0 : \mathbb{P}(Z^x \leq y) = 1\},$$

where $Z^x = Y\mathbb{1}(X \leq x)$. This simple formulation of the monotone frontier function was pointed out by Daouia et al. (2010, Appendix). Note also that the FDH estimator coincides with the maximum of the random variables $Z_i^x = Y_i\mathbb{1}(X_i \leq x)$, $i = 1, \ldots, n$. Moreover, given that the interest is also the estimation of a concept of a partial frontier well inside the sample but near the optimal boundary of Ψ, a natural idea is to define the alternative simple αth frontier function

$$\xi_\alpha(x) := F_{Z^x}^{-1}(\alpha) = \inf\{y \geq 0 : F_{Z^x}(y) \geq \alpha\},$$

where $F_{Z^x}(y) = \mathbb{P}(Z^x \leq y)$. In the context of productivity and efficiency analysis, when the performance of firms is measured in terms of their distance from partial frontiers rather than the full frontier, the use of the αth production frontier $q_\alpha(x)$ as a benchmark can be criticized for its divergence from the optimal frontier as x increases. Instead, Wheelock and Wilson (2008) favored the use of a hyperbolic unconditional variant of $q_\alpha(x)$. Our partial unconditional quantile-type frontier $\xi_\alpha(x)$ provides a more attractive alternative to reduce the vexing defect of the conditional version $q_\alpha(x)$ because of its construction and conceptual simplicity. The benefits of using this new class of partial support curves are demonstrated in Sect. 1.3.1 via two examples in the cases where $F_{Z^x}(\cdot)$ is attracted to the Weibull and Fréchet extreme value type distributions.

A natural estimator of $\xi_\alpha(x)$ is given by the sample quantile

$$\xi_{\alpha,n}(x) := F_{Z^x,n}^{-1}(\alpha) = \inf\{y \geq 0 : F_{Z^x,n}(y) \geq \alpha\},$$

where $F_{Z^x,n}(y) = n^{-1}\sum_{i=1}^{n} \mathbb{1}(Z_i^x \leq y)$. Therefore, the extreme-value theory holds automatically when $\alpha = 1$ and when letting $\alpha = \alpha_n \uparrow 1$ as $n \to \infty$, which is not the case for previous concepts of partial support curves such as those of Cazals et al. (2002), Girard and Jacob (2004), Aragon et al. (2005), Wheelock and

Wilson (2008) and Martins-Filho and Yao (2008). Following Nadaraya (1964), an alternative estimator to $\xi_{\alpha,n}(x)$ is given by the αth quantile

$$\hat{\xi}_\alpha(x) := \hat{F}_{Z^x}^{-1}(\alpha) = \inf\{y \geq 0 : \hat{F}_{Z^x}(y) \geq \alpha\} \tag{1.1}$$

of the kernel-smoothed empirical distribution function $\hat{F}_{Z^x}(y) = n^{-1}\sum_{i=1}^n H\left((y - Z_i^x)/h\right)$ based on a sequence of bandwidths h and an integrated kernel $H(\cdot)$. In the ordinary framework where the order α is a fixed constant in (0,1), Azzalini (1981) established a second-order approximation to the mean squared error of the smoothed quantile $\hat{\xi}_\alpha(x)$ under optimal h, which makes $\hat{\xi}_\alpha(x)$ more efficient than the sample version $\xi_{\alpha,n}(x)$. The smoothed estimator may also be preferable to the sample one for the following additional aspect: the construction of asymptotic confidence intervals for $\xi_\alpha(x)$ using the asymptotic normality of $\xi_{\alpha,n}(x)$ requires the estimation of the derivative $F'_{Z^x}(\xi_\alpha(x))$, whereas smoothing gives a naturally derived estimator of this quantile density function. Nadaraya (1964) has proved under mild conditions the asymptotic normality of $\hat{\xi}_\alpha(x)$ when the order α is fixed in (0,1). In the present chapter, we rather concentrate in Sect. 1.2 on specifying the asymptotic distributions of Nadaraya's estimates $\hat{\xi}_\alpha(x)$ when $\alpha = \alpha_n \to 1$ at different rates as $n \to \infty$, and we verify whether the benefits of smoothing are still valid when considering these extreme quantiles. This does not seem to have been appreciated before in the literature. Theorem 1 characterizes possible limit distributions of $\hat{\xi}_1(x)$ and Theorem 2 discusses moment convergence. When $n(1-\alpha_n)$ is a constant, Theorem 3 shows that $\hat{\xi}_{\alpha_n}(x)$ converges with the same centering and scaling as $\hat{\xi}_1(x)$ to a different extreme value distribution. However, when $n(1-\alpha_n) \to \infty$, Theorem 4 establishes the asymptotic normality of $\hat{\xi}_{\alpha_n}(x)$ as an estimator of $\xi_{\alpha_n}(x)$.

As a matter of fact, in this context where the underlying distribution function $F_{Z^x}(\cdot)$ has a jump at the left-endpoint of its support, we show by using simple arguments that the smoothed maximum $\hat{\xi}_1(x)$ is equal to a deterministic translation of the sample maximum $\xi_{1,n}(x)$ for all $n \geq 1$. Likewise, it turns out that a smoothed quantile of the form $\hat{\xi}_{(n-k+1)/n}(x)$ is within a fixed multiple of the bandwidth of the sample quantile $\xi_{(n-k+1)/n,n}(x)$ for all n large enough. As an immediate consequence, the asymptotic theory of the sample extremes $\xi_{1,n}(x)$ and $\xi_{(n-k+1)/n,n}(x)$ carries over to the smoothed variants.

Our Monte Carlo exercise, provided in Sect. 1.3.2, shows that the smoothed FDH function $\hat{\xi}_1(x)$ is a remarkable bias-corrected estimator of the frontier function $\xi_1(x) < \infty$. It outperforms frankly the sample FDH estimator $\xi_{1,n}(x)$ in terms of both bias and mean-squared error (MSE). Moreover, an explicit expression of the optimal bandwidth is derived in this case by minimizing the asymptotic MSE. Simulations seem to indicate also that, when $F_{Z^x}(\cdot)$ belongs to the maximum domain of attraction of Weibull and the bandwidth is chosen appropriately, the smoothed extreme quantile function $\hat{\xi}_{(n-k+1)/n}(x)$ is superior to the empirical version in terms of MSE although it has a higher positive bias. This result is similar to what happens in the ordinary framework. Although the naive extreme sample quantile $\xi_{(n-k+1)/n,n}(x)$ might not be so efficient, it has the advantage of not requiring the choice of the bandwidth which is not addressed here. In the heavy tailed case, it appears that $\hat{\xi}_{(n-k+1)/n}(x)$ achieves at most the same performance as the empirical version $\xi_{(n-k+1)/n,n}(x)$ in terms of both bias and MSE and so, smoothing seems to be useless in this case.

The limit theorems in Sect. 1.2 are provided in the general setting where the distribution function of Z^x, or equivalently the conditional distribution function of Y given $X \leq x$, is attracted to the three Fisher–Tippett extreme value type distributions, whereas the previous results of e.g. Martins-Filho and Yao (2008) and Daouia et al. (2009) only cover the domain of attraction of a Weibull distribution. These results and their proofs, postponed to the Appendix, are also extensions of some results and techniques described in Daouia et al. (2010, 2009). Illustrations on how the new class of unconditional quantile-based frontiers $\{\xi_\alpha(\cdot), \hat{\xi}_\alpha(\cdot)\}$ differs from the class of conditional quantile-type frontiers $\{q_\alpha(\cdot), \hat{q}_\alpha(\cdot)\}$ are provided in Sect. 1.3 through simulated examples and a real data set.

1.2 Limit Theorems

To simplify the notation we write ξ_α and $\hat{\xi}_\alpha$, respectively, for $\xi_\alpha(x)$ and $\hat{\xi}_\alpha(x)$ throughout this section. We first show that the smooth estimator $\hat{\xi}_1$ of the endpoint ξ_1 has a similar asymptotic behaviour as the maximum $Z_{(n)} = \max\{Z_1, \ldots, Z_n\}$ under the assumption that

(A1) $\int_{-c}^{c} K(u)du = 1$ for some constant $c > 0$.

This is a standard condition in nonparametric estimation, which is satisfied by commonly used density kernels such as Biweight, Triweight, Epanechnikov, etc.

Theorem 1. *Assume that (A1) holds.*

(i) We have $\hat{\xi}_1 = Z_{(n)} + hc$, *for all* $n \geq 1$.
(ii) Suppose there exist $a_n > 0$, $b_n \in \mathbb{R}$, $n \geq 1$ *such that*

$$\mathbb{P}[a_n^{-1}(\hat{\xi}_1 - b_n) \leq z] \longrightarrow G(z) \quad as \quad n \to \infty, \tag{1.2}$$

where G is assumed nondegenerate. Then G has one of the three forms

$$\textit{Fréchet:} \quad G(z) = \Phi_\rho(z) = \begin{cases} 0 & z < 0 \\ \exp\{-z^{-\rho}\} & z \geq 0 \end{cases} \quad \textit{for some} \quad \rho > 0$$

$$\textit{Weibull:} \quad G(z) = \Psi_\rho(z) = \begin{cases} \exp\{-(-z)^{\rho}\} & z < 0 \\ 1 & z \geq 0 \end{cases} \quad \textit{for some} \quad \rho > 0$$

$$\textit{Gumbel:} \quad G(z) = \Lambda(z) = \exp\{-e^{-z}\}, \ z \in \mathbb{R}.$$

It is clear from Theorem 1(i) that (1.2) holds if and only if F_Z belongs to the maximum domain of attraction[2] of an extreme value distribution $G \in \{\Phi_\rho, \Psi_\rho, \Lambda\}$. Then the characterization of a_n and b_n can be easily deduced from the classical theory of limit laws for maxima (see e.g. Resnick 1987) in conjunction with Theorem 1(i). Putting $\bar{F}_Z = 1 - F_Z$, it is well-known that $F_Z \in \mathrm{DA}(\Psi_\rho)$ iff $\xi_1 < \infty$ and[3] $\bar{F}_Z(\xi_1 - \frac{1}{t}) \in \mathrm{RV}_{-\rho}$, $t \to \infty$; in this case we may set $a_n = \xi_1 - \xi_{1-1/n}$ and $b_n = \xi_1 + hc$. Likewise $F_Z \in \mathrm{DA}(\Phi_\rho)$ iff $\bar{F}_Z \in \mathrm{RV}_{-\rho}$; in this case $\xi_1 = \infty$ and (a_n, b_n) can be taken equal to $(\xi_{1-1/n}, hc)$. Finally $F_Z \in \mathrm{DA}(\Lambda)$ iff there exists a strictly positive function g on $\mathbb{R}$ such that $\bar{F}_Z(t + g(t)z)/\bar{F}_Z(t) \to e^{-z}$ as $t \uparrow \xi_1$, for every $z \in \mathbb{R}$; in this case the constants (a_n, b_n) can be taken equal to $(g(\xi_{1-1/n}), \xi_{1-1/n} + hc)$.

On the other hand, by making use of Theorem 1(i), it is easy to show that the convergence in distribution (1.2) of the smoothed maximum $\hat{\xi}_1$ implies the convergence of moments in the three cases $G \in \{\Phi_\rho, \Psi_\rho, \Lambda\}$ under some condition on the left tail of F_Z.

Theorem 2. *Let (A1) and* (1.2) *hold and denote by* $\Gamma^{(k)}$ *the* k*th derivative of the gamma function* Γ.

(i) If $G = \Phi_\rho$ *in* (1.2) *with* $(a_n, b_n) = (\xi_{1-1/n}, hc)$ *and* $\int_{-\infty}^0 |z|^k F_Z(dz) < \infty$ *for some integer* $0 < k < \rho$, *then*

$$\lim_{n\to\infty} \mathbb{E}\{a_n^{-1}(\hat{\xi}_1 - hc)\}^k = \int_{\mathbb{R}} z^k \Phi_\rho(dz) = \Gamma(1 - k/\rho).$$

(ii) If $G = \Psi_\rho$ *with* $(a_n, b_n) = (\xi_1 - \xi_{1-1/n}, \xi_1 + hc)$ *and* $\int_{-\infty}^{\xi_1} |z|^k F_Z(dz) < \infty$ *for some integer* $k > 0$, *then*

$$\lim_{n\to\infty} \mathbb{E}\{a_n^{-1}(\hat{\xi}_1 - \xi_1 - hc)\}^k = \int_{-\infty}^0 z^k \Psi_\rho(dz) = (-1)^k \Gamma(1 + k/\rho).$$

(iii) If $G = \Lambda$ *with* $(a_n, b_n) = (g(\xi_{1-1/n}), \xi_{1-1/n} + hc)$ *and* $\int_{-\infty}^0 |z|^k F_Z(dz) < \infty$ *for some integer* $k > 0$, *then*

$$\lim_{n\to\infty} \mathbb{E}\{a_n^{-1}(\hat{\xi}_1 - b_n)\}^k = \int_{\mathbb{R}} z^k \Lambda(dz) = (-1)^k \Gamma^{(k)}(1).$$

When estimating the endpoint $\xi_1 < \infty$ of $F_Z \in \mathrm{DA}(\Psi_\rho)$, an optimal value of h can be derived by minimizing the asymptotic mean-squared error of $\hat{\xi}_1$. By making use of Theorem 2(ii), it is not hard to check that the optimal bandwidth is given by

---

[2] We write $F_Z \in \mathrm{DA(G)}$ if there exist normalizing constants $a_n > 0$, $c_n \in \mathbb{R}$ such that $a_n^{-1}(Z_{(n)} - c_n) \xrightarrow{d} G$.

[3] A measurable function $\ell : \mathbb{R}_+ \to \mathbb{R}_+$ is regularly varying at ∞ with index γ (written $\ell \in \mathrm{RV}_\gamma$) if $\lim_{t\to\infty} \ell(tx)/\ell(t) = x^\gamma$ for all $x > 0$.

$$h_{opt} = a_n(c\rho)^{-1}\Gamma(1/\rho). \tag{1.3}$$

Next we show that if $\hat{\xi}_1$ converges in distribution, then $\hat{\xi}_{1-k/n}$ converges in distribution as well, with the same centering and scaling, but a different limit distribution.

Theorem 3. *Assume that (A1) holds. If $a_n^{-1}(\hat{\xi}_1 - b_n) \xrightarrow{d} G$ and $a_n^{-1}h \to 0$, then for any integer $k \geq 0$,*

$$a_n^{-1}(\hat{\xi}_{\frac{n-k}{n}} - b_n) \xrightarrow{d} \mathbb{G} \quad as \quad n \to \infty$$

for the distribution function $\mathbb{G}(z) = G(z)\sum_{i=0}^{k}(-\log G(z))^i/i!$.

For the condition $a_n^{-1}h \to 0$ to be satisfied in the case $G = \Phi_\rho$, one only needs to suppose for instance that $\lim_{n\to\infty} h < \infty$. However, for the case $G = \Psi_\rho$, the condition $a_n^{-1}h \to 0$ holds if $h/(\xi_1 - \xi_{1-1/n}) \to 0$. The case $G = \Lambda$ is less flexible since the characterization of the normalization constant $a_n = g(\xi_{1-1/n})$ is not as explicit here than in the cases $G \in \{\Phi_\rho, \Psi_\rho\}$ (see e.g. Resnick (1987, p. 38) for more details). It should be also clear that for $\hat{\xi}_\alpha$ to converge to the extreme-value distribution $\mathbb{G}$, it suffices to choose the sequence $\alpha = \alpha_n \uparrow 1$ such that $n(1-\alpha) = k$, with $k < n$ being an integer, whereas for $\hat{\xi}_\alpha$ to have an asymptotic normal distribution, we show in the next theorem that it suffices to choose $\alpha \to 1$ slowly so that $n(1-\alpha) \to \infty$.

The three Fisher–Tippett extreme-value distributions can be defined as a one-parameter family of types

$$G_\gamma(z) = \begin{cases} \exp\{-(1+\gamma z)^{-1/\gamma}\}, & \gamma \neq 0, \quad 1+\gamma z > 0 \\ \exp\{-e^{-z}\} & \gamma = 0, \quad z \in \mathbb{R}, \end{cases}$$

where γ is the so-called extreme-value index (see, e.g., Beirlant et al. 2004). The heavy-tailed case $F_Z \in \mathrm{DA}(\Phi_\rho)$ corresponds to $\gamma > 0$ and $\rho = 1/\gamma$. For $\gamma = 0$, it is clear that $G_\gamma = \Lambda$. The case $F_Z \in \mathrm{DA}(\Psi_\rho)$ corresponds to $\gamma < 0$ and $\rho = -1/\gamma$. In either case we give in the next theorem asymptotic confidence intervals for high quantiles ξ_α by imposing the extra condition that

(A2) The derivative U' of $U(t) = \xi_{1-1/t}$ exists so that it satisfies $U' \in \mathrm{RV}_{\gamma-1}$.

As pointed out in Dekkers and de Haan (1989), the assumption $U' \in \mathrm{RV}_{\gamma-1}$ on the inverse function $U(t) = (1/(1-F_Z))^{-1}(t)$ is equivalent to $F_Z' \in \mathrm{RV}_{-1-1/\gamma}$ for $\gamma > 0$, $F_Z'(\xi_1 - 1/t) \in \mathrm{RV}_{1+1/\gamma}$ for $\gamma < 0$ and $1/F_Z'$ is Γ-varying for $\gamma = 0$ (for the Γ-Variation, see e.g. Resnick 1987, p.26).

Theorem 4. *Given (A1) and (A2),*

$$\sqrt{k}\left(\hat{\xi}_{\frac{n-k+1}{n}} - \xi_{\alpha_n}\right) / \left(\hat{\xi}_{\frac{n-k+1}{n}} - \hat{\xi}_{\frac{n-2k+1}{n}}\right)$$

is asymptotically normal with mean zero and variance $2^{2\gamma}\gamma^2/(2^\gamma - 1)^2$, *provided that* $\alpha_n \uparrow 1$, $n(1-\alpha_n) \to \infty$ *and the integer part* k *of* $n(1-\alpha_n)$ *satisfies* $hk^{3/2}\{nU'(n/k)\}^{-1} \to 0$ *as* $n \to \infty$.

Note that in the case $\gamma = 0$, the asymptotic variance is understood to be the limit $\{1/\log 2\}^2$ obtained as $\gamma \to 0$. This theorem enables one to construct an asymptotic confidence interval for ξ_{α_n} by replacing the tail index γ in the asymptotic variance with a consistent estimator. One can use for example the moment's estimator introduced by Dekkers et al. (1989) or the recent proposal by Segers (2005). Note also that in frontier and efficiency analysis, econometric considerations often lead to the assumption that the joint density of the random vector $(X, Y) \in \mathbb{R}_+^{p+1}$ has a jump at its support boundary, which corresponds to the case where γ is known and equal to $-1/(p+1)$ as established in Daouia et al. (2010).

1.3 Numerical Illustrations

We provide in this section some modest illustrations in the context of frontier analysis, i.e., when $Z = Z^x$ and $\hat{\xi}_\alpha = \hat{\xi}_\alpha(x)$. In this case, it is important to note that the bandwidth h, the normalizing sequences (a_n, b_n), the extreme value index γ or equivalently the tail index ρ, the order α_n and the sequence k_n should depend on the fixed level $x \in \mathbb{R}_+^p$. We do not enter here into the question of how to choose in an optimal way h, α_n and k_n. Deriving asymptotically optimal values of these parameters is a tedious matter. Using for instance (1.3) calls for selection of subsidiary smoothing parameters (using plug-in methods requires explicit estimation of the spacing a_n and the tail index ρ, which demands optimal selection of the amount of extreme data involved in each estimate, etc). Such complexity is arguably not justified. Instead, we suggest an approximate empirical method as follows. We tune the bandwidth h involved in $\hat{\xi}_\alpha(x) = \hat{F}_{Z^x}^{-1}(\alpha)$ so that approximately a reasonable percentage $\lambda\%$ of the data points $Z_1^x, \ldots, Z_n^x$ fall into the support of $u \mapsto K((z-u)/h)$. In case of kernels with support $[-1, 1]$, as Triweight and Epanechnikov kernels, we use the explicit formula

$$h_x = \frac{\lambda}{200}\left(\max_{i=1,\ldots,n} Z_i^x - \min_{i=1,\ldots,n} Z_i^x\right).$$

Our method itself requires selection of a smoothing parameter λ, but it has the advantage to be very simple to interpret and to implement, particularly in the difficult context of nonparametric curve estimation. The same rule can be applied to the estimator $\hat{q}_\alpha(x) = \hat{F}_{Y^x}^{-1}(\alpha)$ whose computation is similar to $\hat{\xi}_\alpha(x)$. Indeed, similarly to $\hat{F}_{Z^x}(y)$ we have $\hat{F}_{Y^x}(y) = (1/N_x)\sum_{i=1}^{N_x} H\left((y - Y_i^x)/h\right)$, where $N_x = \sum_{i=1}^n \mathbb{1}(X_i \le x)$ and $Y_1^x, \ldots, Y_{N_x}^x$ are the Y_i's such that $X_i \le x$. As a matter of fact, $Y_1^x, \ldots, Y_{N_x}^x$ are the N_x largest statistics of the sample $(Z_1^x, \ldots, Z_n^x)$. Note also that the two families of nonparametric (unconditional and conditional) quantile-based partial frontiers $\hat{\xi}_\alpha(\cdot)$ and $\hat{q}_\alpha(\cdot)$ coincide for $\alpha = 1$, but they differ from one another when $\alpha < 1$. This difference is illustrated through a real data set and two simulated examples in the cases where the support boundary is finite and infinite.

For our practical computations in Sect. 1.3.1, the smooth estimators will be evaluated only for fixed extreme orders α by using a Triweight kernel K and for a grid of values of λ. Section 1.3.2 provides a comparison between the sample extreme frontiers and their smoothed versions via Monte Carlo experiments.

1.3.1 Illustrative Examples on One Sample

1.3.1.1 Case of a Finite Support Boundary

We choose (X, Y) uniformly distributed over the support $\Psi = \{(x, y) | 0 \le x \le 1, 0 \le y \le x\}$. In this case, the true frontier function is $\xi_1(x) = q_1(x) = x$ and the class of conditional quantile-based frontiers $q_\alpha(x) = x(1 - \sqrt{1-\alpha})$ is different from our class of unconditional quantile-type frontiers $\xi_\alpha(x) = \max\{0, x - \sqrt{1-\alpha}\}$, for $\alpha \in (0,1]$. Both partial order-α frontiers are graphed in Fig. 1.1 for some large values of $\alpha = 0.9, 0.95, 0.99$.

Note that in traditional applied econometrics, the distance from the full support frontier is used as a benchmark to measure the production performance of firms.

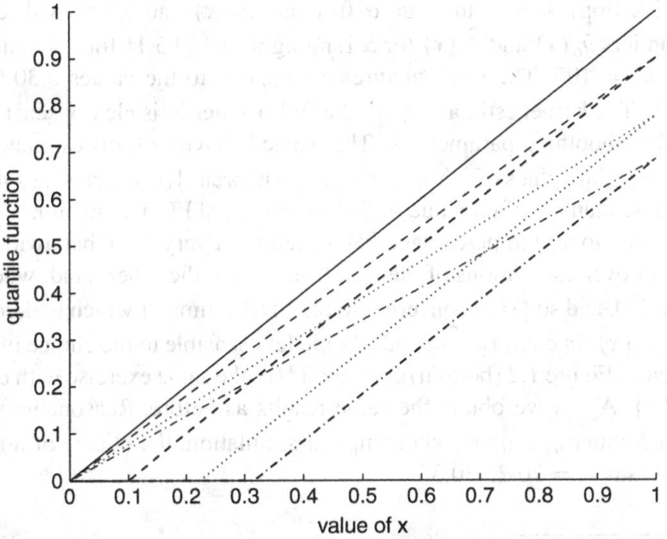

Fig. 1.1 The full frontier in *solid line*. The frontiers $q_\alpha(\cdot)$ in *thin lines* and $\xi_\alpha(\cdot)$ in *thick lines*, with $\alpha = .9$ in *dashdotted lines*, $\alpha = .95$ in *dotted lines* and $\alpha = .99$ in *dashed lines*

The economic efficiency can also be measured in terms of partial frontiers as suggested recently in the econometric literature to avoid non-robustness of the envelopment nonparametric frontier estimators. It is clear that the partial frontiers $\{(x, q_\alpha(x)) : 0 \leq x \leq 1\}$ of the support Ψ diverge from the support boundary as x increases and so, measuring efficiency relative to these curves may result in misleading efficiency measures. To reduce this defect, Wheelock and Wilson (2008) favored the use of a hyperbolic unconditional variant of $q_\alpha(x)$. The new partial frontiers $\xi_\alpha(x)$ parallel the full frontier $\xi_1(x)$ providing thus a simple alternative for measuring partial efficiencies without recourse to the hyperbolic framework. However, this desirable property is valid only for input factors x ranging from the $(1-\alpha)$th quantile of the marginal distribution of X to its endpoint. It is easy to see that the value of $\xi_\alpha(x)$ is zero whenever $\mathbb{P}(X \not\leq x) \geq \alpha$, but the class $\{q_\alpha(x)\}$ does not necessarily take advantage from this drawback of $\{\xi_\alpha(x)\}$. Indeed, given that the interest is in estimating partial frontiers ξ_α and q_α lying close to the full support boundary, the order α shall be selected large enough in such a way that the estimates of ξ_α and q_α capture the shape of the sample's upper boundary without enveloping all the data points[4]. For such a choice of $\alpha \uparrow 1$, the $(1-\alpha)$th quantile of the distribution of X should be very small and so, the shortcoming $\hat{\xi}_\alpha(x_i) = 0$ is expected to hold only for very few observations (x_i, y_i) at the left border of the sample. For these few observations with too small inputs-usage x_i, the estimates $\hat{q}_\alpha(x_i)$ of $q_\alpha(x_i)$ are expected by construction to coincide with the non-robust envelopment FDH estimates (as illustrated below in Figs. 1.2, 1.4 and 1.5), which goes against the concept of partial frontier modeling. Examples are also provided below in Fig. 1.8 where only $\hat{q}_\alpha(\cdot)$ suffers from left border defects, whereas $\hat{\xi}_\alpha(\cdot)$ is clearly the winner.

Figure 1.2 (top) depicts the true α-frontiers $q_\alpha(x)$ and $\xi_\alpha(x)$ with estimated smooth frontiers $\hat{q}_\alpha(x)$ and $\hat{\xi}_\alpha(x)$ for α ranging over $\{0.95, 1\}$ for a simulated data set of size $n = 100$. The three pictures correspond to the values 5,30,60 of the parameter λ. The kernel estimator $\hat{\xi}_1$ of the full frontier ξ_1 is clearly sensible to the choice of the smoothing parameter λ. The worse behavior of this estimator for too large and too small values of λ is explained by Theorem 1(i) which states that $\hat{\xi}_1(x)$ is nothing else than a shifted value of the conventional FDH estimator. The choice of h_x according to a high percentage $\lambda\%$ generates a very large bandwidth, which may result in over-estimations of the true frontier. On the other hand, when $\lambda \downarrow 0$ we have $h_x \downarrow 0$, and so $\hat{\xi}_1(x)$ converges to the FDH estimator which underestimates the frontier $\xi_1(x)$. In contrast, $\hat{\xi}_{.95}$ and $\hat{q}_{.95}$ are less sensible to the choice of λ in this particular case. Figure 1.2 (bottom) corresponds to the same exercise with α ranging over $\{0.99, 1\}$. Again we obtain the same results as before. Reasonable values of the smooth frontier $\hat{\xi}_1$ require, via computer simulation, the choice of a moderate parameter λ, say $\lambda = 20,25,30,35$.

[4]Once a reasonable large value of α is picked out, the idea in practice is then to interpret the observations left outside the αth frontier estimator as highly efficient and to assess the performance of the points lying below the estimated partial frontier by measuring their distances from this frontier in the output-direction.

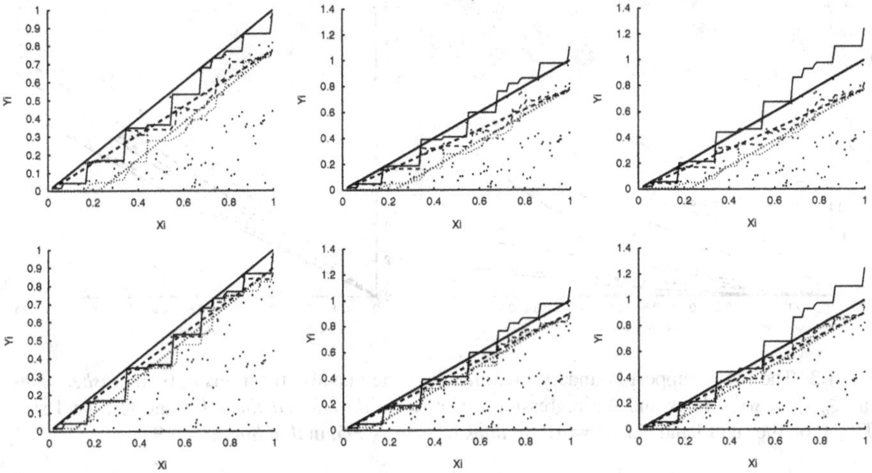

Fig. 1.2 In *thick lines* the frontiers ξ_1 (*solid*), ξ_α (*dotted*) and q_α (*dashed*). In *thin lines* their estimators $\hat{\xi}_1$, $\hat{\xi}_\alpha$ and $\hat{q}_\alpha$. $n = 100$ and from left to right $\lambda = 5,30,60$. From top to bottom $\alpha = .95,.99$

1.3.1.2 Case of an Infinite Support Boundary

We consider the standard case where the distribution of Y given $X = x$ is Pareto of parameter $\beta > 0$, that is

$$\mathbb{P}(Y \le y|X = x) = \begin{cases} 1 - (x/y)^\beta & \text{if} \quad y \ge x \\ 0 & \text{if} \quad y < x. \end{cases}$$

Here we choose X uniform on (0,1). The two partial αth frontiers are then given by

$$q_\alpha(x) = \max\{x, x[(1-\alpha)(1+\beta)]^{-1/\beta}\},$$
$$\xi_\alpha(x) = \max\{x, x^{1+1/\beta}[(1-\alpha)(1+\beta)]^{-1/\beta}\}.$$

Both families of partial frontiers differ following the values of the distribution parameter β. A graphical illustration is displayed in Fig. 1.3. In each picture we superimpose the lower support boundary, the quantile functions $q_\alpha(x)$ and $\xi_\alpha(x)$ and the regression αth quantiles of Y given $X = x$, for a fixed $\beta \in \{1,10\}$ and for two large values of $\alpha = 0.95,0.99$. First note that $\xi_\alpha(\cdot)$ is overall smaller than $q_\alpha(\cdot)$ and that $q_\alpha(\cdot)$ itself is overall smaller than the αth regression quantile function, for any $\alpha \in (0,1)$. Second note that for small values of β (e.g. $\beta = 1$), the use of extreme regression quantiles (e.g. $\alpha = 0.99$) to capture the most efficient firms seems in this particular case to be less justified than the use of $q_\alpha(\cdot)$ and $\xi_\alpha(\cdot)$. Indeed, from an economic point of view, it is not reasonable for optimal dominating firms to be too far in the output direction from the set of relatively inefficient firms (lower

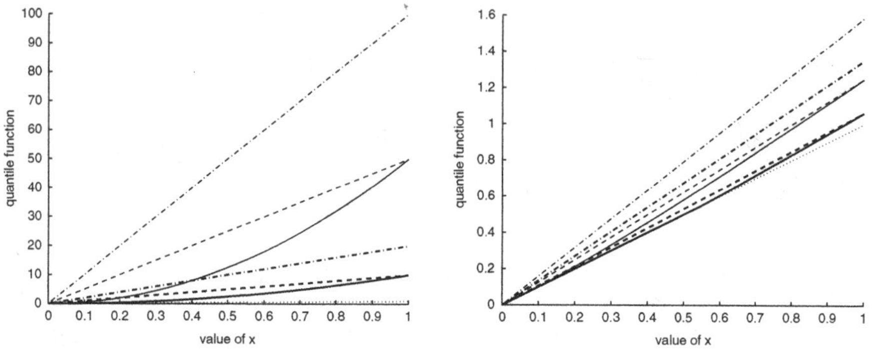

Fig. 1.3 The lower support boundary in *dotted line*, the quantile functions $q_\alpha(\cdot)$ in *dashed lines* and $\xi_\alpha(\cdot)$ in *solid lines* and the regression quantiles in *dashdotted lines*. On the left-hand side $\beta = 1$, on the right-hand side $\beta = 10$. In *thick lines* $\alpha = .95$, in *thin lines* $\alpha = .99$

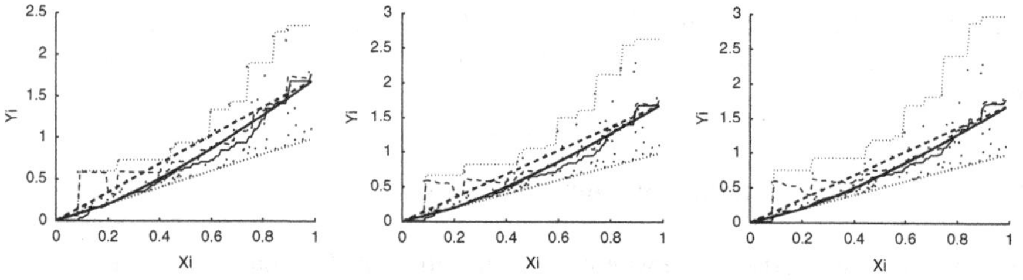

Fig. 1.4 Here $\beta = 3$. In *thick lines*: true lower frontier (*dotted*), $\xi_{.95}$ (*solid*), $q_{.95}$ (*dashed*). In *thin lines*: $\hat{\xi}_1$ (*dotted*), $\hat{\xi}_{.95}$ (*solid*) and $\hat{q}_{.95}$ (*dashed*). $n = 100$ and from left to right $\lambda = 5,30,60$

support boundary). The use of the three types of quantile-based frontiers seems to be more justified from an economic viewpoint for large values of the parameter β (e.g. $\beta = 10$). In particular the parabolic shape of the function $\xi_\alpha(\cdot)$ diminishes as β increases. The three αth quantile functions converge to the linear lower support frontier as $\beta \to \infty$.

For a simulated data set of size $n = 100$ using $\beta = 3$, the true frontiers $q_\alpha(x)$ and $\xi_\alpha(x)$ with the smooth estimators $\hat{q}_\alpha(x)$ and $\hat{\xi}_\alpha(x)$ are graphed in Fig. 1.4 for $\alpha \in \{0.95, 1\}$ and $\lambda = 5,30,60$. We obtain the same conclusions as in the preceding example. Here also, while $\hat{q}_{.95}(x)$ diverges from the extreme smooth frontier $\hat{\xi}_1(x)$ as x increases, the partial frontier $\hat{\xi}_{.95}$ parallels $\hat{\xi}_1(x)$ in much the same way as the partial frontiers $\hat{\xi}_\alpha$ do in Fig. 1.2. In this particular example, it is clear that $\hat{q}_{.95}$ is more attracted by extreme data points with small X_i's. In general, for $\alpha < 1$, the frontiers $\hat{\xi}_\alpha$ are by construction more robust to extremes than $\hat{q}_\alpha$ and are less sensible to the border effects from which the frontiers $\hat{q}_\alpha$ suffer due to the conditioning $X \leq x$.

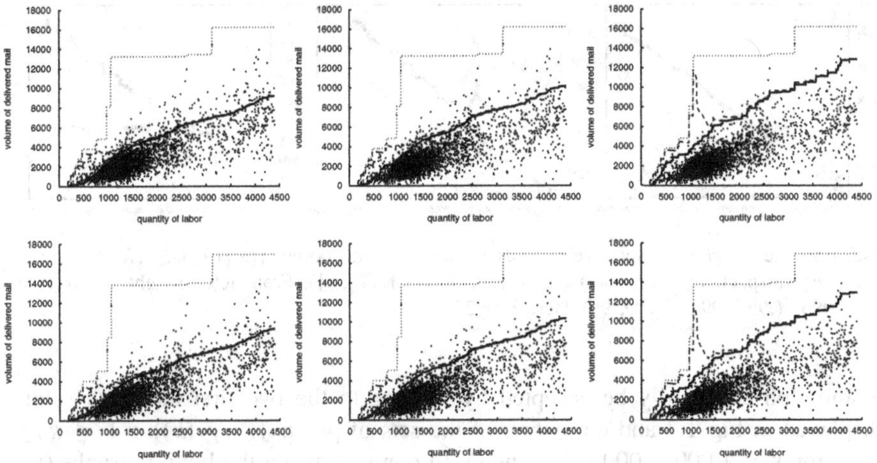

Fig. 1.5 $n = 4{,}000$ French post offices. In *dotted line* $\hat{\xi}_1$, in *solid line* $\hat{\xi}_\alpha$ and in *dashed line* $\hat{q}_\alpha$. From left to right $\alpha = .99, .995, .999$. From top to bottom $\lambda = 25, 35$

1.3.1.3 Frontier Analysis of French Post Offices

To illustrate our methodology, we employ data on 4,000 post offices from France reported in Cazals et al. (2002). These data consist of the cost variable X_i which represents the quantity of labor and the output variable Y_i defined as the volume of delivered mail.

In Fig. 1.5, we provide in each picture a scatterplot of the data and plot the frontiers $\hat{\xi}_1$, $\hat{\xi}_\alpha$ and $\hat{q}_\alpha$, for $\alpha = 0.99$, 0.995 and 0.999, from left to right. From top to bottom, we used $\lambda = 25$ and $\lambda = 35$ in our computations. We first observe that both αth frontiers $\hat{\xi}_\alpha$ and $\hat{q}_\alpha$ are not influenced by the choice of the smoothing parameter λ, whereas the full extreme frontier $\hat{\xi}_1$ changes slightly. We also see that large frontiers $\hat{\xi}_\alpha$ suggest better capability of fitting efficient post offices than large $\hat{q}_\alpha$. It is apparent that the estimates $\hat{\xi}_\alpha$ (solid lines) are less sensitive to the choice of extreme orders α than the frontiers $\hat{q}_\alpha$ are (dashed lines). Figure 1.6 plots the αth frontier estimates $\hat{\xi}_\alpha$ (respectively, $\hat{q}_\alpha$) for the three values $\alpha = 0.99, 0.995, 0.999$ against each other: each panel indicated by 'circles' (respectively, 'dots') compares estimates $\{(\hat{\xi}_{\alpha_1}(X_i), \hat{\xi}_{\alpha_2}(X_i)) : i = 1, \ldots, n\}$ (respectively $\{(\hat{q}_{\alpha_1}(X_i), \hat{q}_{\alpha_2}(X_i)) : i = 1, \ldots, n\}$) for a pair (α_1, α_2) of values for α. Unlike the panels in 'circles', most points fall on or near a straight line for all 'dotted' panels, confirming thus the impression from Fig. 1.5, i.e., while the frontier estimates $\hat{\xi}_\alpha$ obtained with the three extreme values of α are somewhat similar, one sees substantial differences for the frontiers $\hat{q}_\alpha$.

Note that the full extreme frontier $\hat{\xi}_1(x)$ in Fig. 1.5 is far from large partial frontiers $\hat{\xi}_\alpha(x)$ even when α increases (this is not the case for $\hat{q}_\alpha(x)$). This might suggest the heavy-tailed case $F_{Z^x} \in \mathrm{DA}(G_{\gamma_x})$ with $\gamma_x > 0$. However, this

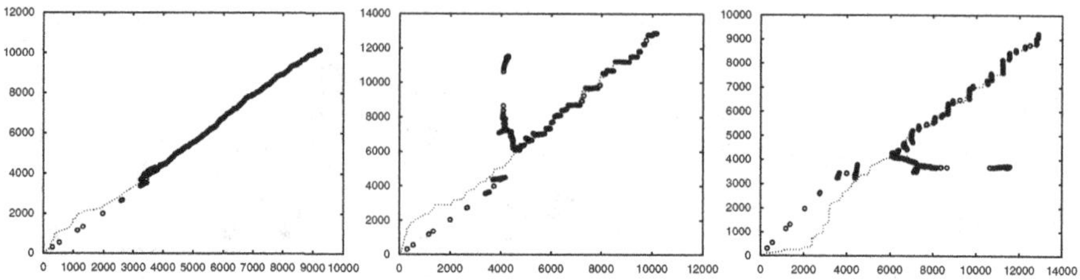

Fig. 1.6 The '*circles*' (respectively, '*dots*') consist of the points $\{(\hat{\xi}_{\alpha_1}(X_i), \hat{\xi}_{\alpha_2}(X_i)) : i = 1, \ldots, n\}$ (respectively $\{(\hat{q}_{\alpha_1}(X_i), \hat{q}_{\alpha_2}(X_i)) : i = 1, \ldots, n\}$). From left to right $(\alpha_1, \alpha_2) =$ (.99,.995), (.995,.999), (.999,.99). Here $\lambda = 25$

assumption can hardly be accepted by looking to the two first moments[5] plots displayed in Fig. 1.7 and consisting of the sets of points $\{(k, \hat{\gamma}_x(k)) : 1 \le k \le N_x\}$, for $x = 1{,}000, 2{,}000$ (top panels). In contrast, when the bottom graphs (for $x = 3{,}000, 4{,}000$) look stable, they correspond to values $\hat{\gamma}_x \ge 0$.

The frontier $\hat{\xi}_1(x)$ being by construction a shifted variant of the FDH estimator, it is more sensitive to extreme values. This frontier is clearly determined by very few outlying post offices. At the opposite, the partial frontier $\hat{\xi}_\alpha(x)$ is more resistant to these outliers than $\hat{\xi}_1(x)$ and $\hat{q}_\alpha(x)$ even for too high values of α. This explains the substantial difference between $\hat{\xi}_\alpha(x)$ and $\hat{\xi}_1(x)$ even when $\alpha = 0.9993, 0.9995$ and 0.9997 as shown in Fig. 1.8.

1.3.2 Monte Carlo Experiments

1.3.2.1 Comparison of the Full Frontier Estimators

Let us first compare the performance of both estimators $\hat{\xi}_1(\cdot)$ and $\xi_{1,n}(\cdot)$ of the full frontier function $\xi_1(\cdot)$. In the particular example described above in Paragraph 1.3.1.1 where $\xi_1(x) = x$, we have $F_{Z^x} \in \mathrm{DA}(\Psi_{\rho_x})$ with $\rho_x = -1/\gamma_x = 2$ and $a_n(x) = x - \max\{0, x - \sqrt{1/n}\}$. For the computation of the smoothed estimator $\hat{\xi}_1(x)$, we use here the true value of the optimal bandwidth $h_{opt}(x) = a_n(x)\rho_x^{-1}\Gamma(\rho_x^{-1})$ derived in (1.3), as well as the plug-in values $h_{opt,1}(x) = \hat{a}_n(x)\rho_x^{-1}\Gamma(\rho_x^{-1})$ and $h_{opt,2}(x) = \hat{a}_n(x)\hat{\rho}_x^{-1}\Gamma(\hat{\rho}_x^{-1})$ obtained by replacing $a_n(x)$ and ρ_x, respectively, with the empirical counterpart $\hat{a}_n(x) = Z^x_{(n)} - Z^x_{(n-1)}$ and the moment's estimator $\hat{\rho}_x = -1/\hat{\gamma}_x$. The computation of $\hat{\rho}_x$ depends on the sample fraction k whose choice is difficult in practice. By definition of the moment's

[5]The moment's estimator $\hat{\gamma}_x$ of the tail index γ_x is defined as $\hat{\gamma}_x = H_n^{(1)} + 1 - \frac{1}{2}\{1 - (H_n^{(1)})^2/H_n^{(2)}\}^{-1}$, with $H_n^{(j)} = (1/k)\sum_{i=0}^{k-1}(\log Z^x_{(n-i)} - \log Z^x_{(n-k)})^j$ for $k < n$ and $j = 1, 2$ (Dekkers et al. 1989).

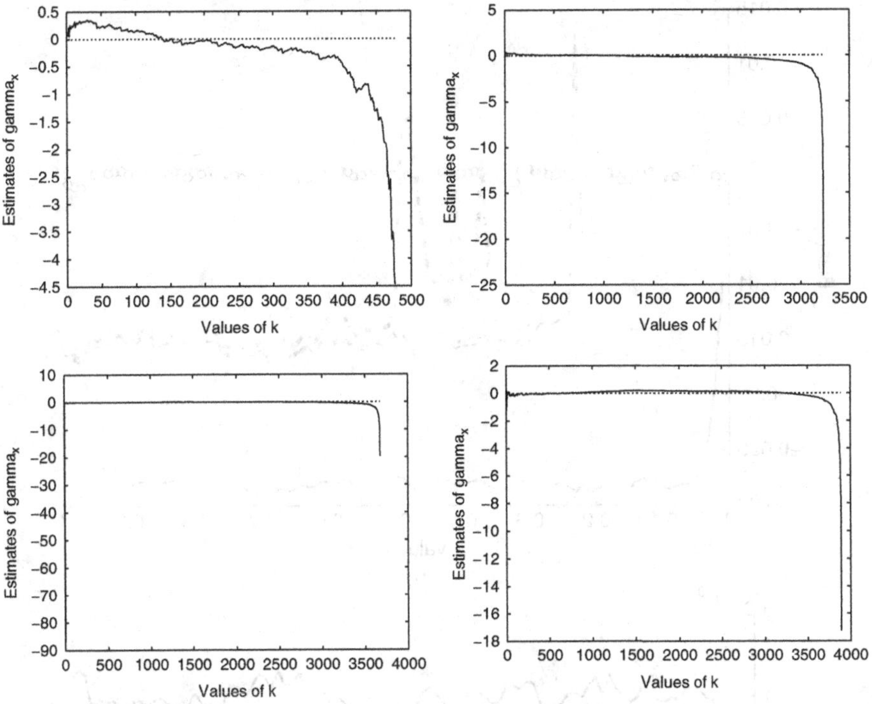

Fig. 1.7 Moments plots for $x = 1{,}000, 2{,}000, 3{,}000, 4{,}000$, respectively from left to right and from top to bottom

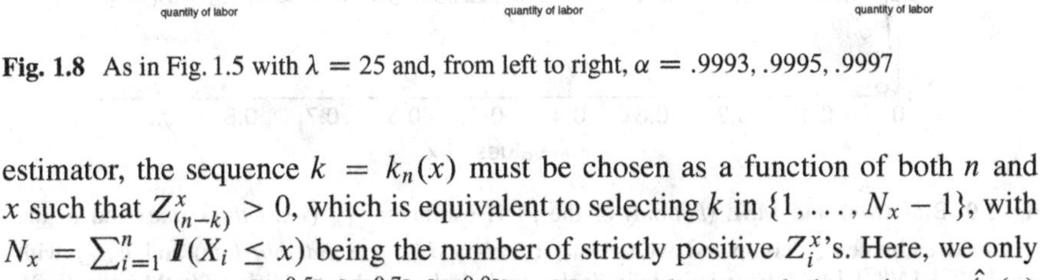

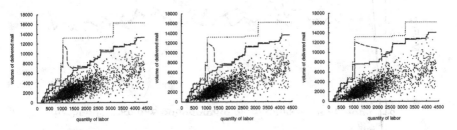

Fig. 1.8 As in Fig. 1.5 with $\lambda = 25$ and, from left to right, $\alpha = .9993, .9995, .9997$

estimator, the sequence $k = k_n(x)$ must be chosen as a function of both n and x such that $Z^x_{(n-k)} > 0$, which is equivalent to selecting k in $\{1, \ldots, N_x - 1\}$, with $N_x = \sum_{i=1}^n \mathbb{1}(X_i \leq x)$ being the number of strictly positive Z_i^x's. Here, we only use the values $k \in \{[N_x^{0.5}], [N_x^{0.7}], [N_x^{0.9}]\}$ to illustrate how much the estimates $\hat{\xi}_1(x)$ based on $h_{opt,2}(x)$ differ from those based on $h_{opt,1}(x)$ and $h_{opt}(x)$.

Figure 1.9 provides the Monte Carlo estimates of the Bias and the Mean-Squared Error (MSE) of $\hat{\xi}_1(x)$ and $\xi_{1,n}(x)$ computed over 2,000 random replications with $n = 1{,}000$ (what is important is not the sample size n itself but the number

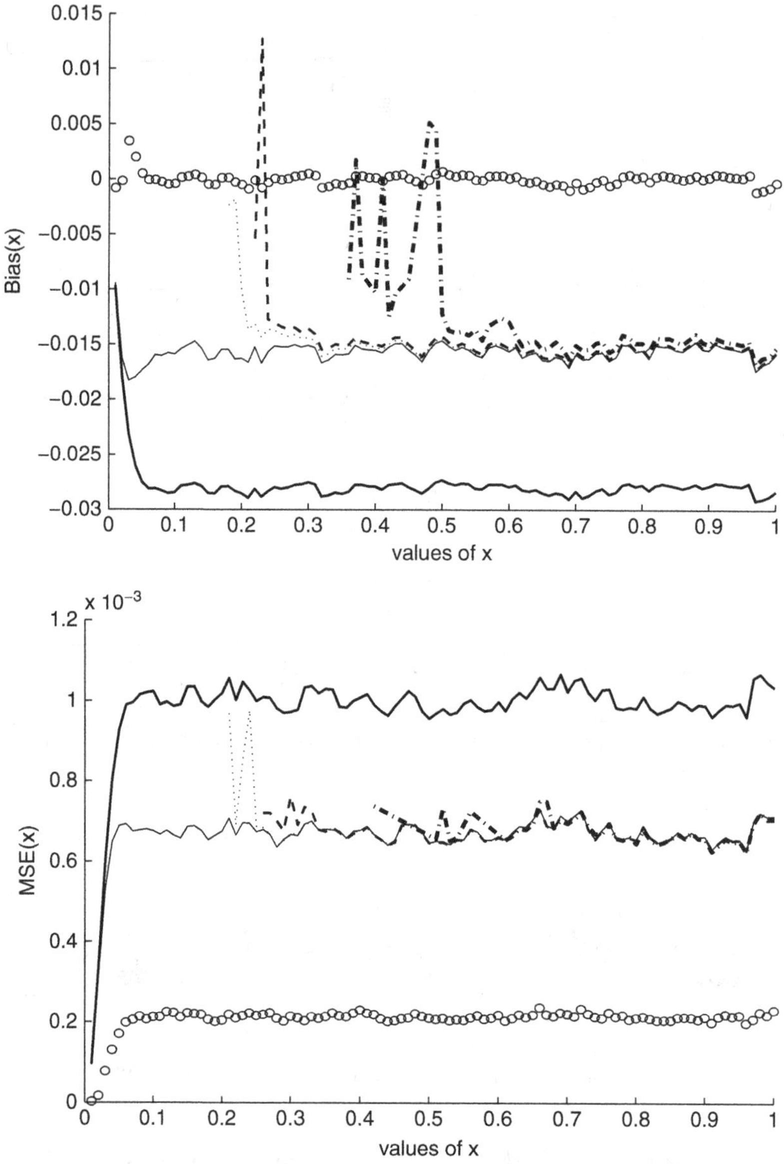

Fig. 1.9 Bias (*top*) and MSE (*bottom*) of the FDH estimator $\xi_{1,n}$ (*solid thick line*) and of the smoothed estimates $\hat{\xi}_1$ computed with : h_{opt} (*circles*), $h_{opt,1}$ (*solid thin line*) and $h_{opt,2}$ with $k = [N_x^{0.5}], [N_x^{0.7}], [N_x^{0.9}]$ (respectively in *dashdotted, dashed and dotted lines*). 2000 Monte-Carlo simulations with $n = 1{,}000$

N_x of observations X_i smaller than or equal to x). The results are displayed for the FDH estimator $\xi_{1,n}(x)$ in solid thick line and for the smoothed version $\hat{\xi}_1(x)$ computed with the true bandwidth $h_{opt}(x)$ in 'circles'. We see clearly that the resulting smoothed maximum $\hat{\xi}_1(x)$ outperforms the sample maximum $\xi_{1,n}(x)$ in terms of both Bias (top picture) and MSE (bottom), for every x.

When only the tail-index ρ_x is supposed to be known and equal to $-1/2$ as is typically the case in the econometric literature on nonparametric frontier estimation, the results for $\hat{\xi}_1(x)$ obtained with the plug-in bandwidth $h_{opt,1}(x)$ are displayed in solid thin line. We observe that the smoothed maximum's performance deteriorates when using the naive spacing $\hat{a}_n(x)$ in place of the theoretical scaling $a_n(x)$, but it remains still appreciably better than the sample maximum's performance in terms of both Bias and MSE, say, for all $x \geq 0.05$ (for too small values of $x < 0.05$, $\hat{\xi}_1(x)$ performs at least as good as $\xi_{1,n}(x)$).

When ρ_x is estimated by $\hat{\rho}_x$, the results for $\hat{\xi}_1(x)$ obtained by using the plug-in bandwidth $h_{opt,2}(x)$ with $k = [N_x^{0.5}], [N_x^{0.7}], [N_x^{0.9}]$, are displayed respectively in dashdotted, dashed and dotted lines. These three lines are graphed only for the values of x where $\hat{\xi}_1(x)$ behaves better than $\xi_{1,n}(x)$ in terms of Bias and MSE. We see that, when N_x is large enough ($x > 0.5$), the three different selected values of k used for the computation of $\hat{\rho}_x$ in $h_{opt,2}(x)$ give very similar results to the "benchmarked" case $h_{opt,1}(x)$ (the idea is that for a properly chosen value of k, both cases $h_{opt,1}(x)$ and $h_{opt,2}(x)$ should approximately yield similar values of $\hat{\xi}_1(x)$). In contrast, the results are all the more sensitive to the choice of k as N_x becomes small : large values of k seem to be needed as N_x decreases in order to get sensible results.

It should also be clear that the minimal value of N_x computed over the 2,000 realizations is given, for instance, by $N_{0.3} = 60$ for $x = 0.3$ and by $N_{0.1} = 1$ for $x = 0.1$. While the estimation of ρ_x from the sample $\{Z^x_{(n-N_x+1)}, \ldots, Z^x_{(n)}\}$ of size N_x as small as $N_{0.3}$ can hardly result in a satisfactory moment's estimate (or any other extreme-value based estimates), it is not even feasible when $N_x < 2$. This is a recurrent problem in extreme-value theory.

Apart from this vexing border defect, we could say in view of the results described above that the smoothed estimator's performance may be improved in terms of Bias and MSE by deriving a more efficient estimate for $a_n(x)$ than the naive spacing $\hat{a}_n(x)$, and by providing an appropriate choice of the sample fraction $k = k_n(x)$ involved in the tail-index estimates.

1.3.2.2 Comparison of Extreme Partial Frontiers

Let us now compare the asymptotically normal estimators $\hat{\xi}_{(n-k+1)/n}$ and $\xi_{(n-k+1)/n,n} = Z^x_{(n-k+1)}$ of the extreme partial frontier $\xi_{(n-k+1)/n}$, for two values of $k = k_n \in \{[n^{0.5}], [n^{0.75}]\}$ and two sample sizes $n \in \{100, 1{,}000\}$.

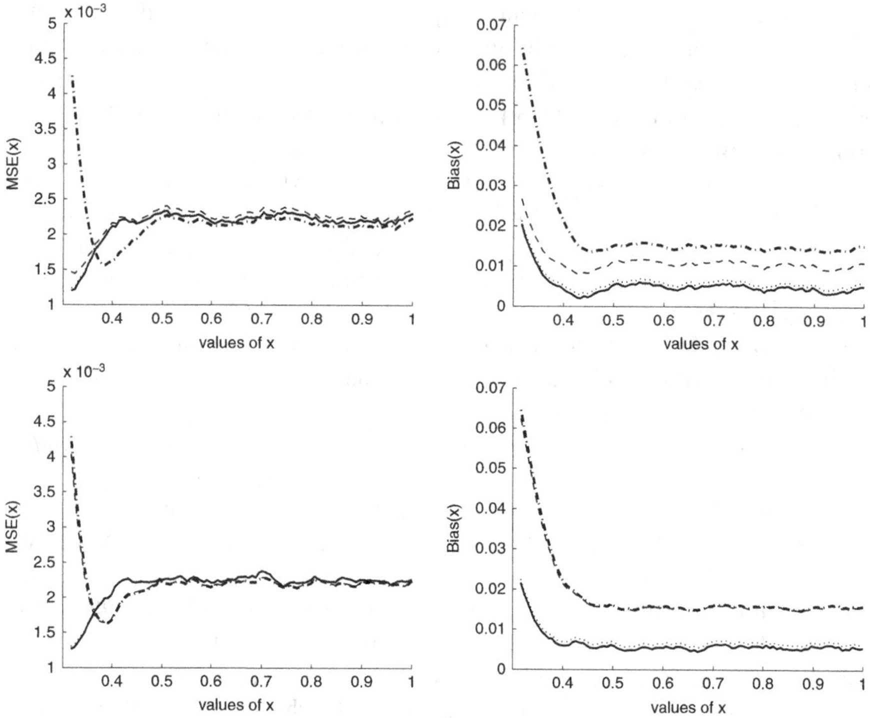

Fig. 1.10 Results for $k = [n^{0.5}]$. Bias (*right panels*) and MSE (*left panels*) of the estimates $\xi_{(n-k+1)/n,n}$ (*solid line*) and $\hat{\xi}_{(n-k+1)/n}$ (*dashdotted, dashed and dotted*, respectively, for $r = 0.5, 1, 1.5$) over 2000 MC simulations, sample size $n = 100$ (*top*) and $n = 1{,}000$ (*bottom*)

Case of a finite frontier: In the scenario of Paragraph 1.3.1.1 above, the derivative of the quantile tail function $U_x(t) := \xi_{1-1/t}(x)$ exists for all $t > x^{-2}$ and is given by $U_x'(t) = t^{\gamma_x - 1}/2 \in \mathrm{RV}_{\gamma_x - 1}$. Hence Condition (A2) holds in this case and the assumption of Theorem 4 that $hk^{3/2}\{nU_x'(n/k)\}^{-1} \to 0$ reduces to the simple condition that $n^{1/2}h \to 0$. For the computation of the smoothed estimator $\hat{\xi}_{(n-k+1)/n}$, we use here $h = n^{-r}$ with three values of $r \in \{0.5, 1, 1.5\}$: the choice $r = 0.5$ (for which $n^{1/2}h \not\to 0$ and $nh \to \infty$) corresponds rather to the asymptotic normality of ordinary smoothed quantiles, whereas the cases with $r > 0.5$ (for which $n^{1/2}h \to 0$ and $nh \not\to \infty$) correspond to the asymptotic normality of extreme smoothed quantiles. The Monte Carlo estimates of the Bias and MSE of the empirical partial frontier $\xi_{(n-k+1)/n,n}$ and the resulting smoothed three variants $\hat{\xi}_{(n-k+1)/n}$ are shown in Fig. 1.10 for $k = [n^{0.5}]$ and in Fig. 1.11 for $k = [n^{0.75}]$.

It may be seen that the smoothed extreme frontier $\hat{\xi}_{(n-k+1)/n}$ achieves its best performance in terms of MSE as h approaches $n^{-1/2}$ and behaves in this case (i.e., for $r \approx 0.5$) better than $\xi_{(n-k+1)/n,n}$, but not by much, as is to be expected from their asymptotic behavior. This appears to be true uniformly in x except for too

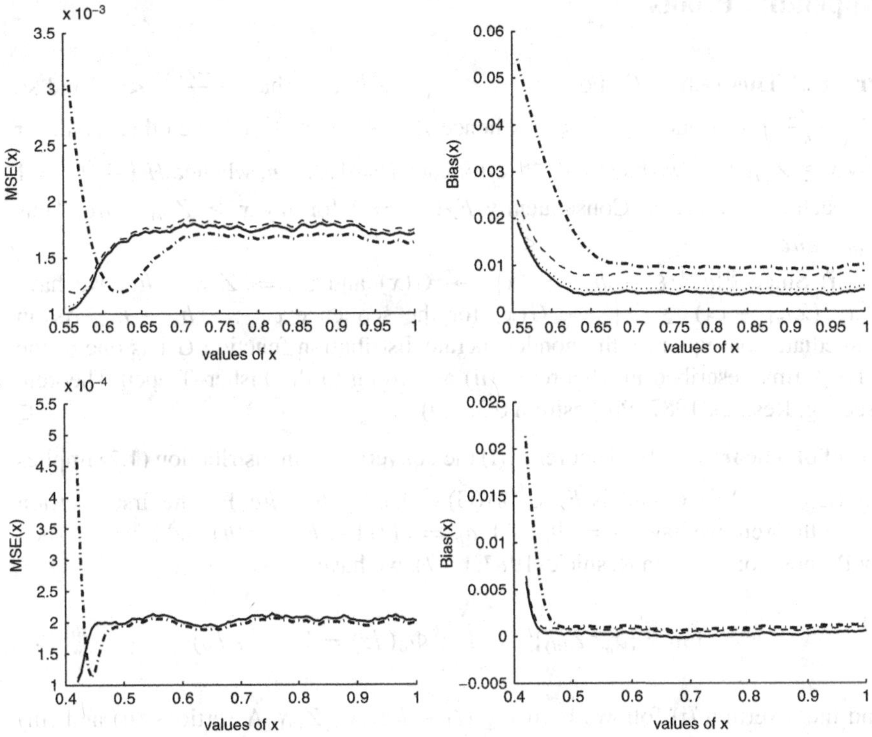

Fig. 1.11 As for Fig. 1.10 but with $k = [n^{0.75}]$

small values : a plausible explanation is that what is important when estimating $\xi_{(n-k+1)/n}(x)$ is not the sample size n itself, but the number of non-null transformed observations $Z_i^x = Y_i \mathbb{I}(X_i \leq x)$, which becomes negligible for x too small and so, smoothing fails in this case.

It is also interesting to note that when $\hat{\xi}_{(n-k+1)/n}$ attains its greatest gains in terms of MSE (as h approaches $n^{-1/2}$), its bias becomes however considerably larger, as is the case for ordinary smoothed quantiles.

Case of an infinite frontier: In the scenario of Paragraph 1.3.1.2 above, where $F_{Z^x} \in \mathrm{DA}(\Phi_{\rho_x})$ with $\rho_x = 1/\gamma_x = \beta$, Condition (A2) holds and the assumption of Theorem 4 that $hk^{3/2}\{nU_x'(n/k)\}^{-1} \to 0$ is equivalent to $hn^{-1/\beta}k^{3/2+1/\beta-1} \to 0$. When $\beta = 3$ as chosen above, this assumption reduces to $n^{1/12}h \to 0$ for $k = [n^{0.5}]$ and to $n^{7/24}h \to 0$ for $k = [n^{0.75}]$. Here, our Monte Carlo experiments are not in favor of the smoothed extreme quantiles and so we do not reproduce the figures for saving place : $\hat{\xi}_{(n-k+1)/n}$ performs at most as well as the sample version $\xi_{(n-k+1)/n,n}$ in terms of both Bias and MSE.

Appendix: Proofs

Proof of Theorem 1. (i) For any $x < Z_{(n)} + hc$, we have $\frac{x-Z_{(n)}}{h} < c$, and so $H\left(\frac{x-Z_{(n)}}{h}\right) < 1$ and $\hat{F}_Z(x) < 1$, whence $Z_{(n)} + hc \le \hat{\xi}_1$. On the other hand, for any $x \ge Z_{(n)} + hc$ we have $\frac{x-Z_i}{h} \ge c$ for each $i = 1, \ldots, n$, whence $H\left(\frac{x-Z_i}{h}\right) = 1$ for each $i = 1, \ldots, n$. Consequently $\hat{F}_Z(x) = 1$ for any $x \ge Z_{(n)} + hc$. Thus $Z_{(n)} + hc \ge \hat{\xi}_1$.

(ii) Since $\mathbb{P}[a_n^{-1}(\hat{\xi}_1 - b_n) \le x] \to G(x)$ and $\hat{\xi}_1 = Z_{(n)} + hc$, we have $\mathbb{P}[a_n^{-1}(Z_{(n)} - c_n) \le x] \to G(x)$ for the sequence $c_n := b_n - hc$. As an immediate consequence, the nondegenerate distribution function G has one of the three forms described in Theorem 1(ii) according to the Fisher–Tippett Theorem (see e.g. Resnick 1987, Proposition 0.3, p.9). □

Proof of Theorem 2. By Theorem 1(i) the convergence in distribution (1.2) implies $a_n^{-1}(Z_{(n)} - c_n) \xrightarrow{d} G$, that is $F_Z \in \text{DA(G)}$ with $c_n = b_n - hc$. For the first assertion of the theorem we have $G = \Phi_\rho$ with $a_n = (1/(1 - F_Z))^{-1}(n)$ and $c_n = 0$. Then by Proposition 2.1(i) in Resnick (1987, p.77), we have

$$\lim_{n\to\infty} \mathbb{E}\{a_n^{-1} Z_{(n)}\}^k = \int_{\mathbb{R}} z^k \Phi_\rho(dz) = \Gamma(1 - k/\rho)$$

and the assertion (i) follows by using $(\hat{\xi}_1 - hc) = Z_{(n)}$. Assertions (ii) and (iii) follow in the same way by an easy application of Proposition 2.1(ii)–(iii) of Resnick (1987). □

Proof of Theorem 3. First note that for any fixed integer $k \ge 0$ we have $k < n$ for all n large enough. Denote by $Z_{(n-k)}$ the $(k+1)$-th largest order statistic and let us show that

$$Z_{(n-k)} - hc < \hat{\xi}_{\frac{n-k}{n}} \le Z_{(n-k)} + hc \quad \text{for all } n \text{ large enough.} \tag{A.1}$$

For any $x \ge Z_{(n-k)} + hc$, we have $\frac{x-Z_{(i)}}{h} \ge c$ for each $i \le n-k$. Then $\hat{F}_Z(x) \ge (1/n)\sum_{i\le n-k} H\left(\frac{x-Z_{(i)}}{h}\right) = \frac{n-k}{n}$. Therefore $Z_{(n-k)} + hc \ge \hat{\xi}_{\frac{n-k}{n}}$. On the other hand, $\hat{F}_Z(Z_{(n-k)} - hc) = (1/n)\sum_{i<n-k} H\left(\frac{Z_{(n-k)}-Z_{(i)}}{h} - c\right) \le \frac{n-k-1}{n} < \frac{n-k}{n} = \hat{F}_Z(\hat{\xi}_{\frac{n-k}{n}})$, which implies $Z_{(n-k)} - hc < \hat{\xi}_{\frac{n-k}{n}}$. Now we can turn to the desired assertion of the theorem. If $a_n^{-1}(\hat{\xi}_1 - b_n) \xrightarrow{d} G$ we have $a_n^{-1}(Z_{(n)} + hc - b_n) \xrightarrow{d} G$ by Theorem 1(i). Then following Theorem 21.18 in van der Vaart (1998, p. 313), we obtain $a_n^{-1}(Z_{(n-k)} + hc - b_n) \xrightarrow{d} H$. Since

$$a_n^{-1}(Z_{(n-k)} + hc - b_n) - 2hca_n^{-1} < a_n^{-1}(\hat{\xi}_{\frac{n-k}{n}} - b_n) \le a_n^{-1}(Z_{(n-k)} + hc - b_n),$$

the desired conclusion follows immediately from the condition $a_n^{-1}h \to 0$. □

Proof of Theorem 4 Let $\sigma = (Z_{(n-k+1)} - Z_{(n-2k+1)})/\sqrt{2k}$. First, we know from Dekkers and de Haan (1989, Theorem 3.1) that $\sigma^{-1}(Z_{(n-k+1)} - \xi_\alpha)$ is asymptotically normal with mean zero and variance $2^{2\gamma+1}\gamma^2/(2^\gamma - 1)^2$, provided that $\alpha \to 1$, $n(1-\alpha) \to \infty$ and k is the integer part of $n(1-\alpha)$. We also know from Dekkers and de Haan (1989, Corollary 3.1) that $(Z_{(n-k+1)} - Z_{(n-2k+1)})/(n/2k)U'(n/2k) \xrightarrow{p} (2^\gamma - 1)/\gamma$ and so, it follows from the condition $h(2k)^{3/2}\{nU'(n/2k)\}^{-1} \to 0$ that $\sigma^{-1}h \xrightarrow{p} 0$. Thus we conclude by making use of (A.1) that $\sigma^{-1}(\hat{\xi}_{\frac{n-k+1}{n}} - \xi_\alpha)$ has the same asymptotic distribution as $\sigma^{-1}(Z_{(n-k+1)} - \xi_\alpha)$. We also have by (A.1) for all n sufficiently large,

$$(Z_{(n-k+1)} - Z_{(n-2k+1)}) - 2hc < \left(\hat{\xi}_{\frac{n-k+1}{n}} - \hat{\xi}_{\frac{n-2k+1}{n}}\right) < (Z_{(n-k+1)} - Z_{(n-2k+1)}) + 2hc.$$

Since $\sigma^{-1}h \xrightarrow{p} 0$, we get $h/(Z_{(n-k+1)} - Z_{(n-2k+1)}) \xrightarrow{p} 0$. Therefore $(\hat{\xi}_{\frac{n-k+1}{n}} - \hat{\xi}_{\frac{n-2k+1}{n}})/(Z_{(n-k+1)} - Z_{(n-2k+1)}) \xrightarrow{p} 1$, which completes the proof by using Slutsky's Lemma. □

Acknowledgements The authors thank an anonymous reviewer for his valuable comments which led to a considerable improvement of the manuscript. This research was supported by the French "Agence Nationale pour la Recherche" under grant ANR-08-BLAN-0106-01/EPI project.

References

Aragon, Y., Daouia, A., & Thomas-Agnan, C. (2005). Nonparametric frontier estimation: a conditional quantile-based approach. *Econometric Theory*, *21*, 358–389.

Azzalini, A. (1981). A note on the estimation of a distribution function and quantiles by a kernel method. *Biometrika*, *68*, 326–328.

Beirlant, J., Teugels, J., Goegebeur, Y., & Segers, J. (2004). Statistics of Extremes: Theory and Applications. Wiley Series in Probability and Statistics. Chichester: Wiley

Cazals, C., Florens, J-P., & Simar, L. (2002). Nonparametric frontier estimation: a robust approach, *Journal of Econometrics*, *106*, 1–25.

Daouia, A., Gardes, L., & Girard, S. (2009). Large Sample Approximation of the Distribution for Smooth Monotone Frontier Estimators, preprint, submitted for publication.

Daouia, A., Florens, J.-P., & Simar, L. (2010). Frontier estimation and extreme value theory. Bernoulli, *16*(4), 1039–1063.

Dekkers, A.L.M., & de Haan, L. (1989). On the estimation of extreme-value index and large quantiles estimation. *The Annals of Statistics*, *17*(4), 1795–1832.

Dekkers, A.L.M., Einmahl, J.H.J., & de Haan, L. (1989). A moment estimator for the index of an extreme-value distribution. *The Annals of Statistics*, *17*(4), 1833–1855.

Deprins, D., Simar, L., & Tulkens, H. (1984). Measuring labor inefficiency in post offices. In M. Marchand, P. Pestieau, & H. Tulkens (Eds.), *The performance of public enterprises: concepts and measurements* (pp. 243–267). Amsterdam: North-Holland.

Girard, S., & Jacob, P. (2004). Extreme values and kernel estimates of point processes boundaries. *ESAIM: Probability and Statistics*, *8*, 150–168.

Martins-Filho, C., & Yao, F. (2008). A smooth nonparametric conditional quantile frontier estimator. *Journal of Econometrics*, *143*, 317–333.

Nadaraya, E. A. (1964). Some new estimates for distribution functions. *Theory of Probability Applications*, *15*, 497–500.

Resnick, S. I. (1987). Extreme values, regular variation, and point processes. New York: Springer.

Segers, J. (2005). Generalized pickands estimators for the extreme value index. *Journal of Statististical Planning and Inference*, *128*, 381–396.

van der Vaart, A. W. (1998). *Asymptotic Statistics*. Cambridge Series in Statistical and Probabilistic Mathematics, 3. Cambridge: Cambridge University Press.

Wheelock, D. C., & Wilson, P. W. (2008). Non-parametric, unconditional quantile estimation for efficiency analysis with an application to Federal Reserve check processing operations. *Journal of Econometrics*, *145*(1-2), 209–225.

Chapter 2
Production Efficiency versus Ownership: The Case of China

Alice Shiu and Valentin Zelenyuk

Abstract In this study, we explore the pattern of efficiency among enterprises in China's 29 provinces across different ownership types in heavy and light industries and across different regions (coastal, central and western). We do so by performing a bootstrap-based analysis of group efficiencies (weighted and non-weighted), estimating and comparing densities of efficiency distributions, and conducting a bootstrapped truncated regression analysis. We find evidence of interesting differences in efficiency levels among various ownership groups, especially for foreign and local ownership, which have different patterns for light and heavy industries.

2.1 Introduction

Extraordinary changes have taken place in China over the past three decades since the adoption of the open door policy. These changes have been exemplified by those seen in China's industrial structure, especially in the radical moves toward non-state ownership. The corporatization of the state sector, the government's encouragement of merger and acquisition activity among state-owned enterprises (SOEs), and the dramatic development of the non-state sector with enormous foreign investments have dominated both the Chinese economy and political debate for the past decade.

A. Shiu (✉)
School of Accounting and Finance, The Hong Kong Polytechnic University, 1 Yuk Choi Road, Hung Hom, Kowloon, Hong Kong
e-mail: afshiu@inet.polyu.edu.hk

V. Zelenyuk
School of Economics and Centre for Efficiency and Productivity Analysis, The University of Queensland, 530, Colin Clark Building (39), St Lucia, Brisbane, QLD4072, Australia
e-mail: v.zelenyuk@uq.edu.au

I. van Keilegom and P.W. Wilson (eds.), *Exploring Research Frontiers in Contemporary Statistics and Econometrics*, DOI 10.1007/978-3-7908-2349-3_2,

The purpose of this chapter is to tackle the timeworn political debate about which type of ownership is more efficient in the Chinese economy and whether it depends on the industry (light or heavy) or the region (central, western or coastal). While the literature includes many studies of productivity in China (see the citations below), none have focused on the relative efficiency of various ownership types for both light and heavy industry combined. This is the issue we attempt to address in our study. Our particular focus is on foreign versus local ownership. While there is little doubt that private ownership should outperform state ownership on average, the situation is not so clear for foreign versus local ownership and whether it depends on the type of industry.

To achieve our goal, we use the most recent census data constructed for Chinese enterprises of different ownership types in 1995. Our methodological approach exploits recent developments in the area of efficiency analysis and is implemented in two stages. The first stage involves the estimation of efficiency scores for individual observations (each province in each type of industry) using the data envelopment analysis (DEA) estimator. In the next stage, we analyze the individual efficiency scores obtained in the first stage using three different methods. The first method is based on the analysis of densities of efficiency distributions for different ownership groups using a kernel density estimator and testing for their equalities using an adaptation of the Li (1996) test. The second method is based on the aggregation method of Färe and Zelenyuk (2003) and investigates group efficiency scores obtained as weighted averages, with the weights representing the economic importance of each observation. Statistical inferences for these group efficiency scores are made via bootstrapping techniques suggested by Simar and Zelenyuk (2007). The third method assumes more of a dependency structure and allows us to analyze the dependency of efficiency scores on hypothetical explanatory variables. Here, we use the truncated regression proposed by Simar and Wilson (2007) in which bootstrapping is used as a means of statistical inference to investigate how the conditional mean of efficiency scores is influenced by explanatory variables such as ownership and regional dummies, as well as by size. These methods yield interesting evidence of performance variations among ownership groups and regions. Remarkably, the pattern of performance for light industry is found to differ from that for heavy industry.

In common with the results of other studies, our results provide robust evidence confirming the expectation that non-state ownership is superior to state ownership in terms of the performance levels achieved. In addition, we confirm our prediction that foreign-owned firms in heavy industry perform distinctly better than their counterparts with other ownership types. Somewhat surprisingly, foreign ownership in light industry appears to be associated with *lower* efficiency, on average, than the other non-state ownership types we consider. This unexpected result can nevertheless be explained by the theory of technology diffusion/adoption, which can be traced back at least as far as the studies of Gerschenkron (1962) and Nelson and Phelps (1966).

Among our other findings, we present evidence of agglomeration effects that are pronounced in light industry but are not particularly marked in heavy industry.

Interestingly, we find no significant difference in average efficiency between light and heavy industries. Overall, apart from confirming a number of previous findings, our study sheds new light on the pattern of productivity in China that will be of interest to researchers and practitioners.

The remainder of this chapter is organized as follows. Section 2.2 briefly discusses our methodology and Sect. 2.3 provides a brief discussion of the data. Section 2.4 reports the empirical results in detail and Sect. 2.5 concludes the chapter.

2.2 Methodology

2.2.1 *Estimation of Efficiency (Stage 1)*

In the first stage of our analysis, we use the data envelopment analysis (DEA) estimator to obtain efficiency scores for each observation. This approach usually assumes that all *decision-making units* (DMUs) within a sample have *access to* the *same technology* for transforming a vector of N inputs, x, into a vector of M outputs, y.[1] We also assume that technology can be characterized by the *technology set*, T, as[2]

$$T = \{(x, y) \in \mathbb{R}_+^N \times \mathbb{R}_+^M : \ x \in \mathbb{R}_+^N \ \textit{can produce}\ y \in \mathbb{R}_+^M\} \qquad (2.1)$$

Note that while our approach requires that all DMUs have *access* to the *same technology*, it also allows for any DMU to be either on or away from the *frontier* of such technology. The distance from each DMU in T to the frontier of T is called the *inefficiency* of each DMU caused by endogenous or exogenous factors specific to that DMU. These endogenous factors could include internal economic incentives influenced by motivation systems, ownership structure, management quality, etc. Exogenous factors might include different demographic or geographic environments, regulatory policies, and so on. Our goal is to *estimate* such inefficiency and analyze its dependency on the hypothesized factors.

Technical efficiency for each DMU $j \in \{1, \ldots, n\}$ is measured using the Farrell (1957)/Debreu-type output-oriented technical efficiency measure

$$TE^j \equiv TE(x^j, y^j) = \max_\theta\{\theta : (x^j, \theta y^j) \in T\}. \qquad (2.2)$$

---

[1]The DEA was originally designed for firm-level analysis, but it has frequently been applied to more aggregated data; see, for example, Färe et al. (1994) and the more recent studies of Kumar and Russell (2002), Henderson and Russell (2005), Henderson and Zelenyuk (2006), and Badunenko et al. (2008).

[2]We assume that the standard regularity conditions of the neo-classical production theory hold (see Färe and Primont (1995) for details).

Obviously, the true T is unobserved, and so we replace it with its *DEA-estimate*, $\widehat{T}$, obtained through the following activity analysis model

$$\widehat{T} = \{(x, y) \in \mathbb{R}_+^N \times \mathbb{R}_+^M : \sum\nolimits_{k=1}^{n} z^k y_m^k \geq y_m, m = 1, \ldots, M,$$

$$\sum\nolimits_{k=1}^{n} z^k x_i^k \leq x_i, \quad i = 1, \ldots, N, \quad z^k \geq 0, \quad k = 1, \ldots n\}, \tag{2.3}$$

where $\{z^k : k = 1, \ldots, n\}$ are the intensity variables over which optimization (2.2) is made. Note that such $\widehat{T}$ is the smallest convex free disposal cone (in (x, y)-space) that contains (or 'envelopes') the input-output data.[3] In our discussions, we focus on the constant returns to scale (CRS) model only for several reasons. First, the CRS model (2.2) has greater discrimination power, making it capable of identifying more inefficiency than non-CRS models. Some of the inefficiency identified under the CRS model will be due to the scale effect (i.e., where a DMU is too small or too large), which will be tested at the second stage by including a proxy for scale. Second, the CRS model compares all DMUs evenly to the same cone, whereas for the non-CRS DEA estimator, a large proportion of DMUs are often in or near the flat regions of the estimated technology and so obtain high or perfect efficiency scores while being quite inefficient from an economic perspective. Third, the CRS model is a natural choice when aggregate (country- or region-level) data are used.

We choose the Farrell efficiency measure over others for two reasons that make it the most popular in practice. This measure has been shown to satisfy a set of attractive mathematical properties that are desirable in an efficiency measure.[4] Moreover, this estimator is fairly easy in terms of computation and allows for straightforward interpretation.

Note that the true efficiency scores from the Farrell measure are bounded between unity and infinity, where unity represents a perfect (technical or technological) *efficiency* score of 100%. On the other hand, $\left(1/TE^j\right)$ would represent the *relative %-level* of the *efficiency* of the jth DMU ($j \in \{1, \ldots, n\}$). By replacing T with $\widehat{T}$ in (2.2), we obtain the DEA estimator of TE^j under the assumptions of CRS, additivity, and free disposability. Applying this estimator will give estimates of the true efficiency scores, $\{TE^j : j = 1, \ldots, n\}$, which we denote as $\{\widehat{TE^j} : j = 1, \ldots, n\}$. These estimated efficiency scores have the same range as the true efficiency scores and, as in many other extreme-value type estimates, are subject to

[3]Alternatively, if we add $\sum_{k=1}^{n} z^k \leq 1$ and $\sum_{k=1}^{n} z^k = 1$ to equation (2.3), then we can model the non-increasing returns to scale (NIRS) or the variable returns to scale (VRS), respectively.

[4]These properties include various forms of *continuity*, (weak) monotonicity, commensurability, homogeneity, and (weak) indication for all technologies satisfying certain regularity conditions (see Russell (1990,1997) for details).

small-sample bias, which nevertheless vanishes asymptotically as the estimates are consistent with their true counterparts.[5]

2.2.2 *Analysis of Efficiency Distributions (Method 1 of Stage 2)*

The aim of the *second* stage of the analysis is to study the dependency of the efficiency scores obtained in the first stage on DMU-specific factors such as ownership structure, regional location, size, etc.

The starting point of our second-stage analysis is to explore the efficiencies within and between groups that might theoretically represent different sub-populations in the population as a whole. For example, state-owned firms have different incentives to other firms which are likely to be reflected in the efficiency distribution of state-owned firms relative to other firms. In particular, we first analyze the *distributions* of efficiency within various groups. Here, we start with estimation and visualization of the densities of corresponding distributions using the *kernel density estimator*. For this, we use the Gaussian kernel, Silverman (1986) reflection method (around unity), to take into account the bounded support of efficiency measure, and Sheather and Jones (1991) method for bandwidth selection. We then apply a version of the Li (1996) test (adapted to the DEA context by Simar and Zelenyuk (2006)) to test the equality of efficiency distributions between various groups of interest.

2.2.3 *Analysis of Aggregate Efficiency Scores (Method 2 of Stage 2)*

We proceed to analyze the various groups by testing the equality of group (aggregate) efficiencies, which is estimated using the weighted and non-weighted averages of the individual efficiency scores for each group. Because the weights used for averaging might be critical here, they must be chosen on the basis of some (more-or-less) objective criterion. We use the weights derived from economic optimization by Färe and Zelenyuk (2003) which were extended to the sub-group case by Simar and Zelenyuk (2007). In summary, our (weighted) group efficiency score for group $l (l = 1, \ldots, L)$ is estimated as

$$\widehat{\overline{TE}}^{l} = \sum_{j=1}^{n_l} \widehat{TE}^{l,j} S^{l,j}, \quad l = 1, \ldots L. \tag{2.4}$$

---

[5]See Korostelev et al. (1995) and Park et al. (2010) for proof of consistency and rates of convergence of the DEA estimator under CRS, and other statistical properties and required assumptions. Also see Kneip et al. (1998, 2008) for related results on VRS.

where the weights are

$$S^{l,j} = py^{l,j}/p\sum\nolimits_{j=1}^{n_l} y^{l,j}, \qquad j = 1,\dots n_l. \tag{2.5}$$

in which p is the vector of output prices. For convenience, we would present the *reciprocals* of the estimated group efficiency scores, i.e., (and the corresponding confidence intervals) to give them meaning in percentage terms.

To make statistical inferences based on these group efficiency scores, we use the bootstrap-based approach suggested by Simar and Zelenyuk (2007); readers are referred to the same study for further details of this method. In summary, the statistic used for testing the null hypothesis that the aggregate efficiencies for any two groups, e.g., A and Z, are equal (i.e., $H_0 : \overline{TE}^A = \overline{TE}^Z$) is given by the relative difference (RD) statistic:

$$\widehat{RD}_{A,Z} = \widehat{\overline{TE}}^A / \widehat{\overline{TE}}^Z \tag{2.6}$$

The null hypothesis will be rejected (at certain level of confidence) in favor of $H_1 : \overline{TE}^A > \overline{TE}^Z$ if $\widehat{RD}_{A,Z} > 1$ (or $H_2 : \overline{TE}^A < \overline{TE}^Z$ if $\widehat{RD}_{A,Z} < 1$) and the bootstrap-estimated confidence interval of $\widehat{RD}_{A,Z}$ does not overlap with unity.

2.2.4 *Regression Analysis of Determinants of Efficiency (Method 3 of Stage 2)*

The last method used in our investigation involves the application of regression analysis to study the dependency between efficiency scores and some expected explanatory variables. Here, we assume and test the following specification

$$TE^j \approx a + Z_j\,\delta + \varepsilon_j, \quad j = 1,\dots,n, \tag{2.7}$$

where a is the constant term, ε_j is statistical noise, and Z_j is a (row) vector of observation-specific variables for DMU j that we expect to influence DMU efficiency score, TE^j, defined in (2.2), through the vector of parameters δ (common for all j) that we need to estimate.

For some time, a practice commonly adopted in the DEA literature was to estimate model (2.7) using the Tobit-estimator. However, Simar and Wilson (2007) illustrate that this approach would be incorrect here and instead propose an approach based on a bootstrapped *truncated regression*, showing that it performs satisfactorily in Monte Carlo experiments. We follow their approach (specifically, their "Algorithm 2") and instead of using the unobserved regressand in (2.7), TE^j, use its bias-corrected estimate, $\widehat{TE}^j_{\mathrm{bc}}$, which is obtained using the heterogeneous parametric bootstrap they propose. Note that because both sides of (2.7) are bounded by unity, the distribution of ε_j is restricted by the condition $\varepsilon_j \geq 1 - a - Z_j\delta$. To simplify the estimation process, we follow Simar and Wilson (2007) by assuming that this distribution is a truncated normal distribution with a mean of zero, unknown variance, and a (left) truncation point determined by $\varepsilon_j \geq 1 - a - Z_j\,\delta$. Formally, our econometric model is given by

$$\widehat{TE}^{j}_{\text{bc}} \approx a + Z_j\,\delta + \varepsilon_j, \quad j = 1, \ldots, n, \tag{2.8}$$

where

$$\varepsilon_j \sim N(0, \sigma_\varepsilon^2), \text{ such that} \quad \varepsilon_j \geq 1 - a - Z_j\delta, \quad j = 1, \ldots, n. \tag{2.9}$$

We then use our data to estimate the model shown in (2.8)–(2.9) by maximizing the corresponding likelihood function with respect to $(a, \delta, \sigma_\varepsilon^2)$. To obtain the bootstrap confidence intervals for the estimates of parameters $(a, \delta, \sigma_\varepsilon^2)$, we use the *parametric* bootstrap for regression that incorporates information on the parametric structure (2.7) and the distributional assumption (2.9). For the sake of brevity, we refer readers to Simar and Wilson (2007) for the details of the estimation algorithm.

2.3 Data

The data used in this chapter are drawn from the *Third National Industrial Census of the People's Republic of China* conducted by the State Statistical Bureau in 1995, which is the latest census for which statistics have been put together and published. The data provided in the census are the only industry-level data available that are categorized by type of ownership. Specifically, the census provides cross-sectional data for Chinese enterprises divided into four ownership types that are aggregated at the *province* level (29 provinces) for light and heavy industries in 1995. The four types of ownership are: (a) state-owned enterprises (SOEs); (b) foreign-funded enterprises (FFEs); (c) township-owned enterprises (TOEs); and (d) collectively-owned enterprises (COEs). Given these data, we have 8 'representative' DMUs for each of the 29 provinces in China: SOEs, FFEs, TOEs, and COEs in the light and heavy industries, respectively.

A brief explanation of the industry sectors is warranted here. "Light industry" refers to the group of industries that produce consumer goods and hand tools. It consists of two categories distinguished from each other according to the materials used. The first category include industries that use farm products as materials while the other category includes industries that use non-farm products as materials.[6]

[6]Some examples of the first category of light industries are food and beverage manufacturing, tobacco processing, and textiles and clothing, and some examples of the second category are the manufacturing of chemicals, synthetic fibers, chemical products, and glass products.

"Heavy industry" refers to industries that produce capital goods and provide materials and technical bases required by various sectors of the national economy.[7] The level of competition among light-industry firms is generally more severe than that among heavy industry participants because there are usually more firms in the former group. Also, because most light-industry firms are non-SOEs, they face hard-budget constraints and are fully responsible for their profits and losses. On the other hand, because most heavy-industry firms are SOEs which are larger and fewer in number, the level of competition between such firms is usually lower than it is among light-industry firms.

To construct the constant returns to scale (CRS) output-oriented activity analysis model for the DEA estimator in the first stage, we use *three inputs* (i.e., total wage, the net value of fixed assets, and the value of intermediate inputs) and *one output* (the gross industrial output of each type of ownership in each province). Some descriptive statistics and a brief discussion of the data are provided in the appendix. Further details can be found in two studies conducted by Shiu (2000, 2001).

2.4 Main Results

2.4.1 Analysis of Densities and Means for Light Industry

After obtaining the DEA estimates of efficiency scores, we use the kernel density estimator to approximate the distributions of the individual efficiency scores for the four ownership groups in each of the light and heavy industry sectors. Statistical tests for the equality of distributions suggested by Li (1996) (and adapted to the DEA context by Simar and Zelenyuk (2006)) are used to test for differences in distributions amongst the ownership groups. Figure 2.1 shows the (estimated) densities of the distributions of the estimated individual efficiency scores for each ownership group in the light industry sector. The estimated densities seem to be relatively divergent among groups. Interestingly, the only ownership group that has a density with an estimated mode of unity is the township-owned enterprises (TOEs). Intuitively, this means that for TOEs, the highest frequency at which the level of efficiency is observed is where one would expect it to be for highly competitive firms: at the 100% level of efficiency. Other groups have estimated modes that are not at unity but are instead at some level of inefficiency, which we view as evidence of some degree of 'pathological' inefficiency. The state-owned enterprises (SOE) group has the most 'inefficient' mode (around 2, i.e., about 50% efficient), making

[7] Heavy industry consists of three branches distinguished according to the purpose of production or how the products are used. They include (1) the mining, quarrying and logging industry that involves the extraction of natural resources; (2) the raw materials industry, which provides raw materials, fuel and power to various sectors of the economy; and (3) the manufacturing industry, which processes raw materials.

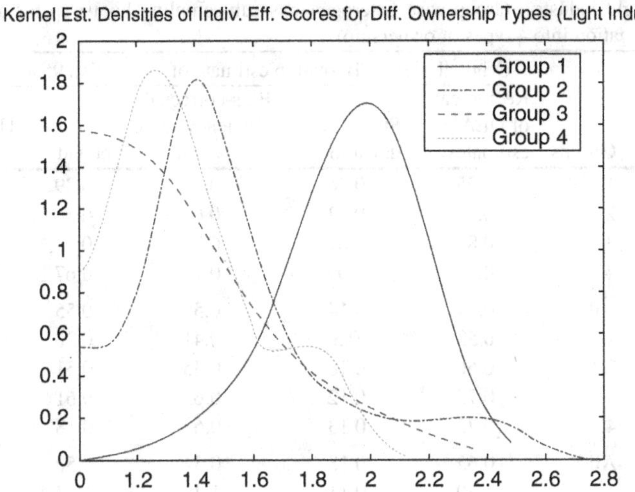

Fig. 2.1 Estimated Densities of Individual Efficiency Scores for Ownership Groups in Light Industry. *Notes*: (i) Groups 1, 2, 3, and 4 refer to SOEs, FFEs, TOEs, and COEs, respectively. (ii) Vertical axis refers to (estimated) probability density function of the distribution of the efficiency scores and horizontal axis refers to efficiency scores

Table 2.1 Simar-Zelenyuk-adapted for DEA Li-test for equality of efficiency distributions across different types of ownership

Null hypothesis	Light industry		Heavy industry		Both industries	
	Test statistic	p-value	Test statistic	p-value	Test statistic	p-value
$f(eff_{SOE}) = f(eff_{FFE})$	10.61	0.00	5.83	0.00	16.6	0.00
$f(eff_{SOE}) = f(eff_{TOE})$	12.24	0.00	0.47	0.47	18.2	0.00
$f(eff_{SOE}) = f(eff_{COE})$	12.39	0.00	0.21	0.76	13.52	0.00
$f(eff_{FFE}) = f(eff_{TOE})$	2.32	0.01	1.24	0.07	1.01	0.08
$f(eff_{FFE}) = f(eff_{COE})$	1.19	0.06	4.52	0.00	−0.13	0.85
$f(eff_{TOE}) = f(eff_{COE})$	0.55	0.34	1.23	0.06	1.32	0.56

Notes: All calculations are done by authors in Matlab using, after adopting from programs used for the work in Simar and Zelenyuk (2006)

it radically different from other groups and the least efficient group. Columns 2 and 3 of Table 2.1 present the results of tests for the equality of distributions between all dyads of ownership groups in the light industry sector; we reject equality for most of the comparisons at the 95% confidence level. The exceptions are the efficiency distributions of foreign-funded enterprises (FFEs) and TOEs versus collectively-owned enterprises (COEs). We also reject equality for FFEs versus COEs at the 10% level (est. p-value of 0.06).

Given the evidence of different efficiency distributions among ownership groups, a further issue that arises is whether this divergence is due to differences in group aggregate efficiency scores and whether these differences are statistically

Table 2.2 The light industry: Group-wise heterogeneous sub-sampling bootstrap for aggregate efficiencies (aggregation into 4 types of ownership)

	Groups	Original: Reciprocal of DEA estimates	Bootstrap estimate of: Standard deviation	Bootstrap estimate of: Bias-corrected efficiency score (reciprocal)	Est. 95% Conf. Int.: Lower bound	Est. 95% Conf. Int.: Upper bound
Weighted (output shares) group efficiencies	1	0.55	0.26	0.43	0.39	0.55
	2	0.71	0.13	0.62	0.57	0.74
	3	0.83	0.07	0.75	0.71	0.88
	4	0.79	0.09	0.71	0.67	0.83
	All	0.69	0.14	0.59	0.55	0.70
Non-weighted group efficiencies	1	0.53	0.3	0.41	0.36	0.53
	2	0.66	0.18	0.55	0.51	0.71
	3	0.75	0.12	0.65	0.61	0.80
	4	0.72	0.13	0.62	0.58	0.77
	All	0.65	0.18	0.54	0.50	0.66
RD statistics for comparing groups in terms of weighted average efficiencies	1 vs. 2[a]	1.30	0.11	1.48	1.23	1.62
	1 vs. 3[a]	1.50	0.18	1.80	1.38	2.00
	1 vs. 4[a]	1.43	0.15	1.70	1.35	1.87
	2 vs. 3[b]	1.16	0.08	1.23	1.02	1.33
	2 vs. 4[b]	1.10	0.06	1.16	1.01	1.24
	3 vs. 4	0.95	0.05	0.93	0.85	1.06
RD statistics for comparing groups in terms of non-weighted average efficiencies	1 vs. 2[a]	1.26	0.11	1.40	1.13	1.54
	1 vs. 3[a]	1.43	0.16	1.67	1.30	1.84
	1 vs. 4[a]	1.38	0.14	1.60	1.27	1.74
	2 vs. 3[c]	1.14	0.08	1.21	0.98	1.31
	2 vs. 4[c]	1.10	0.07	1.15	0.96	1.25
	3 vs. 4	0.96	0.05	0.95	0.84	1.08

Notes: Groups 1, 2, 3, and 4 refer to SOEs, FFEs, TOEs, and COEs, respectively. Also, [a], [b] and [c] indicates the rejection of the null hypothesis of $H_0 : \overline{TE}^A = \overline{TE}^Z$ at 1%, 5% and 10% significance levels, respectively

For convenience, we present *reciprocals* of estimated efficiency scores, i.e., $1/\widehat{\overline{TE}^l}$, $l = 1, 2, 3, 4$ (and the corresponding confidence intervals) so that they have percentage meaning. All calculations are done by authors in Matlab using, after adopting from programs used for the work in Simar and Zelenyuk (2007)

significant The upper part of Table 2.2 lists the weighted efficiency scores for each light industry ownership group in the 29 provinces in 1995. The aggregate efficiency scores are calculated using Färe-Zelenyuk weights, with bias corrected and confidence intervals estimated on the basis of the Simar and Zelenyuk (2007) group-wise-heterogeneous bootstrap-based approach. The second column indicates the ownership groups. The numbers in the 3 and 5 columns represent the *reciprocals* of the original DEA efficiency scores and of the bias-corrected efficiency scores, respectively. (Reciprocals are taken for convenience to show the percentage meaning of the efficiency scores.) The last two columns show the lower and upper bounds of the 95% confidence interval.

The results in the upper part of Table 2.2 indicate that SOE performance is different from non-SOE performance. A relatively large estimated bias in the aggregates of efficiency scores is found among all ownership groups, especially for SOEs (0.55 and 0.43). This indicates that in the light industry sector, the technical efficiency of SOEs varies widely across the provinces.

Furthermore, bootstrap-based tests of the equality of aggregate efficiencies are employed to test for pair-wise comparisons of the aggregate efficiencies of the various sub-groups (see the lower part of Table 2.2). The relative difference (RD) statistics computed for the DEA and bias-corrected aggregate efficiency scores are shown in the third and fifth columns, respectively. If the RD statistic for group A versus group Z is greater than 1 and the confidence interval does not overlap with 1, then the null hypothesis that the aggregate efficiencies of the two groups are equal is rejected in favor of the alternative hypothesis that the aggregate efficiency of group A is worse than that of group Z.[8] The RD statistics suggest that SOEs are operated in a significantly (at the 1% level) less efficient manner than all the other groups. This finding supports the results obtained in our distributional analysis and can be explained by the fact that SOEs are often ill-equipped to meet their business objectives as they tend to use out-of-date capital equipment and usually have no funding available to them for technological upgrades (for more discussion, see Groves et al. 1994; Weitzman and Xu 1994; Zheng et al. 1998; Zhang and Zhang 2001; Dong and Tang 1995; Lin et al. 1998; Huang et al. 1999; Wu 1996, 1998).

Regarding the performance of non-SOEs, it is interesting to find that in the light industry sector, FFEs perform significantly *less* efficiently than COEs and TOEs (at the 5% level for weighted averages and at the 10% level for non-weighted averages). One possible explanation for this result is that the network of bureaucratic restrictions adversely affecting the competitiveness of FFEs offset the benefits gained from the government's preferential policies for foreign investors. Examples include high-profile administrative intervention in the operation of FFEs, the levying of miscellaneous fees of an ambiguous nature, and the imposition of stringent policies. (For more discussion, see ACC 1998; Melvin 1998; Weldon and Vanhonacker 1999; Transparency International 2001). These issues could lead to higher transaction costs being incurred in FFE operations and thereby cancel out certain competitive advantages enjoyed by FFEs over local firms (see, for example, Yeung and Mok 2002). Other reasons that may account for the lower level of efficiency in FFE operations include the large initial investment required and the steep learning curve for foreign investors (e.g., see Wei et al. 2002).

---

[8] E.g., the RD-statistic for comparing the weighted average efficiency scores for groups 1 and 2 was estimated as $\widehat{\overline{TE}^1}/\widehat{\overline{TE}^2} = 1.30$, meaning that group 1 is less efficient than group 2, and this difference is significant, since 95% confidence interval is [1.23, 1.62], not overlapping with 1.

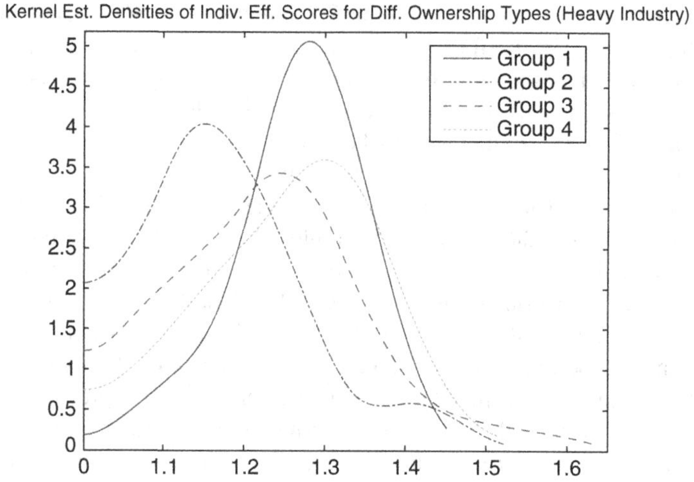

Fig. 2.2 Estimated Densities of Individual Efficiency Scores for Ownership Groups in Heavy Industry. *Notes*: (i) Groups 1, 2, 3, and 4 refer to SOEs, FFEs, TOEs, and COEs, respectively. (ii) Vertical axis refers to (estimated) probability density function of the distribution of the efficiency scores and horizontal axis refers to efficiency scores

2.4.2 Analysis of Densities and Means for Heavy Industry

Figure 2.2 shows the (estimated) densities of individual efficiency distributions of the four ownership groups for the heavy industry sector. The densities appear to be more tightly grouped in the heavy industry sector than those observed for the light industry sector, other than in the case of FFEs, for which we see a clear difference in the density of efficiency relative to that of the other groups. The SOEs group again has less of its distributional mass close to unity, while the FFEs group has more of its distribution close to unity than the other groups. Columns 4 and 5 of Table 2.1 formally support these observations via tests for the equality of distributions between the four groups in the heavy industry sector. Note that the overall situation in the heavy industry sector is somewhat different from what we have seen for the light industry sector. The efficiency distributions cannot be statistically distinguished from each other, the sole exception being the FFEs, for which the distribution appears to be significantly different from those of all the other groups. We also observe significance at the 10% level for TOEs versus COEs (which are not significantly different from each other in the light industry sector).

The results reported in the upper part of Table 2.3 also show that in comparison with the light industry ownership groups, all the ownership groups in the heavy industry sector have relatively small aggregate inefficiency scores and an (absolutely and relatively) lower level of estimated bias. These results suggest that performance varies to a lesser degree among ownership types in the heavy industry sector. This could be explained by the fact that heavy industry operations are more stable

Table 2.3 The heavy industry: Group-wise heterogeneous sub-sampling bootstrap for aggregate efficiencies (aggregation into 4 types of ownership)

	Groups	Original	Bootstrap estimate of		Est. 95% Conf. Int.	
		DEA estimates	Standard deviation	Bias-corrected efficiency score (reciprocal)	Lower bound	Upper bound
Weighted (output shares) group efficiencies	1	0.81	0.06	0.75	0.70	0.82
	2	0.88	0.04	0.83	0.79	0.87
	3	0.87	0.04	0.81	0.78	0.89
	4	0.83	0.06	0.76	0.72	0.85
	All	0.83	0.04	0.77	0.73	0.83
Non-weighted group efficiencies	1	0.79	0.07	0.74	0.68	0.80
	2	0.86	0.04	0.81	0.78	0.86
	3	0.83	0.05	0.75	0.71	0.82
	4	0.79	0.07	0.71	0.67	0.81
	All	0.82	0.04	0.75	0.72	0.81
RD statistics for comparing groups in terms of weighted average efficiencies	1 vs. 2	1.09	0.06	1.09	0.98	1.21
	1 vs. 3	1.08	0.06	1.08	0.97	1.18
	1 vs. 4	1.03	0.05	1.00	0.90	1.09
	2 vs. 3	0.99	0.04	0.98	0.90	1.07
	2 vs. 4	0.95	0.05	0.91	0.83	1.02
	3 vs. 4	0.95	0.04	0.93	0.85	1.03
RD statistics for comparing groups in terms of non-weighted average efficiencies	1 vs. 2[c]	1.08	0.06	1.10	0.99	1.21
	1 vs. 3	1.04	0.05	1.01	0.91	1.10
	1 vs. 4	1.00	0.06	0.95	0.84	1.04
	2 vs. 3[c]	0.96	0.04	0.91	0.83	1.01
	2 vs. 4[a]	0.92	0.06	0.85	0.76	0.98
	3 vs. 4	0.97	0.04	0.94	0.86	1.03

Notes: Groups 1, 2, 3, and 4 refer to SOEs, FFEs, TOEs, and COEs, respectively. Also, [a], [b] and [c] indicates the rejection of the null hypothesis of $H_0 : \overline{\overline{TE}}^A = \overline{\overline{TE}}^Z$ at 1%, 5% and 10% significance levels, respectively

For convenience, we present *reciprocals* of estimated efficiency scores, i.e., $1/\widehat{\overline{TE}^l}, l = 1, 2, 3, 4$ (and the corresponding confidence intervals) so that they have percentage meaning. All calculations are done by authors in Matlab using, after adopting from programs used for the work in Simar and Zelenyuk (2007)

than operations in the light industry sector, which is more dynamic and features larger numbers of firms breaking through and firms lagging behind thereby causing more variation in efficiency. In addition, because heavy industry is more capital-intensive in nature and light industry is more labor-intensive, greater automation in the production process leads to less human-driven inefficiency (such as human mistakes and shirking on the job) in the heavy industry sector. Firms operating in heavy industries therefore tend to operate in a relatively similar manner and are more similar in terms of performance, both of which contribute to less variation in efficiency estimates and, in turn, less estimated bias.

Although it has long been held that SOEs are less efficient than their non-state owned counterparts, our results from the analysis of densities and aggregate efficiencies do not provide strong support for this view in the case of the *heavy* industry sector. Specifically, a comparison of *weighted* aggregate efficiencies between the heavy industry groups using RD statistics indicates *no* statistical difference between them. This result could be attributed to the high level of automation in production activities in the heavy industry sector, a factor which has been discussed in the previous paragraph.

A similar test for the non-weighted efficiency scores confirms the insignificance of the differences between these group efficiencies, other than for the FFEs, which appear to be more efficient than SOEs and TOEs (at about the 10% significance level) and COEs (at about the 1% significance level). This is consistent with our analysis of the distributions for these groups, but contrasts with the results obtained for the light industry sector, where we find that FFEs perform significantly *less* efficiently than COEs and TOEs, while SOEs perform significantly worse than all of the other groups. We explain this difference between the industry sectors in more detail later in this chapter.

2.4.3 Truncated Regression Analysis

The regression analysis method we employ is not simply a generalization of the above analysis because it imposes a particular structure on the dependency between the efficiency of a DMU and the hypothesized explanatory variables. Moreover, the dependent variable (i.e., the efficiency score), does not account for the economic weight (e.g., size) of the observations. Nevertheless, this analysis complements the methods used above in a number of very important respects. In particular, it allows for inferences to be drawn about different factors that simultaneously influence efficiency scores by focusing on the (marginal) effect of each variable. One additional advantage of this approach is that it allows for the effects of continuous variables to be investigated.

Our empirical specification shown on the right-hand side of regression equation (2.8) includes the intercept, dummy variables and one continuous variable. The first dummy variable is the industry indicator (1 for light industry and 0 for heavy industry). The next three dummies – D2, D3, and D4 – represent the DMU ownership type and take the value of 1 if the observation belongs to a group of FFEs, TOEs, and COEs, respectively. Thus, for the sake of convenience in testing, the group of SOEs is taken as the base and so the coefficients on D2, D3, and D4 would estimate the difference in effects between the corresponding group (e.g., FFEs for D2) and the group of SOEs. For example, a negative coefficient on D2 would suggest evidence that FFEs introduce *improvements* relative to SOEs, on average.

The next two dummies – D5 and D6 – represent the regions and are assigned the value of 1 if the observation belongs to the coastal and central regions, respectively.[9] That is, the coefficients on each of these dummies will estimate the difference in effects between their region (e.g., coastal) and the western region, which is taken as the base. The continuous variable on the right-hand side of the regression model is used to control for the size effects (measured as the logarithm of output) of DMUs. The size effect variable is expected to capture at least part of the *agglomeration* effect of the province: the larger the gross output of a particular type of firm in a province in a given industry, the higher we expect the efficiency level to be for this type of ownership. The agglomeration effect is expected to have a positive influence on efficiency for at least two reasons. First, there is a spillover effect derived from the activities of firms that are in the same general industry sector (light or heavy) but are not direct competitors (e.g., shoemakers versus textile producers, etc.). Second, there is also a competition effect between firms producing the same products that is expected to encourage firms to strive for greater efficiency. We expect both effects (the spillover and competition effects) to be 'proxied' by this size control variable, but unfortunately we cannot decompose it into its two components in our data or results because of the aggregate nature of our data.

The results of our bootstrapped (truncated) regression analysis with DEA are presented in Table 2.4.[10] We run several specifications to test the robustness of our conclusions. The results confirm our previous findings, but also shed some additional light on the issue under study. We see consistently strong evidence for the argument that at an aggregate level, non-SOEs of all types of ownership have significantly higher efficiency levels than their SOE counterparts. This evidence is robust in that it is confirmed by all the regression specifications we run. While this result is also consistent with those of many studies and is therefore not surprising, we also provide some interesting new results.

Turning to the pooled models (models 1–4) in which we consider both industries under the same frontier, the greatest efficiency improvement over that of SOEs comes from TOEs and is followed in turn by FFEs and COEs.

The size effect in all four models is found to be significant such that larger output leads to a better (smaller, i.e., closer to unity) efficiency score, on average. This finding supports the hypothesis of a positive spillover effect on efficiency. That is,

---

[9]We follow the categorization used by the State Planning Commission of China: (1) the *Coastal region*, which includes Beijing, Tianjin, Heibei, Liaoning, Shandong, Shanghai, Zhejiang, Jiangsu, Fujian, Guangdong, Hainan, and Guangxi; (2) the *Central region*, which includes Shanxi, Inner Mongolia, Jilin, Heilongjiang, Anhui, Jiangxi, Henan, Hunan, and Hubei; and (3) the *Western region*, which includes Sichuan, Yunnan, Guizhou, Shaanxi, Gansu, Qinghai, Ningxia, Tibet, and Xinjiang.

[10]The significance tests are based on bootstrapped confidence intervals using Algorithm 2 of Simar and Wilson (2007), with 1,000 replications for the bootstrap bias correction of the DEA estimates and 2,000 replications for the bootstrapping of the regression coefficients.

Table 2.4 Result of truncated regression analysis for explaining inefficiency level

	Interpretation of variables	Model 1	Model 2	Model 3	Model 4	Model 5	Model 6	Model 7	Model 8	Model 9	Model 10
Intercept	Constant	2.96^{b}	2.96^{b}	2.96^{b}	2.94^{b}	3.33^{b}	3.29^{b}	1.61^{b}	1.61^{b}	2.45^{b}	1.36^{b}
Industry	Industry Dummy (1 if Light, 0 o.w.)	0.00	–	−0.04	–	Light	Light	Heavy	Heavy	Light	Heavy
D1	Ownership Dummy (1 if SOE, 0 o.w.)	–	–	–	–	–	–	–	–	0.84^{b}	0.25^{b}
D2	Ownership Dummy (1 if FFE, 0 o.w.)	-0.73^{b}	-0.72^{b}	-0.73^{b}	-0.72^{b}	-0.85^{b}	-0.84^{b}	-0.26^{b}	-0.25^{b}	–	–
D3	Ownership Dummy (1 if TOE, 0 o.w.)	-0.90^{b}	-0.89^{b}	-0.89^{b}	-0.88^{b}	-1.28^{b}	-1.27^{b}	-0.13^{b}	-0.13^{b}	-0.42^{b}	0.12^{b}
D4	Ownership Dummy (1 if COE, 0 o.w.)	-0.69^{b}	-0.69^{b}	-0.68^{b}	−0.68	-1.06^{b}	-1.05^{b}	-0.04^{c}	-0.04^{c}	-0.21^{b}	0.21^{b}
D5	Region Dummy (1 if Coastal, 0 o.w.)	0.05	0.05	–	–	0.09	–	0.00	–	–	–
D6	Region Dummy (1 if Central, 0 o.w.)	−0.04	−0.04	–	–	−0.04	–	0.01	–	–	–
Log(y)	Measure of Size	-0.14^{b}	-0.14^{b}	-0.14^{b}	-0.14^{b}	-0.20^{b}	-0.19^{b}	-0.04^{b}	-0.04^{b}	-0.19^{b}	-0.04^{b}
σ^2	Variance of the error term	0.08^{b}	0.08^{b}	0.08^{b}	0.08^{b}	0.10^{b}	0.11^{b}	0.01^{b}	0.01^{b}	0.11^{b}	0.01^{b}

Notes: Dependent variable is "efficiency score" (see (2.8)–(2.9) in the text). Note that ^{c} and ^{b} indicate *significance* at α being 5% and 10%, respectively. Significance tests are based on bootstrapped confidence intervals, using Algorithm 2 of Simar and Wilson (2007), with 1,000 and 2,000 bootstrap replications for bias correction and for confidence intervals, respectively. All calculations are done by authors in Matlab using code of Valentin Zelenyuk, which adopted some earlier codes of Leopold Simar the more activities (total output) performed by a particular type of enterprise in a certain province, the higher the efficiency level is expected to be for that type of enterprise. Notably, the coefficient of the industry dummy is insignificant (and near zero) in Model 1, so we drop it from Model 2 and, as expected, observe almost no change in the estimates. Interestingly, the coefficients on the regional dummies are insignificant in both Model 1 and Model 2, so we drop these dummies from Model 3 and again see almost no change in the coefficients. In Model 4, we drop both the industry dummy and the regional dummies and the coefficients remain almost the same as in the previous three models. These results suggest that, at least on this aggregate level, neither type of industry nor location has a real effect on the level of efficiency. This finding is contrary to the conventional expectation, at least for the coastal region versus the western or even the central region.[11]

More interesting results are revealed when we consider each industry separately. Models 5, 6, and 9 consider light industry alone, while models 7, 8, and 10 consider heavy industry in isolation. There is no qualitative change in most of the results. The region dummies remain insignificant (and almost zero for heavy industry). However, note that the size effect is much more pronounced now for light industry and is much less pronounced in the heavy industry sector relative to what we observed in the pooled models. This suggests that although the agglomeration effect is present in the heavy industry sector, it is much less pronounced than it is in the light industry environment.

Also note that in the heavy industry context, the largest improvement on state ownership comes from FFEs, while the coefficient on the dummy representing the efficiency difference between COEs and SOEs is barely significant. (Recall that in the foregoing analysis, we could not confidently reject the differences between the aggregate group efficiency scores for heavy industry.) On the other hand, we find that in light industry, FFEs make the smallest improvement relative to SOEs (smaller than the other types of ownership), so we use Model 9 to test the efficiency difference between FFEs and other types of ownership in the light industry sector. We see that while SOEs are significantly less efficient than FFEs on average (as was also seen in models 5 and 6), the latter are significantly less efficient than the other (non-state local) ownership groups. Although this result might be somewhat unexpected, it is consistent with the results we obtain using other methods and is robust in this sense. Zelenyuk and Zheka (2006) report a similar result for foreign ownership on a disaggregated level in another transitional country (Ukraine).

[11]However, Zelenyuk (2009) reports Monte Carlo evidence suggesting that the power of the test of the significance of coefficients on dummy variables in the Simar-Wilson (2006) model is very low, even when the true difference is quite substantial from an economic standpoint. It is therefore likely that in some cases, we are simply unable to reject the null hypothesis of equality of efficiencies due to a relatively small sample size, which is clearly not the same as accepting the null hypothesis.

2.5 Concluding Remarks

Over the past three decades, the Chinese economy and its industrial structure have experienced remarkable changes which have been rooted in the reform and open-door policy initiated by Deng Xiaoping in 1978. Although these changes have continued to gain pace over time, their impact has not been uniform across different types of ownership, industries, and regions in China. Given the continued growth of China's economic power since the turn of the new millennium, it is imperative to gain a better understanding of how China has achieved its economic success and how its economy will evolve in the near future.

In this chapter, we investigate efficiency levels and their determinants for different types of ownership, industries and regions in China. The question of the performance of different types of ownership in general, and in China in particular, is a very sensitive issue that often carries political connotations. It goes without saying that great care is required in selecting reliable methods. We employ several recently-developed efficiency analysis methods to examine efficiency variations across different cohorts of Chinese industrial firms. In particular, we employ the latest bootstrap-based estimation procedures involving DEA, aggregation, density estimation and truncated regression. The results obtained in this chapter provide robust statistical evidence that contributes to the ownership-performance debate. While some results support the earlier work of Shiu (2000, 2001), others shed significant new light on the ownership-performance nexus.

We confirm that in comparison with state ownership, all the other types of ownership we consider result in an improvement in performance. This finding is highly robust, is supported by most of the models and methods employed, and is no great surprise. It confirms the results of many other studies that claim modern China is no exception to the economic laws of the free market and related incentives offered by the 'invisible hand' of Adam Smith.

A somewhat unexpected finding that is nevertheless robust is that foreign-owned firms perform worse on aggregate than non-state local enterprises in the light industry sector, but perform slightly better than firms of all other ownership types in the heavy industry sector. To the best of our knowledge, this finding is new to the productivity literature and therefore warrants a greater degree of attention than our other findings.

We consider that the main explanation for this phenomenon stems from the fact that heavy industry, on average, is more capital-intensive than light industry and that purchasing and adopting new technology requires greater financing. As a result, foreign investors in the heavy industry sector, most of which are huge multinational corporations, are likely to have an advantage over local firms in introducing more advanced capital equipment and expensive technologies, both of which lead to better performance.

In light industries, even when foreigners have initial technological and capital advantages, local private firms should be able to absorb, adopt and disseminate such technology according to local specifications more easily and quickly than in heavy industries. On the other hand, because light industries tend to be more labor-intensive than heavy ones, the performance of firms active in the former is more likely to be dependent on local content (culture, traditions, habits, etc.). This is likely to give an advantage to local firms and, given a similar level of technology adoption, should enable them to become more efficient than their foreign counterparts – a prediction we confirm in our study.

Our explanation of the foreign versus local ownership question in the heavy versus light industry puzzle is not entirely new or ad hoc. One theoretical foundation for this explanation is closely related to the technology diffusion argument that goes back at least to the work of Gerschenkron (1962) and Nelson and Phelps (1966), as well as the more recent studies of Grossman and Helpman (1991), Parente and Prescott (1994), Banks (1994), and Helpman and Rangel (1999), in various areas of economics.

2.5.1 *Possible Extensions*

It is worth noting that our results are based on cross-sectional data obtained from the most recently available national census and leave to one side the empirical estimation of changes in efficiency and productivity over time which would be possible with a panel data set. This would be a natural extension of our study and we hope that the work presented in this chapter provides a good foundation for such future research when new census data become available.

Another natural extension to our work would be to use a *non-parametric* truncated regression method, e.g., proposed by Park, Simar and Zelenyuk (2008), which would be possible when more data become available. Yet another interesting extension would be to test for the stochastic dominance of the distributions of efficiency scores of various ownership groups and regions.[12]

Overall, we hope that our study spurs theoretical development of related methodology issues that can improve our work, as well as encourage more of empirical investigations of the current topic using other methods.

Acknowledgements The authors would like to thank Aliaksandr Abrashkevich, Pavlo Demchuk, Natalya Dushkevych, Daniel Henderson, Christopher Parmeter, Ira Howoritz, Vincent Mok, Anatoly Piliawsky, Harry Wu and Xinpeng Xu for their valuable comments. The research support received from the Belgian Government (Belgian Science Policy) through Phase V (No. P5/24) of the "Interuniversity Attraction Pole" is gratefully acknowledged. We also would like to thank the Editor, Paul Wilson for his useful and insightful comments. The authors are entirely responsible for the views expressed and mistakes made in this chapter.

[12] We thank Paul Wilson for this remark.

Appendix

All inputs and outputs used in our activity analysis model for the DEA estimator are measured in units of one hundred million Chinese yuan. Total wage refers to the total remuneration paid to staff and workers during a certain period. This includes wages, salaries and other payments to staff and workers regardless of their source, category and form (in kind or in cash). The net value of fixed assets is calculated as the original value of fixed assets minus depreciation, in which the original value of fixed assets owned by the enterprise is calculated as the price paid at the time the assets were purchased, installed, reconstructed, expanded or subject to technical innovation and transformation. These include expenses incurred in purchasing, packaging, transportation and installation, and so on. The value of intermediate inputs is proxied as the difference between the gross value of industrial output and value added. These are goods that have been processed in one production process and then sold for final processing in another production process. The gross industrial output is the total volume of industrial products sold or available for sale in value terms. It includes the value of finished products and the value of industrial services.

Tables 2.5 and 2.6 show the summary statistics for each ownership type in the heavy and light industry sectors, respectively. See Shiu (2000, 2001) for more information and a discussion of the data set.

Table 2.5 Summary statistics for ownership types (heavy industry)

	Gross industrial output (Hundred million yuan)		Net value of fixed assets (Hundred million yuan)		Total wage (Hundred million yuan)		Intermediate inputs (Hundred million yuan)	
Ownership	Mean	S.D.	Mean	S.D.	Mean	S.D.	Mean	S.D.
SOEs	609.05	401.22	568.77	322.74	65.89	39.96	396.75	267.34
COEs	152.79	258.26	89.19	181.46	6.71	10.36	111.00	189.47
TOEs	197.71	295.94	58.73	75.45	12.69	15.65	147.86	227.69
FFEs	152.79	258.26	89.19	181.46	6.71	10.36	111.00	189.47

Table 2.6 Summary statistics for ownership types (light industry)

	Panel A		Panel B		Panel C		Panel D	
	Gross industrial output (Hundred million yuan)		Net value of fixed assets (Hundred million yuan)		Total wage (Hundred million yuan)		Intermediate inputs (Hundred million yuan)	
Ownership	Mean	S.D.	Mean	S.D.	Mean	S.D.	Mean	S.D.
SOEs	283.51	206.26	167.44	110.05	22.86	15.94	209.46	157.48
COEs	278.80	354.33	86.06	93.20	22.23	21.97	206.38	273.89
TOEs	204.01	336.27	52.84	88.17	11.36	19.00	158.39	263.53
FFEs	216.65	405.08	82.45	134.91	12.22	23.89	169.26	318.28

References

American Chamber of Commerce (ACC). (1998). Business outlook survey.
Badunenko, O., Henderson, D., & Zelenyuk, V. (2008). Technological change and transition: relative contributions to worldwide growth during the 1990's. *Oxford Bulletin of Economics and Statistics*, *70*(4), pp. 461–492.
Banks, R. B. (1994). *Growth and diffusion phenomena*. Berlin: Springer.
Dong, F., & Tang, Z. (1995). *Studies of institutional reforms of chinese state-owned enterprises*. Beijing: The People's Publisher.
Färe, R., Grosskopf, S. Norris, M., & Zhang, Z. (1994). Productivity growth, technical progress, and efficiency change in industrialized countries. *American Economic Review*, *84*(1), 66–83.
Färe, R., & Primont, D. (1995). *Multi-output production and duality: theory and applications*. Boston: Kluwer Academic Publishers.
Färe, R., & Zelenyuk, V. (2003). On aggregate farrell efficiencies. *European Journal of Operations Research*, *146*(3), 615–621.
Farrell, M. J. (1957). The Measurement of Productive Efficiency. *Journal of the Royal Statistical Society*, *120*, 253–281.
Gerschenkron, A. (1962). *Economic backwardness in historical perspective*. Cambridge: Belknap Press of Harvard University Press.
Grossman, G. M., & Helpman, E. (1991). Trade, knowledge spillovers and growth. *European Economic Review*, *35*, 517–26.
Groves, T., Hong, Y., McMillan, J., & Naughton, B. (1994). Autonomy and incentives in chinese state enterprises. *Quarterly Journal of Economics*, *109*, 183–209.
Helpman, E., & Rangel, A. (1999). Adjusting to a new technology: experience and training. *Journal of Economic Growth*, *4*, 359–383.
Henderson, D. J., & Russell, R. R. (2005). Human capital and convergence: a production-frontier approach. *International Economic Review*, *46*, 1167–1205.
Henderson, D. J., & Zelenyuk, V. (2006). Testing for (efficiency) catching-up. *Southern Economic Journal*, *73*(4), 1003–1019.
Huang, H., Cai, F., & Duncan, R. (1999). Reform of state-owned enterprises in china: key measures and policy debate. *ANU*. Canberra: Asia-Pacific Press.
Korostelev, A., Simar, L., & Tsybakov, A. (1995). Efficient estimation of monotone boundaries. *The Annals of Statistics*, *23*(2), 476–489.
Kneip, A., Park, B., & Simar, L. (1998). A note on the convergence of nonparametric DEA estimators for production efficiency scores. *Econometric Theory*, *14*, 783–793.
Kneip, A., Simar, L., & Wilson, P. W. (2008). Asymptotics and consistent bootstraps for DEA estimators in non-parametric frontier models. *Econometric Theory*, *24*, 1663–1697.
Kumar, S., & Russell, R. R. (2002). Technological change, technological catch-up, and capital deepening: relative contributions to growth and convergence. *American Economic Review*, *92*(3), 527– 548.
Li, Q. (1996). Nonparametric testing of closeness between two unknown distributions. *Econometric Reviews*, *15*, 261–274.
Lin, J. Y., Cai, F., & Li, Z. (1998). Competition, policy burdens, and state-owned enterprise reform. *American Economic Review*, *88*(2), 422–427.
Melvin, S. (1998). Business group sizes up shanghai's investment climate. *China Business Review*, *25*(1), 4.
Nelson, R. R., & Phelps, E. S. (1966). Investment in humans, technological diffusion, and economic growth. *American Economic Review*, *56*, 69–75.
Parente, S. L., & Prescott, E. C. (1994). Barriers to technology adoption and development. *Journal of Political Economy*, *102*, 298–321.
Park, B. U., Jeong, S. O., & Simar, L. (2010). Asymptotic distribution of conical-hull estimators of directional edges. *Annals of Statistics*, *38*(3), 1320–1340.

Park, B. U., Simar, L., & Zelenyuk, V. (2008). Local likelihood estimation of truncated regression and its partial derivatives: theory and application. *Journal of Econometrics*, *146*, 185–198.

Russell, R. R. (1990). Continuity of measures of technical efficiency. *Journal of Economic Theory*, *52*, 255–267.

Russell, R. R. (1997). Distance functions in consumer and producer theory in index number theory: essays in honor of Sten Malmquist. *Kluwer Academic Publishers*, 7–90.

Sheather, S. J., & Jones, M. C. (1991). A reliable data based bandwidth selection method for Kernel density estimation. *Journal of the Royal Statistical Society*, *3*, 683–690.

Shiu, A. (2000). Productivity in the Chinese Economy and International Comparisons. Ph.D. *Dissertation*. University of New South Wales, Sydney.

Shiu, A. (2001). Efficiency of chinese enterprises. *Journal of Productivity Analysis*, *18*, 255–267.

Silverman, B. W. (1986). Density Estimation for Statistics and Data Analysis. London: Chapman and Hall.

Simar, L., & Wilson, P. W. (2007). Estimation and inference in two-stage, semi-parametric models of production processes. *Journal of Econometrics*, *136*, 31–64.

Simar, L., & Zelenyuk, V. (2007). Statistical inference for aggregates of farrell-type efficiencies. *Journal of Applied Econometrics*, *22*(7), 1367–1394.

Simar, L., & Zelenyuk, V. (2006). On testing equality of two distribution functions of efficiency scores estimated via DEA. *Econometric Reviews*, *25*(4), 497–522.

State Statistical Bureau, (1995). *The Data of the Third National Industrial Census of the People's Republic of China*. Beijing: State Statistical Bureau.

Transparency International. (2001). Global corruption report (http://www.globalcorruption-report.org/).

Wei, Z., Varela, O., & Hassan, M. K. (2002). Ownership and performance in chinese manufacturing industry. *Journal of Multinational Financial Management*, *12*, 61–78.

Weitzman, M. L., & Xu, C. (1994). Chinese township-village enterprises as vaguely defined cooperatives. *Journal of Comparative Economics*, *18*, 121–145.

Weldon, E., & Vanhonacker, W. (1999). Operating a foreign-invested enterprise in china: challenges for managers and management researchers. *Journal of World Business*, *34*, 94–107.

Wu, Y. (1996). *Productive performance in chinese enterprises – an empirical study*. London: Macmillan Press.

Wu, Y. (1998). Redundancy and firm characteristics in chinese state-owned enterprises. *The Asia Pacific Journal of Economics and Business*, *2*, 33–44.

Yeung, G., & Mok, V. (2002). Government policy and the competitive advantages of foreign-financed firms in Guangdong province of southern china. *Asian Business and Management*, *1*, 227–247.

Zelenyuk, V. (2009). Power of significance test of dummies in simar-wilson two-stage efficiency analysis model. *Applied Economics Letters*, *16*(15), 1493–1497.

Zelenyuk, V., & Zheka, V. (2006). Corporate governance and firm's efficiency: the case of a transitional country, Ukraine. *Journal of Productivity Analysis*, *25*, 143–168.

Zheng, J., Liu, X., & Bigsten, A. (1998). Ownership structure and determinants of technical efficiency: an application of data envelopment analysis to chinese enterprises (1986–1990). *Journal of Comparative Economics*, *26*, 465–484.

Zhang, X. G., & Zhang, S. (2001). Technical efficiency in china's iron and steel industry: evidence from the new census data. *International Review of Applied Economics*, *15*, 199–211.

Chapter 3
Nonparametric Frontier Estimation from Noisy Data

Maik Schwarz, Sébastien Van Bellegem, and Jean-Pierre Florens

Abstract A new nonparametric estimator of production frontiers is defined and studied when the data set of production units is contaminated by measurement error. The measurement error is assumed to be an additive normal random variable on the input variable, but its variance is unknown. The estimator is a modification of the m-frontier, which necessitates the computation of a consistent estimator of the conditional survival function of the input variable given the output variable. In this paper, the identification and the consistency of a new estimator of the survival function is proved in the presence of additive noise with unknown variance. The performance of the estimator is also studied using simulated data.

3.1 Introduction

The modelling and estimation of production functions have been the topic of many research papers on economic activity. A classical formulation of this problem is to consider production units characterized by a vector of inputs $x \in \mathbb{R}_+^p$ producing a

M. Schwarz (✉)
Institut de statistique, biostatistique et sciences actuarielles, Université catholique de Louvain, Voie du Roman Pays 20, 1348 Louvain-la-Neuve, Belgium
e-mail: maik.schwarz@uclouvain.be

S. Van Bellegem
Toulouse School of Economics (GREMAQ), University of Toulouse, Manufacture des Tabacs, Aile J.J. Laffont, 21 Allée de Brienne, 31000 Toulouse, France

Center for Operations Research and Econometrics, Université catholique de Louvain, Voie du Roman Pays 34, 1348 Louvain-la-Neuve, Belgium
e-mail: sebastien.vanbellegem@uclouvain.be

J.-P. Florens
Toulouse School of Economics (GREMAQ), University of Toulouse, Manufacture des Tabacs, Aile J.J. Laffont, 21 Allée de Brienne, 31000 Toulouse, France
e-mail: florens@cict.fr

I. van Keilegom and P.W. Wilson (eds.), *Exploring Research Frontiers in Contemporary Statistics and Econometrics*, DOI 10.1007/978-3-7908-2349-3_3,

vector of outputs $y \in \mathbb{R}^q_+$. The set of production possibilities is denoted by Φ and is a subset of $\mathbb{R}^{p+q}_+$ on which the inputs x can produce the outputs y. Following Shephard (1970), several assumptions are usually imposed on Φ: convexity, free disposability and strong disposability. Free disposability means that if (x, y) belongs to Φ and if x', y' are such that $x' \geqslant x$ and $y' \leqslant y$ then $(x', y') \in \Phi$. Strong disposability requires that one can always produce a smaller amount of outputs using the same inputs.

The boundary of the production set is of particular interest in the efficiency analysis of production units. The efficient frontier in the input space is defined as follows. For all $y \in \mathbb{R}^q_+$, consider the set $\rho(y) = \{x \in \mathbb{R}^p_+ | (x, y) \in \Phi\}$. The radial efficiency boundary is then given by

$$\varphi(y) = \{x \in \mathbb{R}^p_+ \ : \ x \in \rho(y), \theta x \notin \rho(y) \ \forall 0 < \theta < 1\}$$

for all y. Similarly, an efficient frontier in the output space may be defined (e.g. Färe et al. 1985).

In empirical studies, the attainable set Φ is unknown and has to be estimated from data. Suppose a random sample of production units $\mathcal{X}_n = \{(X_i, Y_i) \in \mathbb{R}^{p+q}_+ \ : \ i = 1, \ldots, n\}$ is observed. We assume that each unit (X_i, Y_i) is an independent replication of (X, Y). The joint probability distribution of (X, Y) on $\mathbb{R}^{p+q}_+$ describes the production process. The support of this probability measure is the attainable set Φ, and estimating the efficiency boundary is related to the estimation of the support of (X, Y).

Out of the large literature on the estimation of the attainable set, nonparametric models appeared to be appealing since they do not require restrictive assumptions on the data generating process of $\mathcal{X}_n$. Deprins et al. (1984) have introduced the Free Disposal Hull (FDH) estimator which is defined as

$$\hat{\Phi}_{fdh} = \{(x, y) \in \mathbb{R}^{p+q}_+ \ : \ y \leqslant Y_i, x \geqslant X_i, i = 1, \ldots, n\}$$

and which became a popular estimation method (e.g. De Borger et al. 1994; Leleu 2006). The convex hull of $\hat{\Phi}_{fdh}$, called the Data Envelopment Analysis (DEA), is the smallest free disposal convex set covering the data (e.g. Seiford & Thrall 1990). Among the significant results on this subject, we would like to mention the asymptotic results proved in Kneip et al. (1998) for the DEA and Park et al. (2000) for the FDH.

The consistency of the FDH estimator and other data envelopment techniques is only achieved when the production units are observed without noise, and hence when $\mathbb{P}((X_i, Y_i) \in \Phi) = 1$. However, FDH in particular is very sensitive to the contamination of the data by measurement errors or by outliers (e.g. Cazals et al. 2002; Daouia et al. 2009). Measurement errors are frequently encountered in economic data bases, and therefore there is a need for developing more robust estimation procedures of the production frontier.

In Cazals et al. (2002), a new nonparametric estimator has been proposed to overcome the nonparametric frontier estimation from contaminated samples. When $p = 1$ and under the free disposability assumption, the authors show that the frontier function $\varphi(y)$ can be written as

$$\varphi(y) = \inf\{x \in \mathbb{R}_+ \text{ such that } S_{X|Y\geqslant y}(x) < 1\}, \tag{3.1}$$

where $S_{X|Y\geqslant y}(x) = \mathbb{P}(X \geqslant x|Y \geqslant y)$ denotes the conditional survival function. If $X^1, \ldots, X^m$ are independent replications of $(X|Y \geqslant y)$ for some positive integer m, then a key observation in Cazals et al. (2002) is that the expected minimum input functions

$$\varphi_m(y) := \mathbb{E}(\min\{X^1, \ldots, X^m\}|Y \geqslant y) \quad m = 1, 2, 3, \ldots \tag{3.2}$$

are such that

$$\varphi_m(y) := \int_0^\infty \{S_{X|Y\geqslant y}(u)\}^m \, du \tag{3.3}$$

and $\varphi_m(y)$ converges point-wise to the frontier $\varphi(y)$ as m tends to infinity (assuming the existence of $\varphi_m(y)$ for all m). The functions $\varphi_m(y)$ are estimated in Cazals et al. (2002) from nonparametric estimators of the conditional survival function $S_{X|Y\geqslant y}$. The empirical survival function is defined by $\hat{S}_{X,Y}(x, y) = n^{-1} \sum_i \boldsymbol{I}(X_i \geqslant x, Y_i \geqslant y)$ and the empirical version of $S_{X|Y\geqslant y}$ is thus given by

$$\hat{S}_{X|Y\geqslant y} = \frac{\hat{S}_{X,Y}(x, y)}{\hat{S}_Y(y)}, \tag{3.4}$$

where $\hat{S}_Y(y) = n^{-1} \sum_i \boldsymbol{I}(Y_i \geqslant y)$. Cazals et al. (2002) have studied the asymptotic properties of the frontier estimator

$$\hat{\varphi}_{m,n}(y) := \int_0^\infty \left\{\hat{S}_{X|Y\geqslant y}(u)\right\}^m du \tag{3.5}$$

which is called the m-frontier estimator. They argue that this estimator is less sensitive to extreme values or noise in the sample of production units than FDH or DEA-type estimators.

In this chapter, we slightly amend this claim and show that when the noise level on the data does not vanish as the sample size n grows, then the m-estimator is no longer asymptotically consistent. When the noise level is too high, we show that consistency may be recovered when a robust estimate of the conditional survival function is plugged in the integral in (3.5). By "robust estimate", we refer here to an estimator of $S_{X|Y\geqslant y}$ that is consistent even in the presence of a non-vanishing noise in the sample.

In this chapter, a new robust estimator of the survival function is studied when the inputs X are contaminated by an additive error. We show the consistency of the estimator under the assumption that we only have partial information on the distribution of the error. More precisely, we assume that the additive noise is a zero-mean Gaussian random variable with an unknown variance.

The paper is organized as follows. In Sect. 3.2, we give an overview of existing methods to nonparametrically estimate a density from noisy observations when the distribution of the noise is partially unknown. In Sect. 3.3, we define a new estimator of the survival function in the univariate case, when the data are contaminated by an additive Gaussian random noise with an unknown variance. We prove the asymptotic consistency of our estimator. Finite sample properties are also considered through Monte Carlo simulations. In Sect. 3.4, we define and illustrate on data a new robust m-frontier estimator that is defined similarly to the estimator in (3.5), except that our robust estimator of the conditional survival function is plugged in the integral. The consistency of the robust m-frontier estimator is also established theoretically in this section. The last section summarizes the results of this chapter and suggests future directions of research.

3.2 Density Estimation from Noisy Observations

Estimating the distribution of a real random variable X from a noisy sample is a standard problem in nonparametric statistics. The usual setting is to assume independent and identically distributed (iid) observations from a random variable Z such that $Z = X + \varepsilon$, where ε represents an additive error independent of X. Many research papers focus on the accurate estimation of the cumulative distribution function (cdf) of X under the assumption that the cdf of ε is known. The additive measurement error implies that the density of Z, if it exists, is the convolution between the density of ε and the one of X:

$$f^Z(z) = f^\varepsilon \star f^X(z) := \int_{-\infty}^{\infty} f^\varepsilon(t) f^X(z-t)\mathrm{d}t \ .$$

Based on this result, most estimators of f^X studied in the literature use the Fourier transform of the densities since the Fourier coefficients of the convolution are the product of the coefficients:

$$\psi^Z(\ell) = \psi^\varepsilon(\ell)\psi^X(\ell), \quad \ell \in \mathbb{Z}$$

where $\psi^U(\ell) := \mathbb{E}\{\exp(i\ell U)\}$ denotes the ℓ-th Fourier coefficient of a density f^U. A usual estimator of $\psi^Z(\ell)$ is (a functional of) the empirical characteristic function of the sample $(Z_1, \ldots, Z_n)$, i.e.

$$\hat{\psi}^Z(\ell) := \frac{1}{n}\sum_{i=1}^{n} \exp(i\ell Z_i), \quad \ell \in \mathbb{Z} \ .$$

From this estimator and under the condition that f_ℓ is known and nonzero, the standard estimators are based on the inverse Fourier transform of $\hat{\psi}^Z(\ell)/\psi^\varepsilon(\ell)$ (e.g. Carroll & Hall 1988; Fan 1991). Alternative estimators have also been studied in the literature, for instance in the wavelet domain (Pensky & Vidakovic, 1999; Johnstone et al., 2004; Bigot & Van Bellegem, 2009).

The exact knowledge of the cdf of the error is however not realistic in many empirical studies. If we want to relax the condition that the cdf of the error is known, one major obstacle is that the cdf of X is no longer identifiable. To circumvent this problem, at least three research directions can be found in the literature.

A first approach assumes that an independent sample from the measurement error ε is available in addition to the sample of Z. From the independent observation of ε, the density f^{ε} is identified and so is the target density f^X. In a first step, a nonparametric estimator of f^{ε} can be constructed from the sample of ε's . This estimator is then used in the construction of the estimator of f^X (Neumann, 2007; Johannes & Schwarz, 2009; Johannes et al., 2010). If this approach may be realistic for a set of practical situations (e.g. in some problems in biostatistics and astrophysics), it is hardly applicable in production frontier estimation.

A second approach is to assume various sampling processes. Li & Vuong (1998) suppose that repeated measurements for one single value of X are available, such as $Z_j = X + \varepsilon_j$ for $j = 1, \ldots, m$. Assuming further that X, ε_1, and ε_2 are mutually independent, $\mathbb{E}(\varepsilon_j) = 0$, and that the characteristic functions of X and ε are non-zero everywhere, they show how the latter characteristic functions can be expressed as functions of the joint characteristic function of (Z_1, Z_2). From this representation it follows that the cumulative distribution function (cdf) of both X and ε can be identified from the observation of the pair (Z_1, Z_2). The joint characteristic function of (Z_1, Z_2) can be estimated from a sample of (Z_1, Z_2) and is then used to derive an estimator of f^X. The characteristic functions of X and ε, denoted by ψ^X and ψ^{ε}, can then be computed using the above-mentioned representation. Delaigle et al. (2008) have also considered this setting and present modified kernel estimators which, if the number of repeated measurements is large enough, can perform as well as they would under known error distribution.

A related situation is when there are repeated measurements of X in a multilevel model. In Neumann (2007) it is assumed that $Z_{ij} = X_i + \varepsilon_{ij}$ for $j = 1, \ldots, N$ and $i = 1, \ldots, n$ are observed (see also Meister et al. 2010). In this sampling process, the identification of the cdf of X is ensured by a condition on the zero-sets of the characteristic functions of X and ε. Let $\mathcal{Z} = (Z_{i1}, \ldots, Z_{iN})'$, $\psi^{\mathcal{Z}}$ its characteristic function, and $\hat{\psi}^{\mathcal{Z}}$ the empirical characteristic function of $\mathcal{Z}$. A consistent estimator of the density of X is obtained by minimizing the discrepancy

$$\int_{\mathbb{R}^n} \left|\psi^X(t_1 + \ldots + t_n)\psi^{\varepsilon}(t_1)\cdots\psi^{\varepsilon}(t_n) - \hat{\psi}_n^{\mathcal{Z}}(t_1, \ldots, t_n)\right| h(t_1, \ldots, t_n)\mathrm{d}t_1 \ldots \mathrm{d}t_n$$

over certain classes of possible characteristic functions ψ^X and ψ^{ε} of X and ε, respectively.

Repeated measurements of multilevel sampling appear in some economic situations, for instance when production units are observed over time (a case considered e.g. in Park et al. 2003; Daskovska et al. 2010).

A third approach to recover the identification of X in spite of the noise ε is to assume that the cdf of ε is only *partially unknown*. A realistic case for practical purposes is to assume that ε is normally distributed, but the variance of ε is unknown. Of course the cdf of X is not identified in this setting, and it is necessary to restrict the class of cdfs of X in order to recover identification.

Several recent research papers have proposed identification restrictions on the class of X given a partial knowledge about the cdf of the noise. Butucea & Matias (2005) assume that the error density, is "s-exponential" meaning that its Fourier transform, ψ^{ε}, satisfies

$$b\exp(-|u|^{s}) \leqslant |\psi^{\varepsilon}(u)| \leqslant B\exp(-|u|^{s})$$

for some constants b, B, s and $|u|$ large enough. In their approach the error density is supposed to be known up to its scale σ (called "noise level"). As for the density f^X, both polynomial and exponential decay of its Fourier transform are shown to lead to a fully identified model. To define an estimator, let $\psi^{\varepsilon}_{\sigma}$ be the Fourier transform of (σf^{ε}). The key to the estimation of σ is the observation that the function $|F(\tau,u)| = |\psi^Z(u)|/|\psi^{\varepsilon}_{\tau}(u)|$ diverges as $u \to \infty$ when $\tau > \sigma$ and that it converges to 0 otherwise. Let $\hat{F}(\tau,u_n) = |\hat{\psi}^Z(u_n)|/|\psi^{\varepsilon}_{\tau}(u_n)|$. Then Butucea & Matias (2005) show that

$$\hat{\sigma}_n = \inf\{\tau > 0 : |\hat{F}(\tau,u_n)| \geqslant 1\}$$

yields a consistent estimator of σ for some well balanced sequence u_n. This estimator is then used to deconvolve the empirical density of Z and to get an estimator of the density of X. Some extensions are proposed in Butucea et al. (2008), where the error density is assumed to have a stable symmetric distribution with $\psi^{\varepsilon}(u) = \exp(-|\gamma u|^{s})$ in which γ represents some known scale parameter and s is an unknown index, called the self-similarity index.

A similar setting is considered in Meister (2006). In this chapter, the error is supposed to be normally distributed with an unknown variance parameter. Identification is recovered by assuming that ψ^X lies in $\{\psi : c_1|u|^{-\beta} \leqslant |\psi(u)| \leqslant c_2|u|^{-\beta}$ for all $u \gg 0\}$ for some strictly positive constants c_1, c_2.

In Meister (2007), it is assumed that ψ^{ε} is known on some arbitrarily small interval $[-\nu, \nu]$ and that it belongs to some class

$$\mathcal{G}_{\mu,\nu} = \{f \text{ is a density such that } \|f\|_{\infty} \leqslant C, |\psi^{f}(t)| \geqslant \mu\ \forall |t| \geqslant \nu\}.$$

The target density f^X is assumed to belong to

$$\mathcal{F}_{S,C,\beta} = \Big\{f \text{ is a density such that } \int_{-S}^{S} f(u)\mathrm{d}u = 1 \text{ and}$$
$$\int |\psi^{f}(t)|^2(1+t^2)^{\beta}\mathrm{d}t \leqslant C\Big\},$$

that is to the class of densities with compact support that are uniformly bounded in the Sobolev norm. The direct empirical access to ψ^X via Fourier deconvolution is restricted to the interval $[-\nu, \nu]$. However, it is shown using a Taylor expansion that ψ^X is uniquely determined by its restriction to $[-\nu, \nu]$, and therefore identified everywhere.

Because the deconvolution of the density of Z is solved via the Fourier transform, most of the assumptions on X or ε recalled above are expressed in terms of their characteristic functions. They appear to be *ad hoc* assumptions, as they do not have any obvious economic interpretation. In Schwarz & Van Bellegem (2010), an identification theorem is proved on the target density under assumptions that are not expressed in the Fourier domain. It is instead assumed that the measurement error ε is normally distributed with an unknown variance parameter, and that f^X lies in the class of densities that vanish on a set of positive Lebesgue measure. This restriction on the class of target densities is reasonable for our purpose of frontier estimation, in which it is structurally assumed that the density of X (or the conditional density of $(X|Y \geqslant y)$) is zero beyond the frontier. Since this is a natural assumption in the setting of frontier estimation, we use this framework in the next section in order to estimate a survival function from noisy data.

3.3 A New Estimator of the Survival Function from Noisy Observations

3.3.1 *Identification of the Survival Function*

Suppose we observe a sample $\{Z_1, \ldots, Z_n\}$ of n independent replications of Z from the model

$$Z = X + \varepsilon \,, \tag{3.6}$$

where ε is a $N(0, \sigma^2)$ random variable, independent from X, and with an unknown variance σ^2. As explained in the previous section, the probability density of Z is the convolution $\phi_\sigma \star f^X$, where f^X is the probability density of X and ϕ_σ denotes the Normal density with standard error σ. The following theorem, quoted from Schwarz & Van Bellegem (2010), defines a set of identified probability distributions f^X for model (3.6). The survival function S^X of X will hence be identified on that set from the observation of Z.

Theorem 3.3.1. *Define the following set of probability distributions:*

$$\mathcal{P}_0 := \{P \text{ distribution } : \exists \text{ Borel set } A \text{ such that } |A| > 0 \text{ and } P(A) = 0\},$$

where $|A|$ denotes the Lebesgue measure of A. The model defined by (3.6) is identifiable for the parameter space $\mathcal{P}_0 \times (0, \infty)$. In other words, for any two probability measures $P^1, P^2 \in \mathcal{P}_0$ and $\sigma_1, \sigma_2 > 0$, we have that $\phi_{\sigma_1} \star P^1 = \phi_{\sigma_2} \star P^2$ implies $P^1 = P^2$ and $\sigma_1 = \sigma_2$.

3.3.2 A Consistent Estimator

From model (3.6), we also observe after a straightforward calculation that the survival function of Z, denoted by S^Z, also follows a convolution formula:

$$S^Z(z) = \phi_\sigma \star S^X(z)$$

where S^X is the survival function of the variable X and ϕ_σ denotes the density function of a $N(0, \sigma^2)$ random variable.

Our estimator of S^X is approximated in a sieve as follows. For any integers $k, D > 0$, define $\Delta^{(k,D)} := \{\delta \in \mathbb{R}^k \ : \ 0 \leqslant \delta_1 \leqslant \ldots \leqslant \delta_k \leqslant D\}$ and for $\delta \in \Delta^{(k,D)}$ define

$$S_\delta(t) := \frac{1}{k} \sum_{j=1}^{k} \mathbb{1}(\delta_j > t) \; . \tag{3.7}$$

For any $\delta \in \Delta^{(k,D)}$, denote by P_δ the probability distribution corresponding to the survival function S_δ. The choice of the approximating function is performed minimizing the contrast function

$$\gamma(S, \zeta; T) := \int_{-\infty}^{\infty} \left|(\phi_\zeta \star S)(t) - T(t)\right| h(t) \mathrm{d}t,$$

where h is some strictly positive probability density ensuring the existence of the integral.

We are now in position to define our estimator of the survival function. Let $(k_n)_{n\in\mathbb{N}}$ and $(D_n)_{n\in\mathbb{N}}$ be two positive, divergent sequence of integers. The estimator $(S_{\hat{\delta}(n)}, \hat{\sigma}_n)$ is defined by

$$(\hat{\delta}(n), \hat{\sigma}_n) := \underset{\substack{\delta \in \Delta^{(k_n, D_n)} \\ \sigma \in [0, D_n]}}{\arg\min}\; \gamma(S_\delta, \sigma; \hat{S}_n^Z) \; , \tag{3.8}$$

where $\hat{S}_n^Z := n^{-1} \sum_{k=1}^n \mathbb{1}(Z_k > t)$ is the empirical survival function of Z. Note that the argmin is attained because it is taken over a compact set of parameters. Though, it is not necessary unique. If it is not, an arbitrary value among the possible solutions may be chosen.

Theorem 3.3.2. *The estimator* $(S_{\hat{\delta}(n)}, \hat{\sigma}_n)$ *is consistent in the sense that*

$$P^X_{\hat{\delta}_n} \xrightarrow{\mathcal{L}} P^X \quad \text{and} \quad \hat{\sigma}_n \to \sigma$$

almost surely as $n \to \infty$*, where* $\xrightarrow{\mathcal{L}}$ *denotes weak convergence of probability measures.*

Table 3.1 The inputs simulated in this experiment are uniformly distributed over [1, 2]. For each sample size and noise level, we compute the mean of $\sigma - \hat{\sigma}_n$ from $B = 2{,}000$ replications (the standard deviation is given between parentheses)

	True σ		
n	1	2	5
100		1.30	−1.08
		(1.05)	(0.51)
200	0.91	0.07	−0.38
	(3.84)	(0.45)	(0.45)
500	0.37	0.06	0.14
	(0.30)	(0.44)	(0.49)

The proof of this result is based on some technical lemmas and can be found in the appendix below.

To illustrate the estimator, we now present the result of a Monte Carlo experiment. The estimator of the standard deviation σ of the noise is of particular interest. In the following experiment, we consider two designs for the input X. One is uniformly distributed over [1, 2], and the other is a mixture $U[1,2] + Exp(1)$. In both cases the density of X is zero below 1, and in the second case the support of X is not bounded to the right. For various true values of σ, we calculate the estimators $(\hat{\delta}(n), \hat{\sigma}_n)$ for sample sizes $n = 100$, 200 and 500. No particular optimization over the value of k (appearing in (3.7)) is provided, except that we increase k as the sample size increases. For the considered sample sizes, we set $k = 10\,n^{1/2}$. The minimization of the contrast function is calculated using the algorithm `optim` in the R software. For this algorithm, we have chosen the initial values of δ_j to be equispaced values over the interval [0, 3] and the initial value of σ is the empirical standard deviation of the sample $Z_1, \ldots, Z_n$.

Tables 3.1 and 3.2 show the result of the Monte Carlo simulation using $B = 2{,}000$ replications of each design. The mean and standard deviation of $\sigma - \hat{\sigma}_n$ over the B replications are displayed. Some results are not reported for very small sizes, because a stability problem has been observed, especially in the mixture case. In these cases, the `optim` algorithm did not often converge (a similar phenomenon has been observed using the `nlm` algorithm). It also has to be mentioned that the stability is very sensitive to the choice of k and to the choice of initial values for δ and σ. For larger sample sizes, or larger values of the noise, the results overall improve with the sample size.

3.4 Robust *m*-Frontier Estimation in the Presence of Noise

3.4.1 *Inconsistency of the m-Frontier Estimator*

Let us now consider our initial problem of consistently estimating the production frontier $\varphi(y)$ from a sample of production units (X_i, Y_i), where X_i is the input

Table 3.2 The inputs simulated in this experiment are a mixture $U[1,2] + Exp(1)$. For each sample size and noise level, we compute the mean of $\sigma - \hat{\sigma}_n$ from $B = 2,000$ replications (the standard deviation is given between parentheses)

	True σ		
n	1	2	5
100		2.84	−0.92
		(7.80)	(7.15)
200		−0.49	−0.49
		(6.32)	(5.92)
500	1.78	0.029	0.014
	(5.90)	(4.88)	(6.69)

and Y_i is the output. To simplify the discussion, we assume that the dimension of the input and the output are $p = q = 1$.

In the introduction we have recalled the definition of the m-frontier estimator in equation (3.5). Compared to the FDH or DEA estimator, this nonparametric frontier estimator provides a more robust estimator of the frontier in the presence of noise. In (Cazals et al., 2002, Theorem 3.1) it is also proved that for any interior point y in the support of the distribution Y and for any $m \geqslant 1$, it holds that

$$\hat{\varphi}_{m,n}(y) \to \varphi_m(y) \quad \text{almost surely as } n \to \infty \tag{3.9}$$

where $\varphi_m(y)$ is the expected minimum input function of order m given in equation (3.2).

When the input of the production units is contaminated by an additive error, the actually observed inputs are

$$Z_i = X_i + \varepsilon_i, \quad \varepsilon_i \sim N(0, \sigma^2)$$

instead of X_i, for some positive, unknown variance parameter σ^2. If σ^2 does not vanish asymptotically, the limit appearing in (3.9) is no longer given by the expected minimum input function (3.2). Instead we get

$$\hat{\varphi}_{m,n}(y) \to \mathbb{E}\left(\min\{Z_1, \ldots, Z_m\} | Y \geqslant y\right) \quad \text{almost surely as } n \to \infty .$$

The expectation appearing on the right hand side is not (3.2) because the support of the variable Z is the whole real line. Therefore, the m-frontier estimator does not converge to the desired target function, due to the non-vanishing error variance. Note that this is in contrast with the approach of Hall & Simar (2002) or Simar (2007). In the two latter references, the noise level is assumed to be asymptotically negligible.

The inconsistency of the m-frontier estimator is illustrated in Figs. 3.1 and 3.2. The true production frontier in this simulation is given by $\varphi(y) = \sqrt{y}$ and is represented by the dotted line. We have simulated 200 production inputs from model

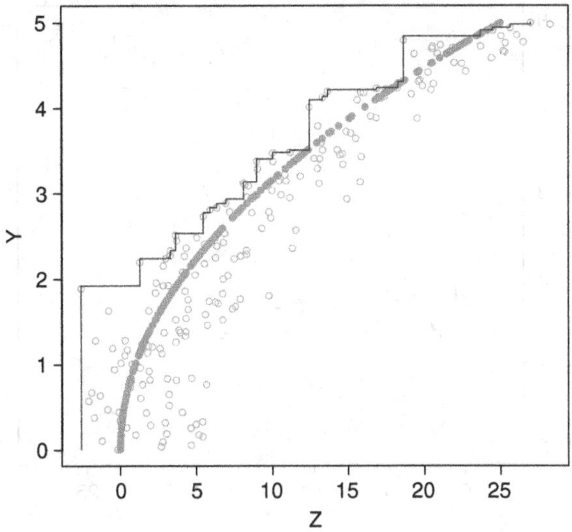

Fig. 3.1 The *gray circles* are the simulated production units and the *dotted line* is the true production frontier. The *solid line* is the Free Disposal Hull (FDH) estimator of the frontier

$X_i = Y_i^2 + E_i$, where $E_i \sim Exp(1)$. The production inputs are then contaminated by an additive noise, so that the observed inputs are $Z_i = X_i + \varepsilon_i$ instead of X_i, where ε_i are independently generated from a zero mean normal variable with standard error $\sigma = 2$.

The FDH estimator computed in Fig. 3.1 is known to be inconsistent in this situation, because it is constructed under the assumption that all production units are in the production set Φ with probability one. Figure 3.2 shows the m-frontier of Cazals et al. (2002) for $m = 1$ and 50 respectively (cf. 3.5). As discussed in Cazals et al. (2002), an appropriate choice of m is delicate and, as far as we know, there is no automatic procedure to select it from the data. If m is too low, the m-frontier is not a good estimator of the production function. In the theory of Cazals et al. (2002), m is an increasing parameter with respect to the sample size. For large values of y, the estimator is above the true frontier.

For larger values of m, as shown in Fig. 3.2, the estimator is close to the FDH estimator. Because the value of m increases with n in theory, the two estimators will be asymptotically close. This illustrates the inconsistency of the m-frontier in the case where the noise on the data is not vanishing with increasing sample size.

3.4.2 Robust m-Frontier Estimation

In order to recover the consistency of the m-frontier, we need to plug-in a consistent estimator of the conditional survival function in (3.3). The construction of the

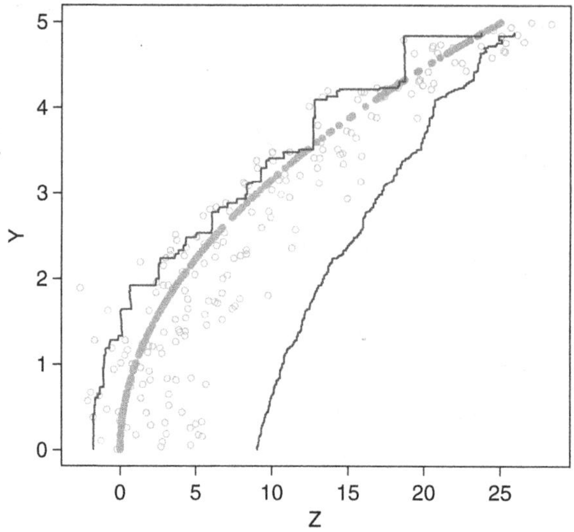

Fig. 3.2 Using the same data as in Fig. 3.1, the two *solid lines* are the m-frontier estimator with $m = 1$ and $m = 50$ respectively

estimator is easy from the above results if we assume that the additive noise to the inputs is independent from the input X and the output Y. Let y be a point in the output domain where the support of Y is strictly positive. Restricting the data set to $(Z_i | Y_i \geqslant y)$, we can construct the empirical conditional survival function $\hat{S}_{Z|Y \geqslant y}$ using the usual nonparametric estimator (3.4). Note that this estimator does not require any regularization parameter such as a bandwidth. In analogy to (3.8), we also define

$$(\hat{\delta}(n), \hat{\sigma}_n) := \underset{\substack{\delta \in \Delta^{(k_n, D_n)} \\ \sigma \in [0, D_n]}}{\arg\min}\ \gamma(S_\delta, \sigma; \hat{S}_{Z|Y \geqslant y}) . \tag{3.10}$$

The final robust m-frontier estimator is then given by

$$\hat{\varphi}^{rob}_{m,n}(y) = \int_0^\infty \left\{ S_{\hat{\delta}(n)}(u) \right\}^m \mathrm{d}u . \tag{3.11}$$

Note that this integral is easy to compute since $S_{\hat{\delta}(n)}$ is a step function. The following result establishes the consistency of this new estimator under a condition on the parameter m.

Proposition 3.4.1. *Suppose we observe production units $\{(Z_i, Y_i); i = 1, \ldots, n\}$ in which the univariate inputs are such that $Z_i = X_i + \varepsilon_i$, where ε_i models a measurement error that is independent from X_i and Y_i, normally distributed with* *zero mean and unknown variance σ^2. Consider the robust m-frontier estimator given by equations (3.10) and (3.11) and let m_n be a strictly divergent sequence of positive integers such that*

$$\{S_{\hat{\delta}(n)}(\varphi(y))\}^{m_n} \to 1 \tag{3.12}$$

almost surely as $n \to \infty$. Then $\hat{\varphi}^{rob}_{m_n,n}(y) \to \varphi(y)$ almost surely as $n \to \infty$.

This result illustrates well the role of the parameter $m = m_n$, which has to tend to infinity at an appropriate rate as $n \to \infty$ in order to achieve consistency of the robust frontier estimator. Indeed, if m_n is bounded by some $M > 0$, Fatou's Lemma implies that almost surely

$$\lim_{n\to\infty} \hat{\varphi}^{rob}_{m_n,n}(y) \geqslant \int_0^\infty \left\{S_{X|Y\geqslant y}(u)\right\}^M du = \varphi(y) + \int_{\varphi(y)}^\infty \left\{S_{X|Y\geqslant y}(u)\right\}^M du.$$

Except for the trivial case where the true conditional survival function is the indicator function of the interval $(-\infty, \varphi(y))$, the last integral on the right hand side is strictly positive. This shows that the robust estimator asymptotically overestimates the true frontier $\varphi(y)$ if m_n does not diverge to infinity.

On the other hand, if m_n increases too fast in the sense that the condition in (3.12) does not hold, then $\hat{\varphi}^{rob}_{m_n,n}(y)$ may asymptotically underestimate the true frontier $\varphi(y)$ as one can see decomposing the integral from (3.11) into

$$\int_0^\infty \left\{S_{\hat{\delta}(n)}(u)\right\}^{m_n} du = \int_0^{\varphi(y)} \left\{S_{\hat{\delta}(n)}(u)\right\}^{m_n} du + \int_{\varphi(y)}^\infty \left\{S_{\hat{\delta}(n)}(u)\right\}^{m_n} du.$$

The second integral on the right hand side tends to 0 almost surely for $n \to \infty$ as we explain in the proof of Proposition 3.4.1. As for the first one, the integrand converges to a non-negative monotone function S with $S(\varphi(y)) < 1$, and hence the integral may tend to a limit that is smaller than the true frontier $\varphi(y)$. However, this need not be the case, and thus the condition in (3.12) is sufficient but not necessary.

Summarizing the above discussion, the sufficient condition in (3.12) implicitly defines an appropriate rate at which m_n may diverge to infinity such that the new robust frontier estimator is consistent. This rate depends on characteristics of the true conditional survival function, and we do not know at present how to choose it in an adaptive way. Nevertheless, the simulations show that even for finite samples, large choices of m do not deteriorate the performance of the robust estimator.

The estimator is computed for each possible value of y. In practice, it is not necessary to estimate the standard deviation of the noise for each y. We can first estimate the noise level using the marginal data set of inputs only, and use the techniques developed in Sect. 3.4. We then use this estimated value in (3.10) even as an initial parameter of the `optim` algorithm, or as a fixed, known parameter of the noise standard deviation.

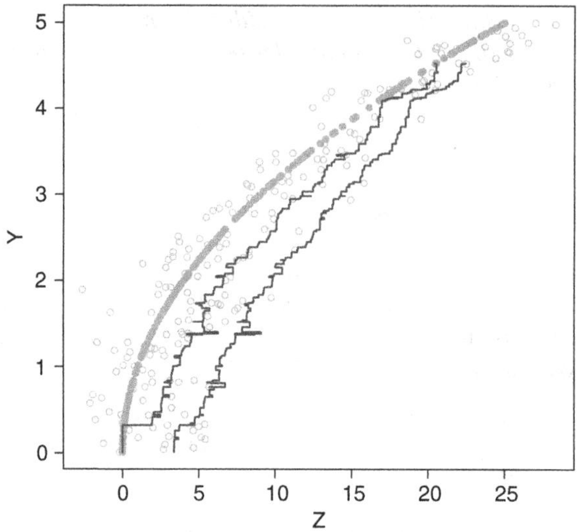

Fig. 3.3 Using the same data as in Fig. 3.1, the two *solid lines* are the robust m-frontier estimator with $m = 1$ and $m = 50$ respectively

Figure 3.3 shows the estimator on the simulated data of Fig. 3.1. As for the standard m-frontier, the robust m-frontier with $m = 1$ is not a satisfactory estimator. The interesting fact about the robust m-frontier is that it does not deteriorate the frontier estimation for large values of m. For the sake of comparison with Figs. 3.2 and 3.3 also displays the robust m-frontier estimator with $m = 50$. This estimator does not cross the true production frontier and does not converge to the FDH estimator.

3.5 Conclusion and Further Research

One original idea in this paper is to consider stochastic frontier estimation when the data generating process has an additive noise on the inputs. The noise is not assumed to vanish asymptotically. In this situation, the m-frontier estimator introduced by Cazals et al. (2002) is still a valuable tool in robust frontier estimation, but it requires to plug-in a consistent estimator of the conditional survival function in order to be consistent itself.

Constructing this consistent estimator is a deconvolution problem. We have solved this problem in this chapter. An important feature of our results is that the noise level is not known, and therefore needs to be estimated from a cross section of production units.

Measurement errors are frequently encountered in empirical economic data, and the new robust estimator is designed to be consistent in this setting. The rate of convergence of the estimator is however unknown. This study might be of interest for future research in efficiency analysis.

As it was suggested by a referee, one might also be interested in the case where the measurement error is in the output rather than in the input variable. We would like to end this paper by explaining how the above methods can be transferred to this problem and where the limitations are. In this setting, in contrast to Sect. 3.4, the inputs X_i are directly observed, but only a contaminated version

$$W_i = Y_i + \eta_i, \quad \eta_i \sim N(0, \sigma^2) \tag{3.13}$$

of the true output variables Y_i is observed, with η_i independent from X_i and Y_i. Let us briefly discuss the case where both the input and the output spaces are one-dimensional, i.e. $p = q = 1$. As the frontier function $\varphi : \mathbb{R}_+ \to \mathbb{R}_+$ given in (3.1) is strictly increasing, its inverse function $\varphi^{-1} : \mathbb{R}_+ \to \mathbb{R}_+$ exists. The efficiency boundary can be described by either of the functions φ and φ^{-1}. Estimating φ^{-1} is thus equivalent to estimating φ itself. The inverse frontier function can be written as

$$\varphi^{-1}(x) = \inf\{y \in \mathbb{R}_+ \ : \ F_{Y|X \leqslant x}(y) = 1\},$$

where $F_{Y|X \leqslant x}$ denotes the conditional distribution function of Y given $X \leqslant x$. To apply the robust m-frontier methodology we therefore need to estimate the conditional distribution function $F_{Y|X \leqslant x}$. From the model (3.13), one can easily show that the estimation of $F_{Y|X \leqslant x}$ is again a deconvolution problem, and recalling that $F_{Y|X \leqslant x} = 1 - S_{Y|X \leqslant x}$, we can define

$$(\hat{\delta}(n), \hat{\sigma}_n) := \operatorname*{arg\,min}_{\substack{\delta \in \Delta^{(k_n, D_n)} \\ \sigma \in [0, D_n]}} \gamma(S_\delta, \sigma; \hat{S}_{W|X \leqslant x}) \ \text{ and } \ \hat{F}_n := 1 - S_{\hat{\delta}(n)}$$

in analogy to Sect. 3.4.2. $\hat{F}_n$ is the deconvolving estimator of the conditional distribution function $F_{Y|X \leqslant x}$. We proceed by defining the robust m-frontier estimator of φ^{-1} as

$$\hat{\varphi}^{-1}_{m,n}(x) := A - \int_0^A \left\{\hat{F}_n(u)\right\}^m du,$$

where $A > 0$ is some constant fixed in advance. Let m_n be a strictly divergent sequence such that $\{\hat{F}_n(\varphi(x))\}^{m_n} \to 1$ almost surely as $n \to \infty$. In analogy to Proposition 3.4.1, it can be shown that for such a sequence, $\hat{\varphi}^{-1}_{m_n,n}(x)$ is consistent if $A > \varphi^{-1}(x)$. Otherwise, $\hat{\varphi}^{-1}_{m_n,n}(x)$ tends to A almost surely. This suggests the following adaptive choice of A. First, one computes the estimator with some arbitrary initial value of A. If the result is close to A, recompute it repeatedly for increasing values of A until a value smaller than A is obtained.

This estimator is thus robust with respect to noise in the output variable, but note that it is not obvious how to generalize this procedure to a multi-dimensional setting. Moreover, it is not clear how one could cope with a situation with error in both variables. These questions could be subject to further investigation.

Appendix: Proofs

Proof of Theorem 3.3.2

In order to show the consistency of the robust frontier estimator, we first need to prove two lemmas.

Lemma 3.5.1. *The estimator* $(S_{\hat{\delta}(n)}, \hat{\sigma}_n)$ *satisfies*

$$\gamma(S_{\hat{\delta}(n)}, \hat{\sigma}_n; \hat{S}_n^Z) \to 0 \quad \textit{as } n \to \infty.$$

Proof. By the triangle inequality, we have, for any $(S', \sigma') \in \mathbb{C} \times \mathbb{R}^+$,

$$\begin{aligned}
\gamma(S_{\hat{\delta}(n)}, \hat{\sigma}_n; \hat{S}_n^Z) &= \min_{\substack{\delta \in \Delta^{(k_n, D_n)} \\ \tilde{\sigma} \in [0, D_n]}} \gamma(S_\delta, \tilde{\sigma}; \hat{S}_n^Z) \\
&\leqslant \min_{\substack{\delta \in \Delta^{(k_n, D_n)} \\ \sigma \in [0, D_n]}} \gamma(S_\delta, \sigma; S^X \star \phi_\sigma) + \gamma(S^X, \phi_\sigma; \hat{S}_n^Z).
\end{aligned} \tag{A.1}$$

Let $\eta > 0$ and $T > 0$ be such that $\int_T^\infty S^X(x)\, dx \leqslant \eta/2$. For n sufficiently large, we have $\sigma \leqslant D_n$ and there is $\delta \in \Delta^{(k_n, D_n)}$ with $\int_0^T |(S_\delta - S^X)(x)|\, dx \leqslant \eta/2$, such that $\int_{\mathbb{R}} |(S_\delta - S^X)(x)|\, dx \leqslant \eta$. It follows that the first term on the right hand side of (A.1) is a null sequence, because

$$\gamma(S_\delta, \sigma; S^X \star \phi_\sigma) \leqslant \|(S_\delta - S^X) \star \phi_\sigma\|_{L^1} \leqslant \|S_\delta - S^X\|_{L^1} \|\phi_\sigma\|_{L^1} \leqslant \eta.$$

The second term is also a null sequence by virtue of Glivenko-Cantelli's and Lebesgue's Theorem. □

Lemma 3.5.2. *The estimator* $S_{\hat{\delta}(n)}$ *defined by (3.8) satisfies*

$$(P_{\hat{\delta}(n)} \star \phi_{\hat{\sigma}_n}) \xrightarrow{\mathcal{L}} P^Z$$

almost surely as $n \to \infty$.

Proof. The survival function S^Z is continuous everywhere as it can be written as a convolution with some normal density. Therefore, the convergence

$$\hat{S}_n^Z(x) \xrightarrow{n \to \infty} S^Z(x) \quad \text{a.s.}$$

holds for every $x \in \mathbb{R}$. Hence, by Lebesgue's theorem,

$$\gamma(S^X, \sigma; \hat{S}_n^Z) \xrightarrow{n\to\infty} 0 \quad \text{a.s.}$$

The triangle inequality, together with Lemma 3.5.1, implies

$$\gamma(S_{\hat{\delta}(n)}, \hat{\sigma}_n; S^Z) \leqslant \gamma(S_{\hat{\delta}(n)}, \hat{\sigma}_n; \hat{S}_n^Z) + \gamma(S^X, \sigma; \hat{S}_n^Z) \xrightarrow{n\to\infty} 0 \quad \text{a.s.}$$

A continuity argument implies

$$(S_{\hat{\delta}(n)} \star \phi_{\hat{\sigma}_n})(x) \xrightarrow{n\to\infty} S^Z(x) \quad \text{a.s.}$$

for every $x \in \mathbb{R}$, which is in fact weak convergence and hence concludes the proof. □

Our proof of consistency also needs the following two lemmas. The first one is quoted from Schwarz & Van Bellegem (2010), the second one is an immediate consequence of Lemma 3.4 from the same article.

Lemma 3.5.3. *Let Q_n be a sequence of probability distributions and σ_n a sequence of positive real numbers. Suppose further that $(Q_n \star N(0, \sigma_n))_{n\in\mathbb{N}}$ converges weakly to some probability distribution. Then, there exist an increasing sequence $(n_k)_{k\in\mathbb{N}}$, a probability distribution Q_∞, and a constant $\sigma_\infty \geqslant 0$ such that*

$$Q_{n_k} \xrightarrow{\mathcal{L}} Q_\infty \quad \textit{and} \quad \sigma_{n_k} \to \sigma_\infty$$

as $n \to \infty$.

Lemma 3.5.4. *A weakly convergent sequence of probability distributions that have all their mass on the positive axis has its limit in $\mathcal{P}_0$.*

We are now in position to prove the consistency theorem.

Proof of Theorem 3.3.2. For probability distributions P, P' and positive real numbers σ, σ', define the distance $\Delta(P, \sigma; P', \sigma') := d(P, P') + |\sigma - \sigma'|$, where $d(\cdot,\cdot)$ denotes a distance that metrizes weak convergence, e.g. the Lévy distance. The theorem is hence equivalent to

$$\Delta(P_{\hat{\delta}(n_k)}, \hat{\sigma}_{n_k}; P^X, \sigma) \xrightarrow{n\to\infty} 0$$

almost surely. The proof is obtained by contradiction. Suppose that there is some $d > 0$ and an increasing sequence $(n_k)_{k\in\mathbb{N}}$ such that

$$\Delta(P_{\hat{\delta}_{n_k}}, \hat{\sigma}(n_k); P^X, \sigma) > d$$

for all $k \in \mathbb{N}$.

By Lemma 3.5.2, we know that the distributions given by $(S_{\hat{\delta}(n)} \star \phi_{\hat{\sigma}_n})$ converge almost surely weakly to P^Z. Lemma 3.5.3 implies that there is a distribution P_∞, some $\sigma_\infty \geqslant 0$, and a sub-sequence $(n'_k)_{k\in\mathbb{N}}$ such that almost surely

$$P_{\hat{\delta}(n'_k)} \xrightarrow{\mathcal{L}} P_\infty \quad \text{and} \quad \hat{\sigma}_{n'_k} \to \sigma_\infty,$$

which implies the almost sure point-wise convergence of $S_{\hat{\delta}_{n'_k}}$ to S_∞. Fatou's lemma then implies

$$\gamma(S_\infty, \sigma_\infty; S^Z) \leqslant \liminf_{k\to\infty} \gamma(S_{\hat{\delta}(n'_k)}, \hat{\sigma}_{n'_k}; S^Z) = 0 \quad \text{a.s.},$$

where the last equality holds because of Lemma 3.5.2. Hence, $\gamma(S_\infty, \sigma_\infty; S^Z) = 0$, and using continuity again, we conclude that $S_\infty \star \phi_{\sigma_\infty} = S^X \star \phi_\sigma$. Or equivalently, in terms of distributions, $P_\infty \star \phi_{\sigma_\infty} = P^X \star \mathcal{N}_\sigma$. As all the distributions $P_{\hat{\delta}(n'_k)}$ have their mass on the positive axis, Lemma 3.5.4 implies that $P_\infty \in \mathcal{P}_0$, and hence that $P_\infty = P^X$ and $\sigma_\infty = \sigma$, which contradicts the assumption and thus concludes the proof. □

Proof of Proposition 3.4.1

We begin the proof by plugging-in the sequence m_n into the robust estimator and by splitting up the integral occurring in (3.11) into

$$\int_0^\infty \left\{S_{\hat{\delta}(n)}(u)\right\}^{m_n} du = \int_0^{\varphi(y)} \left\{S_{\hat{\delta}(n)}(u)\right\}^{m_n} du + \int_{\varphi(y)}^\infty \left\{S_{\hat{\delta}(n)}(u)\right\}^{m_n} du =: A_n + B_n$$

with obvious definitions for A_n and B_n. We have that $B_n \to 0$ almost surely as n tends to infinity. To see this, let $t_n := \varphi(y) \vee \sup\{t \in \mathbb{R} \;:\; S_{\hat{\delta}(n)}(t) = 1\}$ and decompose B_n further into

$$\int_{\varphi(y)}^\infty \{S_{\hat{\delta}(n)}(u)\}^{m_n} du = \int_{\varphi(y)}^{t_n} 1\, du + \int_{t_n}^\infty \{S_{\hat{\delta}(n)}(u)\}^{m_n} du. \tag{A.1}$$

Firstly, $t_n \to \varphi(y)$ as $n \to \infty$ because of the consistency of $S_{\hat{\delta}(n)}$. Therefore, the first integral on the right hand side of (A.1) tends to 0 as $n \to \infty$. Secondly, $S_{\hat{\delta}(n)}$ is non-increasing and strictly smaller than 1 on (t_n, ∞) for every $n \in \mathbb{N}$. As the sequence $S_{\hat{\delta}(n)}$ is further surely point-wise convergent on $\mathbb{R}$, the other integral of the decomposition in (A.1) also tends to 0.

It remains to show that $A_n \to \varphi(y)$ almost surely as $n \to \infty$. Since $S_{\hat{\delta}(n)}$ is non-increasing and $S_{\hat{\delta}(n)}(0) = 1$, we have that $s_n \leqslant \varphi(y)$. On the other hand, $s_n \geqslant \varphi(y)\,\{S_{\hat{\delta}(n)}(\varphi(y))\}^{m_n}$, which proves the result by virtue of the assumption. □

Acknowledgements This work was supported by the "Agence National de la Recherche" under contract ANR-09-JCJC-0124-01 and by the IAP research network nr P6/03 of the Belgian Government (Belgian Science Policy). Comments from Ingrid Van Keilegom and an anonymous referee were most helpful to improve the final version of the manuscript. The usual disclaimer applies.

References

Bigot, J., & Van Bellegem, S. (2009). Log-density deconvolution by wavelet thresholding. *Scandinavian Journal of Statistics, 36*, 749–763.

Butucea, C., & Matias, C. (2005). Minimax estimation of the noise level and of the deconvolution density in a semiparametric deconvolution model. *Bernoulli, 11*, 309–340.

Butucea, C., Matias, C., & Pouet, C. (2008). Adaptivity in convolution models with partially known noise distribution. *Electronic Journal of Statistics, 2*, 897–915.

Carroll, R., & Hall, P. (1988). Optimal rates of convergence for deconvolving a density. *Journal of the American Statistical Association, 83*, 1184–1186.

Cazals, C., Florens, J. P., & Simar, L. (2002). Nonparametric frontier estimation: a robust approach. *Journal of Econometrics, 106*, 1–25.

Daouia, A., Florens, J., & Simar, L. (2009). *Regularization in nonparametric frontier estimators* (Discussion Paper No. 0922). Université catholique de Louvain, Belgium: Institut de statistique, biostatistique et sciences actuarielles.

Daskovska, A., Simar, L., & Van Bellegem, S. (2010). Forecasting the Malmquist productivity index. *Journal of Productivity Analysis, 33*, 97–107.

De Borger, B., Kerstens, K., Moesen, W., & Vanneste, J. (1994). A non-parametric free disposal hull (FDH) approach to technical efficiency: an illustration of radial and graph efficiency measures and some sensitivity results. *Swiss Journal of Economics and Statistics, 130*, 647–667.

Delaigle, A., Hall, P., & Meister, A. (2008). On deconvolution with repeated measurements. *The Annals of Statistics, 36*, 665–685.

Deprins, D., Simar, L., & Tulkens, H. (1984). Measuring labor inefficiency in post offices. In M. Marchand, P. Pestieau, & H. Tulkens (Eds.), *The performance of public enterprises: Concepts and measurements* (pp. 243–267). Amsterdam: North-Holland.

Fan, J. (1991). On the optimal rate of convergence for nonparametric deconvolution problems. *The Annals of Statistics, 19*, 1257–1272.

Färe, R., Grosskopf, S., & Knox Lovell, C. (1985). *The measurements of efficiency of production* (Vol. 6). New York: Springer.

Hall, P., & Simar, L. (2002). Estimating a changepoint, boundary, or frontier in the presence of observation error. *Journal of the American Statistical Association, 97*, 523-534.

Johannes, J., & Schwarz, M. (2009). *Adaptive circular deconvolution by model selection under unknown error distribution* (Discussion Paper No. 0931). Université catholique de Louvain, Belgium: Institut de statistique, biostatistique et sciences actuarielles.

Johannes, J., Van Bellegem, S., & Vanhems, A. (2010). Convergence rates for ill-posed inverse problems with an unknown operator. *Econometric Theory*. (forthcoming)

Johnstone, I., Kerkyacharian, G., Picard, D., & Raimondo, M. (2004). Wavelet deconvolution in a periodic setting. *Journal of the Royal Statistical Society Series B, 66*, 547–573.

Kneip, A., Park, B., & Simar, L. (1998). A note on the convergence of nonparametric DEA estimators for production efficiency scores. *Econometric Theory, 14*, 783–793.

Leleu, H. (2006). A linear programming framework for free disposal hull technologies and cost functions: Primal and dual models. *European Journal of Operational Research, 168*, 340–344.

Li, T., & Vuong, Q. (1998). Nonparametric estimation of the measurement error model using multiple indicators. *Journal of Multivariate Analysis, 65*, 139–165.

Meister, A. (2006). Density estimation with normal measurement error with unknown variance. *Statistica Sinica*, *16*, 195–211.

Meister, A. (2007). Deconvolving compactly supported densities. *Mathematical Methods in Statistics*, *16*, 195–211.

Meister, A., Stadtmüller, U., & Wagner, C. (2010). Density deconvolution in a two-level heteroscedastic model with unknown error density. *Electronic Journal of Statistics*, *4*, 36–57.

Neumann, M. H. (2007). Deconvolution from panel data with unknown error distribution. *Journal of Multivariate Analysis*, *98*, 1955–1968.

Park, B. U., Sickles, R. C., & Simar, L. (2003). Semiparametric efficient estimation of AR(1) panel data models. *Journal of Econometrics*, *117*, 279-311.

Park, B. U., Simar, L., & Weiner, C. (2000). The FDH estimator for productivity efficiency scores: asymptotic properties. *Econometric Theory*, *16*, 855–877.

Pensky, M., & Vidakovic, B. (1999). Adaptive wavelet estimator for nonparametric density deconvolution. *The Annals of Statistics*, *27*, 2033–2053.

Schwarz, M., & Van Bellegem, S. (2010). Consistent density deconvolution under partially known error distribution. *Statistics and Probability Letters*, *80*, 236–241.

Seiford, L., & Thrall, R. (1990). Recent developments in DEA: The mathematical programming approach to frontier analysis. *Journal of Econometrics*, *46*, 7–38.

Shephard, R. W. (1970). *Theory of cost and production functions*. Princeton, NJ: Princeton University Press.

Simar, L. (2007). How to improve the performances of DEA/FDH estimators in the presence of noise? *Journal of Productivity Analysis*, *28*, 183–201.

Chapter 4
Estimating Frontier Cost Models Using Extremiles

Abdelaati Daouia and Irène Gijbels

Abstract In the econometric literature on the estimation of production technologies, there has been considerable interest in estimating so called cost frontier models that relate closely to models for extreme non-standard conditional quantiles (Aragon et al. Econ Theor 21:358–389, 2005) and expected minimum input functions (Cazals et al. J Econometrics 106:1–25, 2002). In this paper, we introduce a class of extremile-based cost frontiers which includes the family of expected minimum input frontiers and parallels the class of quantile-type frontiers. The class is motivated via several angles, which reveals its specific merits and strengths. We discuss nonparametric estimation of the extremile-based costs frontiers and establish asymptotic normality and weak convergence of the associated process. Empirical illustrations are provided.

4.1 Introduction

In the analysis of productivity and efficiency, for example of firms, the interest lies in estimating a production frontier or cost function. Among the basic references in economic theory are Koopmans (1951), Debreu (1951) and Shephard (1970). The activity of a production unit (e.g. a firm) is characterized via a set of outputs, $y \in \mathbb{R}_+^q$ that is produced by a set of inputs $x \in \mathbb{R}_+^p$. The set of attainable points can be characterized as

---

A. Daouia
Toulouse School of Economics (GREMAQ), University of Toulouse, Manufacture des Tabacs, Aile J.J. Laffont, 21 Allée de Brienne, 31000 Toulouse, France
e-mail: daouia@cict.fr

I. Gijbels (✉)
Department of Mathematics and Leuven Statistics Research Center, Katholieke Universiteit Leuven, Celestijnenlaan 200B, Box 2400, 3001 Leuven (Heverlee), Belgium
e-mail: Irene.Gijbels@wis.kuleuven.be

I. van Keilegom and P.W. Wilson (eds.), *Exploring Research Frontiers in Contemporary Statistics and Econometrics*, DOI 10.1007/978-3-7908-2349-3_4,

$$\Psi = \{(y, x) \in \mathbb{R}_+^{q+p} | \ y \text{ can be produced by } x\}.$$

This set can be described mathematically by its sections. In the input space one has the input requirement sets $C(y) = \{x \in \mathbb{R}_+^p | (y, x) \in \Psi\}$, defined for all possible outputs $y \in \mathbb{R}_+^q$. The radial (or input-oriented) efficiency boundary is then given by $\partial C(y)$, the boundary of the input requirement set. In the case of univariate inputs $\partial C(y) = \min C(y)$, the input-efficiency function, also called the frontier cost function. From an economic point of view, a monotonicity assumption on this function is reasonable, meaning that higher outputs go along with a higher minimal cost. Different other assumptions can be made on Ψ such as for example free disposability, i.e. if $(y, x) \in \Psi$ then $(y', x') \in \Psi$ for any $x' \geq x$ and $y' \leq y$ (the inequalities here have to be understood componentwise); or convexity, i.e., every convex combination of feasible production plans is also feasible. See Shephard (1970) for more information and economic background.

In this paper we will focus the presentation on the input orientation[1], where we want to estimate the minimal cost frontier in the case of univariate inputs. To our disposal are observations $\mathcal{X}_n = \{(Y_i, X_i) | \ i = 1, \cdots, n\}$ generated by the production process defined through for example the joint distribution of a random vector (Y, X) on $\mathbb{R}_+^q \times \mathbb{R}_+$, where the q-dimensional vector Y represents the outputs and the second variable X is the single input. Let $(\Omega, \mathcal{A}, \mathbb{P})$ be the probability space on which both Y and X are defined. In the case where the production set Ψ is equal to the support of the joint distribution of (Y, X), a probabilistic way for defining the cost frontier is as follows. The cost function $\partial C(y)$ is characterized for a given set of outputs y by the lower boundary of the support of the conditional distribution of X given $Y \geq y$, i.e.

$$\varphi(y) := \inf\{x \geq 0 | \bar{F}_y(x) < 1\} \equiv \partial C(y), \tag{4.1}$$

where $\bar{F}_y(x) = 1 - F_y(x)$, with $F_y(x) = \mathbb{P}(X \leq x | Y \geq y)$ being the conditional distribution function of X given $Y \geq y$, for y such that $\mathbb{P}(Y \geq y) > 0$. The frontier function φ is monotone nondecreasing, which corresponds to the free disposability property of the support Ψ. When the support boundary $\partial C(\cdot)$ is not assumed to be monotone, $\varphi(\cdot)$ is in fact the largest monotone function which is smaller than or equal to the lower boundary $\partial C(\cdot)$. See Cazals et al. (2002) for this formulation and a detailed discussion on the concept of frontier cost function.

There is a vast literature on the estimation of frontier functions from a random sample of production units $\mathcal{X}_n$. There have been developments along two main approaches: the deterministic frontier models which suppose that with probability one, all the observations in $\mathcal{X}_n$ belong to Ψ, and the stochastic frontier models where random noise allows some observations to be outside of Ψ.

[1]The presentation for the output orientation, where we want to estimate the maximal production frontier in the case of univariate outputs, is a straightforward adaptation of what is done here.

In deterministic frontier models, two different nonparametric methods based on envelopment techniques have been around. The free disposal hull (FDH) technique and the data envelopment analysis (DEA) technique. Deprins et al. (1984) introduced the FDH estimator that relies only on the free disposability assumption on Ψ. The DEA estimator initiated by Farrell (1957) and popularized as linear programming estimator by Charnes et al. (1978), requires stronger assumptions, it relies on the free disposability assumption and the convexity of Ψ. Such a convexity assumption is widely used in economics, but it is not always valid. Because of the additional assumption of convexity the FDH estimator is a more general estimator than the DEA estimator. The asymptotic distribution of the FDH estimator was derived by Park et al. (2000) in the particular case where the joint density of (Y, X) has a jump at the frontier and by Daouia et al. (2010) in the general setting. The asymptotic distribution of the DEA estimator was derived by Gijbels et al. (1999). Today, most statistical theory of these estimators is available. See Simar and Wilson (2008), among others.

In stochastic frontier models, where noise is allowed, one often imposes parametric restrictions on the shape of the frontier and on the data generating process to allow identification of the noise from the cost frontier and subsequently estimation of this frontier. These parametric methods may lack robustness if the distributional assumptions made do not hold.

Since nonparametric deterministic frontier models rely on very few assumptions, they are quite appealing. Moreover, the FDH estimator of the frontier cost function can simply be viewed as a plug-in version of (4.1) obtained by just replacing the conditional distribution function by its empirical analog $\widehat{F}_y(x)$ resulting into

$$\hat{\varphi}(y) = \inf\{x \in \mathbb{R}_+ | \hat{F}_y(x) > 0\} = \min_{i:Y_i \geq y} X_i \ ,$$

with $\hat{F}_y(x) = \sum_{i=1}^{n} \mathbb{1}(X_i \leq x, Y_i \geq y) / \sum_{i=1}^{n} \mathbb{1}(Y_i \geq y)$.

The FDH estimator, as well as the DEA estimator, however are very sensitive to outlying observations. In the literature two robust nonparametric estimators of (partial) cost frontiers have been proposed to deal with this sensitivity. Cazals et al. (2002) introduced the concept of expected minimal cost of order $m \in \{1, 2, 3, ...\}$. It is defined as the expected minimal cost among m firms drawn in the population of firms exceeding a certain level of outputs. More precisely, for a given level of outputs y, the cost function of order m is given by

$$\varphi_m(y) = E[\min(X_1^y, \cdots, X_m^y)] = \int_0^{\infty} \{\bar{F}_y(x)\}^m dx,$$

where $(X_1^y, \cdots, X_m^y)$ are m independent identically distributed random variables generated from the distribution of X given $Y \geq y$. Its nonparametric estimator is defined by

$$\hat{\varphi}_{m,n}(y) = \int_0^{\infty} \{1 - \hat{F}_y(x)\}^m dx.$$

The estimator $\hat{\varphi}_{m,n}(y)$ does not envelop all the data points, and so it is more robust to extreme values than the FDH estimator $\hat{\varphi}(y)$. By choosing m appropriately as a function of the sample size n, $\hat{\varphi}_{m,n}(y)$ estimates the cost function $\varphi(y)$ itself while keeping the asymptotic properties of the FDH estimator.

A second approach to deal with the sensitivity to outlying observations was proposed by Aragon et al. (2005). They consider extreme quantiles of the conditional distribution of X given $Y \geq y$. Such non-standard conditional quantiles provide another concept of a partial cost frontier as an alternative towards the order-m partial cost frontier introduced by Cazals et al. (2002). The duality between expected minimum input frontiers and quantile-type cost frontiers has been investigated by Daouia and Gijbels (2011).

In this paper we introduce a new class of extremile-based cost frontiers which includes the class of order-m expected minimum input frontiers. The class also parallels the class of quantile partial cost frontiers in the sense that it is related to the mean of a random variable rather than the median (or quantile more generally).

The chapter is organized as follows. In Sect. 4.2 we introduce the class of extremile-based cost frontier functions, and discuss the relation with the order-m partial cost functions and the quantile-type cost functions. Some basic properties of the new class of frontier functions are provided. Section 4.3 is devoted to nonparametric estimation of an extremile-based cost frontier, and studies the asymptotic properties of the estimators. An empirical study on a simulation model and on a real data example is provided in Sect. 4.4. Section 4.5 concludes.

4.2 The Extremile-Based Cost Function

Consider a real $\gamma \in (0,1)$ and let K_γ be a measure on $[0,1]$ whose distribution function is

$$K_\gamma(t) = \begin{cases} 1-(1-t)^{s(\gamma)} & \text{if } 0 < \gamma \leq \frac{1}{2} \\ t^{s(1-\gamma)} & \text{if } \frac{1}{2} \leq \gamma < 1 \end{cases}$$

where

$$s(\gamma) = \frac{\log(1/2)}{\log(1-\gamma)} \geq 1 \quad \text{for} \quad \gamma \in [0, 1/2].$$

Define the score function $J_\gamma(\cdot)$ to be the density of the measure K_γ on $(0,1)$.

Definition 4.3. *The extremile-based cost function of order γ denoted by $\xi_\gamma(y)$ is the real function defined on $\mathbb{R}^q_+$ as*

$$\xi_\gamma(y) = \mathbb{E}\left[XJ_\gamma\left(F_y(X)\right)|Y \geq y\right]$$

where we assume the existence of this expectation.

As a matter of fact, the partial cost function $\xi_\gamma(y)$ coincides with the γth extremile (see Daouia and Gijbels 2009) of the conditional distribution of X given $Y \geq y$. The following proposition is a basic property of extremiles.

Proposition 4.4. *If* $\mathbb{E}(X|Y \geq y) < \infty$*, then* $\xi_\gamma(y)$ *exists for all* $\gamma \in (0,1)$.

Proof. Following Daouia and Gijbels (2009, Proposition 1 (ii)), the γth extremile exists provided that the underlying distribution has a finite absolute mean which corresponds here to $\mathbb{E}(X|Y \geq y) < \infty$. □

From an economic point of view, the quantity X of inputs-usage is often assumed to be bounded or at least to have a finite mean and so, in this case, the γth cost function is well defined for any order γ in $(0,1)$ and all $y \in \mathbb{R}^q_+$ such that $\mathbb{P}(Y \geq y) > 0$.

More specifically, the extremile function $\xi_\gamma(y)$ is proportional to conditional probability-weighted moments:

$$\xi_\gamma(y) = \begin{cases} s(\gamma)\mathbb{E}\left[X\,\{\bar{F}_y(X)\}^{s(\gamma)-1}\,|Y \geq y\right] & \text{for } \; 0 < \gamma \leq \frac{1}{2} \\ s(1-\gamma)\mathbb{E}\left[X\,\{F_y(X)\}^{s(1-\gamma)-1}\,|Y \geq y\right] & \text{for } \; \frac{1}{2} \leq \gamma < 1. \end{cases}$$

In the special case where $\gamma \leq 1/2$ with $s(\gamma)$ being a positive integer, $\xi_\gamma(y)$ equals the expectation of the minimum of $s(\gamma)$ independent random variables $(X_1^y, \ldots, X_{s(\gamma)}^y)$ generated from the distribution of X given $Y \geq y$. Whence

$$\xi_\gamma(y) = \mathbb{E}\left[\min\left(X_1^y, \ldots, X_{s(\gamma)}^y\right)\right] = \varphi_{s(\gamma)}(y).$$

Thus the class of our conditional extremiles includes the family of expected minimum input functions introduced by Cazals et al. (2002). Likewise, if $\gamma \geq 1/2$ with $s(1-\gamma) = 1, 2, \ldots$ we have $\xi_\gamma(y) = \mathbb{E}\left[\max\left(X_1^y, \ldots, X_{s(1-\gamma)}^y\right)\right]$, where $X_1^y, \ldots, X_{s(1-\gamma)}^y$ are independent random variables generated from the distribution of X given $Y \geq y$.

Proposition 4.5. *If the conditional distribution of* X *given* $Y \geq y$ *has a finite mean, then it can be characterized by the subclass* $\{\xi_\gamma(y) : s(\gamma) = 1, 2, \ldots\}$ *or* $\{\xi_\gamma(y) : s(1-\gamma) = 1, 2, \ldots\}$.

Proof. This follows from the well known result of Chan (1967) which states that a distribution with finite absolute first moment can be uniquely defined by its expected maxima or expected minima. □

The non-standard conditional distribution of X given $Y \geq y$ whose $\mathbb{E}(X|Y \geq y) < \infty$ is uniquely defined by its discrete extremiles. This means that no two such non-standard distributions with finite means have the same expected minimum input functions.

Of particular interest is the left tail $\gamma \leq 1/2$ where the partial γth cost function has the following interpretation

$$\mathbb{E}\left[\min\left(X_1^y,\ldots,X_{[s(\gamma)]+1}^y\right)\right] \leq \xi_\gamma(y) \leq \mathbb{E}\left[\min\left(X_1^y,\ldots,X_{[s(\gamma)]}^y\right)\right]$$

where $[s(\gamma)]$ denotes the integer part of $s(\gamma)$ and $X_1^y, X_2^y, \ldots$ are iid random variables of distribution function F_y. In other words, we have $\varphi_{[s(\gamma)]+1}(y) \leq \xi_\gamma(y) \leq \varphi_{[s(\gamma)]}(y)$ for $\gamma \leq 1/2$. Hence $\xi_\gamma(y)$ benefits from a similar "benchmark" interpretation as expected minimum input functions. For the manager of a production unit working at level (x, y), comparing its inputs-usage x with the benchmarked value $\xi_\gamma(y)$, for a sequence of few decreasing values of $\gamma \searrow 0$, could offer a clear indication of how efficient its firm is compared with a fixed number of $(1 + [s(\gamma)])$ potential firms producing more than y.

Yet, there is still another way of looking at $\xi_\gamma(y)$. Let $\mathcal{X}_\gamma^y$ be a random variable having cumulative distribution function

$$F_{\mathcal{X}_\gamma^y} = \begin{cases} 1 - \{\bar{F}_y\}^{s(\gamma)} & \text{if } 0 < \gamma \leq \frac{1}{2} \\ \{F_y\}^{s(1-\gamma)} & \text{if } \frac{1}{2} \leq \gamma < 1. \end{cases}$$

Proposition 4.6. *We have $\xi_\gamma(y) = \mathbb{E}(\mathcal{X}_\gamma^y)$ provided this expectation exists.*

Proof. Since $\mathbb{E}|\mathcal{X}_\gamma^y| = \mathbb{E}(\mathcal{X}_\gamma^y) < \infty$, we have $\mathbb{E}(\mathcal{X}_\gamma^y) = \int_0^1 F_{\mathcal{X}_\gamma^y}^{-1}(t)dt$ in view of a general property of expectations (see Shorack 2000, p.117). On the other hand, it is easy to check that $\xi_\gamma(y) = \int_0^1 J_\gamma(t) F_y^{-1}(t)dt = \int_0^1 F_y^{-1}(t)dK_\gamma(t) = \int_0^1 F_{\mathcal{X}_\gamma^y}^{-1}(t)dt$. □

This allows to establish how our class of extremile-based cost functions is related to the family of quantile-based cost functions defined by Aragon et al. (2005) as

$$Q_\gamma(y) = F_y^{-1}(\gamma) := \inf\{x \in \mathbb{R}_+ |\ F_y(x) \geq \gamma\} \quad \text{for} \quad 0 < \gamma < 1.$$

Indeed, while $\xi_\gamma(y)$ equals the mean of the random variable $\mathcal{X}_\gamma^y$, it is easy to see that the quantile function $Q_\gamma(y)$ coincides with the median of the same variable $\mathcal{X}_\gamma^y$. Consequently the γth extremile-based cost function is clearly more tail sensitive and more efficient than the γth quantile-based cost function. The latter means that the (asymptotic) variance for the extremile-based cost function estimator is smaller than the (asymptotic) variance of the γth quantile-based cost function. Recall that for many population distributions (such as e.g. a normal distribution) the sample mean has a smaller asymptotic variance than the sample median, when both are estimating the same quantity. See for example Serfling (1980).

One way of defining $\xi_\gamma(y)$, with $0 \leq \gamma \leq 1$, is as the following explicit quantity.

Proposition 4.7. *If* $\mathbb{E}(\mathcal{X}_\gamma^y) < \infty$, *we have*

$$\xi_\gamma(y) = \begin{cases} \varphi(y) + \int_{\varphi(y)}^{\infty} \{\bar{F}_y(x)\}^{s(\gamma)}\, dx & \text{for } 0 \le \gamma \le \frac{1}{2} \\ \varphi(y) + \int_{\varphi(y)}^{\infty} \left(1 - \{F_y(x)\}^{s(1-\gamma)}\right) dx & \text{for } \frac{1}{2} \le \gamma \le 1. \end{cases} \quad (4.2)$$

Proof. We have $\xi_\gamma(y) = \mathbb{E}(\mathcal{X}_\gamma^y)$ by Proposition 4.6 and $\mathbb{E}(\mathcal{X}_\gamma^y) = \int_0^\infty \left\{1 - F_{\mathcal{X}_\gamma^y}(x)\right\} dx = \varphi(y) + \int_{\varphi(y)}^\infty \left\{1 - F_{\mathcal{X}_\gamma^y}(x)\right\} dx$ by a general property of expectations (Shorack 2000, p.117). □

This explicit expression is very useful when it comes to proposing an estimator for $\xi_\gamma(y)$. Obviously, the central extremile-based cost function $\xi_{1/2}(y)$ reduces to the regression function $\mathbb{E}(X|Y \ge y)$. The conditional extremile $\xi_\gamma(y)$ is clearly a continuous and increasing function in γ and it maps $(0, 1)$ onto the range $\{x \ge 0 | 0 < F_y(x) < 1\}$. The left and right endpoints of the support of the conditional distribution function $F_y(\cdot)$ coincide respectively with the lower and upper extremiles $\xi_0(y)$ and $\xi_1(y)$ since $s(0) = \infty$. Hence the range of $\xi_\gamma(y)$ is the entire range of X given $Y \ge y$.

Of interest is the limiting case $\gamma \downarrow 0$ which leads to access the full cost function $\varphi(y) = \xi_0(y)$. Although the limit frontier function $\varphi(\cdot)$ is monotone nondecreasing, the partial cost function $\xi_\gamma(\cdot)$ itself is not necessarily monotone. To ensure the monotonicity of $\xi_\gamma(y)$ in y, it suffices to assume, as it can be easily seen from Proposition 4.7, that the conditional survival function $\bar{F}_y(x)$ is nondecreasing in y. As pointed out by Cazals et al. (2002), this assumption is quite reasonable from an economic point of view since the chance of spending more than a cost x does not decrease if a firm produces more.

The next proposition provides an explicit relationship between the γth quantile and extremile type cost functions at $\gamma \downarrow 0$. Let $\mathrm{DA}(W_\rho)$ denote the minimum domain of attraction of the Weibull extreme-value distribution

$$W_\rho(x) = 1 - \exp\{-x^\rho\} \quad \text{with support} \quad [0, \infty), \quad \text{for some} \quad \rho > 0,$$

i.e., the set of distribution functions whose asymptotic distributions of minima are of the type of W_ρ. According to Daouia et al. (2010), if there exists a sequence $\{a_n > 0\}$ such that the normalized minima $a_n^{-1}(\hat{\varphi}(y) - \varphi(y))$ converges to a non-degenerate distribution, then the limit distribution function is of the type of W_ρ for a positive function $\rho = \rho(y)$ in y.

Proposition 4.8. *Suppose* $\mathbb{E}(X|Y \ge y) < \infty$ *and* $F_y(\cdot) \in DA(W_{\rho(y)})$. *Then*

$$\frac{\xi_\gamma(y) - \varphi(y)}{Q_\gamma(y) - \varphi(y)} \sim \Gamma(1 + \rho^{-1}(y))\{\log 2\}^{-1/\rho(y)} \quad as \quad \gamma \downarrow 0,$$

where $\Gamma(\cdot)$ *denotes the gamma function.*

Proof. This follows immediately by applying Proposition 2 (ii) in Daouia and Gijbels (2009) to the distribution of $-X$ given $Y \geq y$. □

Consequently, as $\gamma \downarrow 0$, the quantile curve $Q_\gamma(\cdot)$ is closer to the true cost frontier $\varphi(\cdot)$ than is the extremile curve $\xi_\gamma(\cdot)$ following the value of the tail index ρ. In most situations described so far in the econometric literature on frontier analysis, it is assumed that there is a jump of the joint density of (Y, X) at the frontier: this corresponds to the case where the tail index $\rho(y)$ equals the dimension of data $(1+q)$ according to Daouia et al. (2010). It was shown in that paper that $\beta(y) = \rho(y) - (1+q)$, where $\beta(y)$ denotes the algebraic rate with which the joint density decreases to 0 when x approaches the point at the frontier function. Since a jump of the joint density at the frontier implies that $\beta(y) = 0$, it follows that $\rho(y) = 1 + q$ in that case. In such situations, $Q_\gamma(\cdot)$ is asymptotically closer to $\varphi(\cdot)$ than is $\xi_\gamma(\cdot)$ when $q \leq 2$, but $\xi_\gamma(\cdot)$ is more spread than $Q_\gamma(\cdot)$ when $q > 2$.

On the other hand, the score function $J_\gamma(\cdot)$ being monotone increasing for $\gamma \geq 1/2$ and decreasing for $\gamma \leq 1/2$, the conditional extremile $\xi_\gamma(y)$ depends by construction on all feasible values of X given $Y \geq y$ putting more weight to the high values for $\gamma \geq 1/2$ and more weight to the low values for $\gamma \leq 1/2$. Therefore $\xi_\gamma(y)$ is sensible to the magnitude of extreme values for any order $\gamma \in (0,1)$. In contrast, the conditional quantile $Q_\gamma(y)$ is determined solely by the tail probability (relative frequency) γ, and so it may be unaffected by desirable extreme values whatever the shape of tails of the underlying distribution. On the other hand, when $Q_\gamma(y)$ breaks down at $\gamma \downarrow 0$ or $\gamma \uparrow 1$, the γth conditional extremile, being an L-functional, is more resistant to extreme values. Hence, $\xi_\gamma(y)$ steers an advantageous middle course between the extreme behaviors (insensitivity and breakdown) of $Q_\gamma(y)$.

4.3 Nonparametric Estimation

Instead of estimating the full cost function, an original idea first suggested by Cazals et al. (2002) and applied by Aragon et al. (2005) to quantiles is rather to estimate a partial cost function lying near $\varphi(y)$. Thus the interest in this section will be in the estimation of the extremile function $\xi_\gamma(y)$ for $\gamma \leq 1/2$. The right tail (i.e. $\gamma \geq 1/2$) can be handled in a similar way and so is omitted. Results below are easily derived by means of L-statistics theory applied to the dimensionless transformation $Z^y = X\mathbb{1}(Y \geq y)$ of the random vector $(Y, X) \in \mathbb{R}_+^{q+1}$. Let $\bar{F}_{Z^y} = 1 - F_{Z^y}$ be the survival function of Z^y. It is easy to check that $\bar{F}_y(X) = \bar{F}_{Z^y}(Z^y)/\mathbb{P}(Y \geq y)$. Then

$$\begin{aligned} \xi_\gamma(y) &= \mathbb{E}\left[Z^y J_\gamma\left(F_y(X)\right)\right]/\mathbb{P}(Y \geq y) \quad \text{for} \quad 0 < \gamma < 1 \\ &= \frac{s(\gamma)}{\mathbb{P}(Y \geq y)}\mathbb{E}\left[Z^y\left\{\frac{\bar{F}_{Z^y}(Z^y)}{\mathbb{P}(Y \geq y)}\right\}^{s(\gamma)-1}\right] = \frac{\xi_{Z^y}(\gamma)}{\{\mathbb{P}(Y \geq y)\}^{s(\gamma)}} \quad \text{for} \quad 0 < \gamma \leq \frac{1}{2} \end{aligned} \tag{4.3}$$

where $\xi_{Z^y}(\gamma) = s(\gamma)\mathbb{E}\left[Z^y\left\{\bar{F}_{Z^y}(Z^y)\right\}^{s(\gamma)-1}\right] = \int_0^1 F_{Z^y}^{-1} dK_\gamma$ is by definition the ordinary γth extremile of the random variable Z^y. Therefore it suffices to replace $\mathbb{P}(Y \geq y)$ by its empirical version $\hat{\mathbb{P}}(Y \geq y) = (1/n)\sum_{i=1}^n \boldsymbol{1}(Y_i \geq y)$ and $\xi_{Z^y}(\gamma)$ by a consistent estimator to obtain a convergent estimate of the conditional extremile $\xi_\gamma(y)$.

As shown in Daouia and Gijbels (2009), a natural estimator of the ordinary extremile $\xi_{Z^y}(\gamma)$ is given by the L-statistic generated by the measure K_γ:

$$\hat{\xi}_{Z^y}(\gamma) = \sum_{i=1}^n \left\{K_\gamma\left(\frac{i}{n}\right) - K_\gamma\left(\frac{i-1}{n}\right)\right\} Z_{(i)}^y, \tag{4.4}$$

where $Z_{(1)}^y \leq Z_{(2)}^y \leq \cdots \leq Z_{(n)}^y$ denote the order statistics generated by the sample $\{Z_i^y = X_i \boldsymbol{1}(Y_i \geq y) : i = 1, \cdots, n\}$. It is easy to see that the resulting estimator of the γth cost function $\xi_\gamma(y)$, given by

$$\hat{\xi}_\gamma(y) = \hat{\xi}_{Z^y}(\gamma)/\{\hat{\mathbb{P}}(Y \geq y)\}^{s(\gamma)}$$

coincides with the empirical conditional extremile obtained by replacing $F_y(x)$ in expression (4.2) with its empirical version $\hat{F}_y(x)$, i.e.,

$$\hat{\xi}_\gamma(y) = \int_0^\infty \left\{1 - \hat{F}_y(x)\right\}^{s(\gamma)} dx = \hat{\varphi}(y) + \int_{\hat{\varphi}(y)}^\infty \left\{1 - \hat{F}_y(x)\right\}^{s(\gamma)} dx. \tag{4.5}$$

This estimator converges to the FDH input efficient frontier $\hat{\varphi}(y)$ as γ decreases to zero. In particular, when the power $s(\gamma)$ is a positive integer $m = 1, 2, \ldots$ we recover the estimator $\hat{\varphi}_{m,n}(y)$, of the expected minimum input function of order m proposed by Cazals et al. (2002). See Sect. 4.1. The following theorem summarizes the asymptotic properties of $\hat{\xi}_\gamma(y)$ for a fixed order γ.

Theorem 4.10. *Assume that the support of* (Y, X) *is compact and let* $\gamma \in (0, 1/2]$.

(i) For any point $y \in \mathbb{R}_+^q$ *such that* $P(Y \geq y) > 0$, $\hat{\xi}_\gamma(y) \overset{a.s.}{\to} \xi_\gamma(y)$ *as* $n \to \infty$, *and* $\sqrt{n}\left(\hat{\xi}_\gamma(y) - \xi_\gamma(y)\right)$ *has an asymptotic normal distribution with mean zero and variance* $\mathbb{E}\left[\mathbb{S}_\gamma(y, Y, X)\right]^2$, *where* $\mathbb{S}_\gamma(y, Y, X) = \frac{s(\gamma)}{\{\mathbb{P}(Y \geq y)\}^{s(\gamma)}} \int_0^\infty \{\mathbb{P}(X > x, Y \geq y)\}^{s(\gamma)-1} \boldsymbol{1}(X > x, Y \geq y)dx - \frac{s(\gamma)\xi_\gamma(y)}{\mathbb{P}(Y \geq y)} \boldsymbol{1}(Y \geq y)$.

(ii) For any subset $\mathcal{Y} \subset \mathbb{R}_+^q$ *such that* $\inf_{y \in \mathcal{Y}} P(Y \geq y) > 0$, *the process* $\sqrt{n}\left(\hat{\xi}_\gamma(\cdot) - \xi_\gamma(\cdot)\right)$ *converges in distribution in the space of bounded functions on* $\mathcal{Y}$ *to a q-dimensional zero mean Gaussian process indexed by* $y \in \mathcal{Y}$ *with covariance function*

$$\Sigma_{k,l} = \mathbb{E}\left[\mathbb{S}_\gamma(y^k, Y, X)\, \mathbb{S}_\gamma(y^l, Y, X)\right].$$

Proof. Let $m = s(\gamma)$. When $m = 1, 2, \ldots$ the two results (i)–(ii) are given respectively by Theorem 3.1 and Appendix B in Cazals et al. (2002). In fact, it is not hard to verify that the proofs of these results remain valid even when the trimming parameter m is not an integer. □

The conditional distribution function F_y even does not need to be continuous, which is not the case for the empirical conditional quantiles $\hat{Q}_\gamma(y) = \hat{F}_y^{-1}(\gamma)$ whose asymptotic normality requires at least the differentiability of F_y at $Q_\gamma(y)$ with a strictly positive derivative (see Aragon et al. (2002) for the pointwise convergence and Daouia and Gijbels (2009) for the functional convergence).

Next we show that if the FDH estimator $\hat{\varphi}(y)$ converges in distribution, then for a specific choice of γ as a function of n, $\hat{\xi}_\gamma(y)$ estimates the true full cost function $\varphi(y)$ itself and converges in distribution as well to the same limit as $\hat{\varphi}(y)$ and with the same scaling.

Theorem 4.11. *Suppose the support of* (Y, X) *is compact. If* $a_n^{-1}(\hat{\varphi}(y) - \varphi(y)) \xrightarrow{d} W_{\rho(y)}$, *then* $a_n^{-1}\left(\hat{\xi}_{\gamma_y(n)}(y) - \varphi(y)\right) \xrightarrow{d} W_{\rho(y)}$ *provided*

$$\gamma_y(n) \le 1 - \left\{1 - \frac{1}{n\hat{\mathbb{P}}(Y \ge y)}\right\}^{\frac{\log(2)}{(\beta+1)\log(Cn)}} \quad or \quad \gamma_y(n) \le 1 - \exp\left\{\frac{(1 + o(1))\log(1/2)}{(\beta + 1)n\log(Cn)\mathbb{P}(Y \ge y)}\right\},$$

with $\beta > 0$ *such that* $a_n n^\beta \to \infty$ *as* $n \to \infty$, *and* C *being a positive constant.*

Proof. We have $a_n^{-1}\left(\hat{\xi}_\gamma(y) - \varphi(y)\right) = a_n^{-1}(\hat{\varphi}(y) - \varphi(y)) + a_n^{-1}\left(\hat{\xi}_\gamma(y) - \hat{\varphi}(y)\right)$. Let $N_y = \sum_{i=1}^n \mathbb{1}(Y_i \ge y) = \sum_{i=1}^n \mathbb{1}(Z_i^y > 0)$. It is easily seen from (4.5) that

$$\left(\hat{\xi}_\gamma(y) - \hat{\varphi}(y)\right) = \sum_{j=1}^{N_y}\left\{\frac{N_y - j}{N_y}\right\}^{s(\gamma)}\left(Z^y_{(n-N_y+j+1)} - Z^y_{(n-N_y+j)}\right).$$

The support of (Y, X) being compact, the range of Z^y is bounded and so $\left(\hat{\xi}_\gamma(y) - \hat{\varphi}(y)\right) = O\left(n\left\{1 - \frac{1}{N_y}\right\}^{s(\gamma)}\right)$. Then, for the term $a_n^{-1}\left(\hat{\xi}_\gamma(y) - \hat{\varphi}(y)\right)$ to be $o_p(1)$ as $n \to \infty$, it is sufficient to choose $\gamma = \gamma_y(n)$ such that $\left\{1 - \frac{1}{N_y}\right\}^{s(\gamma_y(n))} = O\left(n^{-(\beta+1)}\right)$ or equivalently $\left\{1 - \frac{1}{N_y}\right\}^{s(\gamma_y(n))} \le (Cn)^{-(\beta+1)}$ with $C > 0$ being a constant and $\beta > 0$ is such that $a_n^{-1}n^{-\beta} = o(1)$ as $n \to \infty$. Whence the condition $s(\gamma_y(n)) \ge \frac{(\beta+1)}{\log\left(1-\frac{1}{N_y}\right)}\frac{\log(Cn)}{\log(1/2)}$, or equivalently, $\gamma_y(n) \le 1 - \left\{1 - \frac{1}{N_y}\right\}^{\frac{\log 2}{(\beta+1)\log(Cn)}}$. Since $\log\left(1 - \frac{1}{N_y}\right) \sim -\frac{1}{N_y} \sim -\frac{1}{n\mathbb{P}(Y\ge y)}$ as $n \to \infty$, with probability 1, it suffices to assume that $s(\gamma_y(n)) \ge \frac{(\beta+1)n\log(Cn)\mathbb{P}(Y\ge y)}{\log(2)(1+o(1))}$, or equivalently, $\gamma_y(n) \le 1 - \exp\left\{\frac{(1+o(1))\log(1/2)}{(\beta+1)n\log(Cn)\mathbb{P}(Y\ge y)}\right\}$. □

Note that the condition of Theorem 4.11 on the order $\gamma_y(n)$ is also provided in the proof in terms of $s(\gamma_y(n))$ and reads as follows:

$$s(\gamma_y(n)) \geq \frac{(\beta+1)\log(Cn)}{\log\left(1-\frac{1}{n\widehat{\mathbb{P}}(Y\geq y)}\right)\log(1/2)} \quad \text{or} \quad s(\gamma_y(n)) \geq \frac{(\beta+1)n\log(Cn)\mathbb{P}(Y\geq y)}{\log(2)\,(1+o(1))}. \tag{4.6}$$

Note also that in the particular case considered by Cazals et al. (2002) where the joint density of (Y, X) is strictly positive at the upper boundary and the frontier function $\varphi(y)$ is continuously differentiable in y, the convergence rate a_n satisfies $a_n^{-1} \sim (n\ell_y)^{1/\rho(y)}$ with $\rho(y) = 1 + q$ and $\ell_y > 0$ being a constant (see Park et al. 2000). In this case, the condition $a_n n^{\beta} \to \infty$ reduces to $\beta > 1/(1+q)$.

It should be clear that the main results of Cazals et al. (2002) are corollaries of our Theorems 4.10 and 4.11. Indeed, when the real parameter $s(\gamma) \in [1, \infty)$ in our approach is taken to be a positive integer $m = 1, 2, \ldots$, we recover Theorems 3.1 and 3.2 of Cazals et al. (2002). However, we hope to have shown that the sufficient condition $m_y(n) = O(\beta n \log(n)\mathbb{P}(Y \geq y))$ of Cazals et al. (2002, Theorem 3.2) on the trimming parameter $m_y(n) \equiv s(\gamma_y(n))$ is somewhat premature and should be replaced by (4.6).

Alternative estimators of the conditional extremile $\xi_\gamma(y)$ can be constructed from expression (4.3). Instead of the sample extremile (4.4), one may estimate the ordinary extremile $\xi_{Z^y}(\gamma)$ by

$$\tilde{\xi}_{Z^y}(\gamma) = \frac{1}{n}\sum_{i=1}^{n} J_\gamma\left(\frac{i}{n+1}\right) Z^y_{(i)}.$$

This estimator which is in fact first-order equivalent with $\hat{\xi}_{Z^y}(\gamma)$ (see Daouia and Gijbels 2009) leads to the alternative estimator $\tilde{\xi}_\gamma(y)$ of $\xi_\gamma(y)$ defined as $\tilde{\xi}_\gamma(y) = \tilde{\xi}_{Z^y}(\gamma)/\{\widehat{\mathbb{P}}(Y \geq y)\}^{s(\gamma)}$. In the particular case considered by Cazals et al. (2002) where $s(\gamma)$ is only a positive integer, the statistic

$$\xi^*_{Z^y}(\gamma) = \frac{s(\gamma)}{n}\sum_{i=1}^{n-s(\gamma)+1}\left(\prod_{j=1}^{s(\gamma)-1}\frac{(n-i+1-j)}{(n-j)}\right) Z^y_{(i)}$$

is an unbiased estimator of the ordinary extremile $\xi_{Z^y}(\gamma)$ with the same asymptotic normal distribution as $\hat{\xi}_{Z^y}(\gamma)$ and $\tilde{\xi}_{Z^y}(\gamma)$ (Daouia and Gijbels 2009). This provides an attractive estimator $\xi^*_\gamma(y) = \xi^*_{Z^y}(\gamma)/\{\widehat{\mathbb{P}}(Y \geq y)\}^{s(\gamma)}$ for the order-$s(\gamma)$ expected minimum input function $\xi_\gamma(y) \equiv \varphi_{s(\gamma)}(y)$.

4.4 Empirical Illustration

Let N_y be the number of the Y_i observations greater than or equal to y, i.e. $N_y = \sum_{i=1}^n \mathbb{1}(Y_i \geq y)$, and, for $j = 1, \ldots, N_y$, denote by $X_{(j)}^y$ the jth order statistic of the X_i's such that $Y_i \geq y$. It is then clear that $X_{(j)}^y = Z_{(n-N_y+j)}^y$ for each $j = 1, \ldots, N_y$, and the estimator $\hat{\xi}_\gamma(y)$ can be easily computed as

$$\hat{\xi}_\gamma(y) = \frac{\hat{\xi}_{Z^y}(\gamma)}{\{n^{-1}N_y\}^{s(\gamma)}} = X_{(1)}^y + \sum_{j=1}^{N_y-1} \left(1 - \frac{j}{N_y}\right)^{s(\gamma)} \left\{X_{(j+1)}^y - X_{(j)}^y\right\}.$$

This estimator always lies above the FDH $\hat{\varphi}(y) = X_{(1)}^y$ and so is more robust to extremes and outliers. Moreover, being a linear function of the data, $\hat{\xi}_\gamma(y)$ suffers less than the empirical γth quantile $\hat{Q}_\gamma(y)$ to sampling variability or measurement errors in the extreme values $X_{(j)}^y$. The quantile-based frontier only depends on the frequency of tail costs and not on their values. Consequently, it could be too liberal (insensitive to the magnitude of extreme costs $X_{(j)}^y$) or too conservative following the value of γ. In contrast, putting more weight to high and low observations in the input-orientation, the extremile-based frontier is always sensible to desirable extreme costs. Nevertheless, being a linear function of all the data points (L-statistic), it remains resistant in the sense that it could be only attracted by outlying observations without enveloping them.

We first apply Theorem 4.11 in conjunction with these sensitivity and resistance properties to estimate the optimal cost of the delivery activity of the postal services in France. The data set contains information about 9,521 post offices (Y_i, X_i) observed in 1994, with X_i being the labor cost (measured by the quantity of labor which represents more than 80% of the total cost of the delivery activity) and the output Y_i is defined as the volume of delivered mail (in number of objects). See Cazals et al. (2002) for more details. Here, we only use the $n = 4,000$ observations with the smallest inputs X_i to illustrate the extremile-based estimator $\hat{\xi}_{\gamma_y(n)}(y)$ of the efficient frontier $\varphi(y)$. The important question of how to pick out the order $\gamma_y(n)$ in practice can be addressed as follows.

We know that the condition of Theorem 4.11 provides an upper bound on the value of $\gamma_y(n)$. Remember also that in most situations described so far in the econometric literature on frontier analysis, the joint density of (Y, X) is supposed to be strictly positive at the frontier. In this case, the upper bound for $\gamma_y(n)$ is given by

$$\gamma_{(C)} = 1 - \left\{1 - \frac{1}{N_y}\right\}^{\frac{(1+q)\log(2)}{(2+q)\log(Cn)}},$$

where the number of outputs q equals here 1 and the positive constant C should be selected so that $\log(Cn) \neq 0$, i.e., $C > 1/n$. The practical question now is how

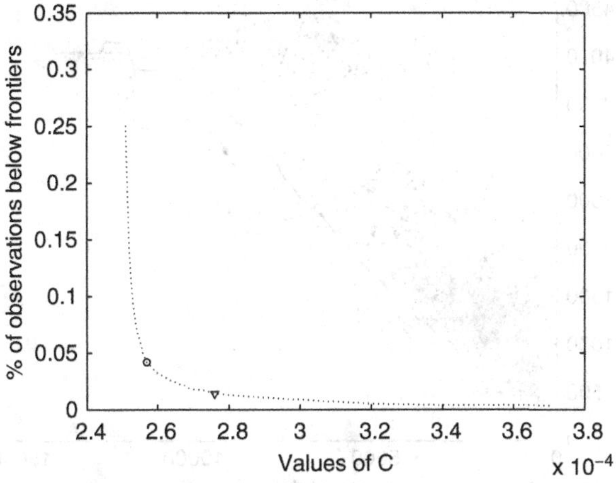

Fig. 4.1 Evolution of the percentage of observations below the frontier $\hat{\xi}_{\gamma(C)}$ with C

to choose $C > 0.00025$ in such a way that $\hat{\xi}_{\gamma(C)}$ provides a reasonable estimate of the frontier function φ. This can be achieved by looking to Fig. 4.1 which indicates how the percentage of points below the curve $\hat{\xi}_{\gamma(C)}$ decreases with the constant C. The idea is to choose values of C for which the frontier estimator $\hat{\xi}_{\gamma(C)}$ is sensible to the magnitude of desirable extreme post offices and, at the same time, is robust to outliers (or at least not being drastically influenced by outliers as is the case for the FDH estimator).

The evolution of the percentage in Fig. 4.1 has clearly an "L" structure. This deviation should appear whatever the analyzed data set due to both sensitivity and resistance properties of extremiles. The percentage falls rapidly until the circle, i.e., for $C \leq 0.000257$. This means that the observations below the frontiers $\{\hat{\xi}_{\gamma(C)} : C < 0.000257\}$ are not really extreme and could be interior observations to the cloud of data points. So it is not judicious to select $C < 0.000257$. In contrast, the percentage becomes very stable from the triangle on (i.e. $C \geq 0.000276$), where precisely 1.4% of the 4000 observations are left out. This means that these few 1.4% observations are really very extreme in the input-direction and could be outlying or perturbed by noise. Although the frontier $\hat{\xi}_{\gamma(C)}$, for $C \geq 0.000276$, is resistant to these suspicious extremes, it can be severely attracted by them due to its sensitivity. This suggests to choose $C < 0.000276$. Thus, our strategy leads to the choice of a constant C ranging over the interval $[0.000257, 0.000276)$ where the decrease of the percentage is rather moderate.

The two extreme (lower and upper) choices of the frontier estimator $\hat{\xi}_{\gamma(C)}$ are graphed in Fig. 4.2, where the solid line corresponds to the lower bound $C_\ell = 0.000257$ and the dotted line corresponds to the upper bound $C_u = 0.000276$. The frontier estimator $\hat{\xi}_{\gamma(C)}$ in dashed line corresponds to the medium value $C_m = (C_\ell + C_u)/2$. The obtained curves are quite satisfactory.

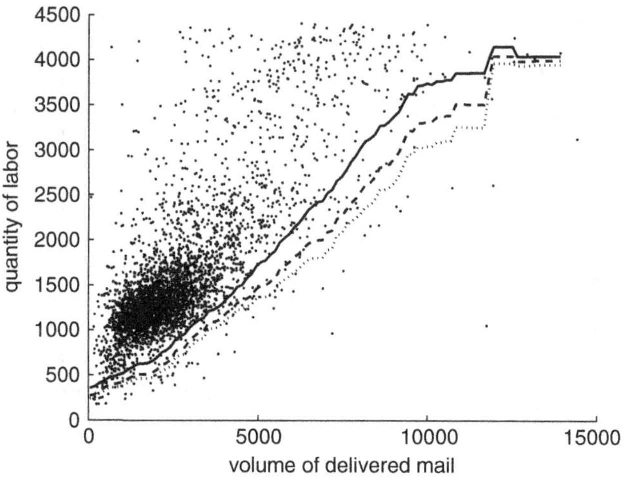

Fig. 4.2 $\hat{\xi}_{\gamma(C_\ell)}$ in *solid line*, $\hat{\xi}_{\gamma(C_u)}$ in *dotted line* and $\hat{\xi}_{\gamma(C_m)}$ in *dashed line*

Let us now test this strategy on a data set of 100 observations simulated following the model

$$Y = \exp(-5 + 10X)/(1 + \exp(-5 + 10X))\exp(-U),$$

where X is uniform on $(0, 1)$ and U is exponential with mean $1/3$. Five outliers indicated as "*" in Fig. 4.3, right-hand side, are added to the cloud of data points (here $n = 105$). The picture on the left-hand side of Fig. 4.3 provides the evolution of the percentage of observations below the frontier $\hat{\xi}_{\gamma(C)}$ with C. This percentage falls rapidly until the circle (i.e., for $C \leq 0.0122$) and then becomes very stable suggesting thus the value 0.0122 for the constant C. The resulting estimator $\hat{\xi}_{\gamma(.0122)}$ and the true frontier φ are superimposed in Fig. 4.3, right-hand side. The frontier estimator $\hat{\xi}_{\gamma(.0122)}$ (in solid line) has a nice behavior: it is somewhat affected by the five outliers, but remains very resistant.

We did the same exercise without the five outliers. The results are displayed in Fig. 4.4. The percentage of observations below the extremile-based frontiers becomes stable from the circle on (i.e., for $C \geq 0.0167$) and so it is enough to choose the value 0.0167 for the constant C. One can also select C in the interval $[C_\ell = 0.0167, C_u = 0.0244)$ which corresponds to the range of points between the circle and the triangle. As expected, in absence of outliers, both estimators $\hat{\xi}_{\gamma(C_\ell)}$ (solid line) and $\hat{\xi}_{\gamma(C_u)}$ (dashed line) are very close from the FDH frontier (i.e., the largest step and nondecreasing curve enveloping below all observations). However, as desired, here also $\hat{\xi}_{\gamma(C_\ell)}$ and $\hat{\xi}_{\gamma(C_u)}$ capture the shape of the efficient boundary of the cloud of data points without enveloping the most extreme observations.

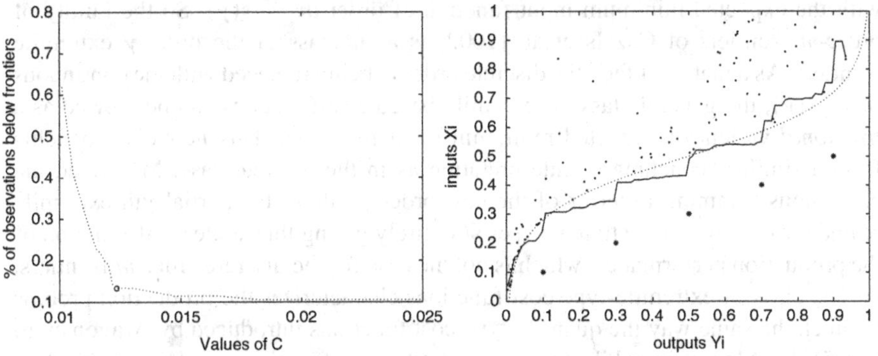

Fig. 4.3 *Left-hand side*, the percentage curve. *Right-hand side*, the frontiers φ and $\hat{\xi}_{\gamma(.0122)}$ superimposed (in *dotted* and *solid lines* respectively). Five outliers included as "*"

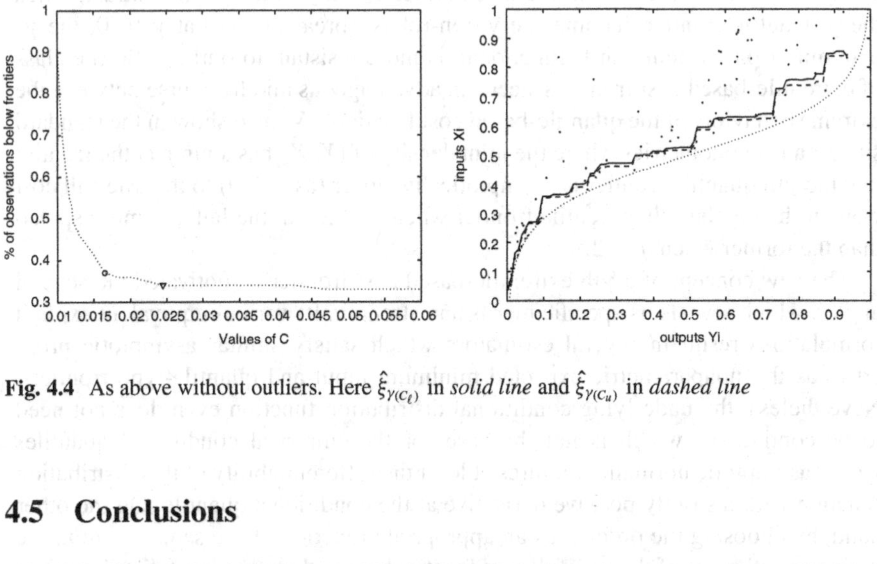

Fig. 4.4 As above without outliers. Here $\hat{\xi}_{\gamma(C_\ell)}$ in *solid line* and $\hat{\xi}_{\gamma(C_u)}$ in *dashed line*

4.5 Conclusions

Instead of estimating the full cost frontier we rather propose in this paper to estimate a boundary well inside the production set Ψ but near its optimal frontier by using extremiles of the same non-standard conditional distribution considered by Cazals et al. (2002) and Aragon et al. (2005). The extremile cost function of order $\gamma \in (0, 1)$ is proportional to a specific conditional probability-weighted moment. It defines a natural concept of a partial cost frontier instead of the m-trimmed frontier suggested by Cazals et al. (2002). The concept is attractive because the "trimming" is continuous in terms of the transformed index $s(\gamma)$, where $s(\gamma) \in [1, \infty)$, whereas $m \in \{1, 2, \ldots\}$. In the particular case where $s(\gamma)$ is discrete (i.e. $s(\gamma) = 1, 2, \ldots$), the corresponding γth extremile-based function coincides with the expected minimum input function of order $m = s(\gamma)$. So the family of order-m frontiers of Cazals et al. (2002) is a subclass of the order-γ extremile frontiers. As a matter of fact, the discrete order m being replaced with the continuous index $s(\gamma)$, the general class of extremile-based cost functions can be viewed as a *fractional* variant of expected minimum input functions. This new class benefits from a similar "benchmark" interpretation as in the discrete case. Moreover, the continuous trimming in terms of the new order γ allows the partial γth extremile boundaries to cover the attainable set Ψ entirely giving thus a clear information of the production performance, which is not the case for the discrete order-m frontiers.

The class of extremile-type cost functions characterizes the production process in much the same way the quantile-type cost functions introduced by Aragon et al. (2005) do. Moreover, while the γth quantile-type function can be expressed as the median of a specific power of the underlying conditional distribution, the γth extremile-type function is given by its expectation. Being determined solely by the tail probability γ, the γth quantile-based cost frontier may be unaffected by desirable extreme observations, whereas the γth extremile-based cost frontier is always sensible to the magnitude of extremes for any order γ. In contrast, when the γ-quantile frontier becomes very non-robust (breaks down) at $\gamma \downarrow 0$, the γ-extremile frontier being an L-functional, is more resistant to outliers. So the class of extremile-based cost frontiers steers an advantageous middle course between the extreme behaviors of the quantile-based cost frontiers. We also show in the standard situation in econometrics where the joint density of (Y, X) has a jump at the frontier that the γth quantile frontier is asymptotically closer (as $\gamma \downarrow 0$) to the true full cost frontier than is the γth extremile frontier when $q \leq 2$, but the latter is more spread than the former when $q > 2$.

The new concept of a γth extremile-based cost frontier is motivated via several angles, which reveals its specific merits and strength. Its various equivalent explicit formulations result in several estimators which satisfy similar asymptotic properties as the nonparametric expected minimum input and quantile-type frontiers. Nevertheless, the underlying conditional distribution function even does not need to be continuous, which is not the case for the empirical conditional quantiles whose asymptotic normality requires at least the differentiability of this distribution function with a strictly positive derivative at the conditional quantile. On the other hand, by choosing the order γ as an appropriate function of the sample size n, we derive an estimator of the true full cost frontier having the same limit distribution as the conventional FDH estimator. Combining the sensitivity and resistance properties of this frontier estimator with the theoretical conditions on the order $\gamma = \gamma(n)$, we show how to pick out in practice reasonable values of $\gamma(n)$. Our empirical rule is illustrated through a simulated and a real data set providing remarkable results. It should be clear that, unlike the approaches of Cazals et al. (2002) and Daouia and Simar (2007), the conditional extremile approach is not extended here to the full multivariate case (multi-inputs and multi-outputs). This problem is worth investigating.

Acknowledgements This research was supported by the French "Agence Nationale pour la Recherche" under grant ANR-08-BLAN-0106-01/EPI project (Abdelaati Daouia) and the Research Fund KULeuven (GOA/07/04-project) and by the IAP research network P6/03, Federal Science Policy, Belgium (Irène Gijbels).

References

Aragon, Y., Daouia, A., & Thomas-Agnan, C. (2005). Nonparametric frontier estimation: a conditional quantile-based approach. *Econometric Theory*, *21*, 358–389.

Cazals, C., Florens, J. P., & Simar, L. (2002). Nonparametric frontier estimation: a robust approach. *Journal of Econometrics*, *106*, 1–25.

Chan, L. K. (1967). On a characterization of distributions by expected values of extreme order statistics. *American Mathematical Monthly*, *74*, 950–951.

Charnes, A., Cooper, W. W., & Rhodes, E. (1978). Measuring the inefficiency of decision making units. *European Journal of Operational Research*, *2*, 429–444.

Daouia, A., Florens, J.-P., & Simar, L. (2008). Functional convergence of quantile-type frontiers with application to parametric approximations. *Journal of Statistical Planning and Inference*, *138*, 708–725.

Daouia, A., Florens, J.-P., & Simar, L. (2010). Frontier estimation and extreme value theory. Bernoulli, *16*(4), 1039–1063.

Daouia, A., & Gijbels, I. (2009). Extremiles, manuscript.

Daouia, A., & Gijbels, I. (2011). Robustness and inference in nonparametric partial-frontier modeling. Journal of Econometrics, *161*, 147–165.

Daouia, A., & Simar, L. (2007). Nonparametric efficiency analysis: a multivariate conditional quantile approach. *Journal of Econometrics*, *140*(2), 375–400.

Debreu, G. (1951). The coefficient of resource utilization. *Econometrica*, *19*(3), 273–292.

Deprins, D., Simar, L., & Tulkens, H. (1984). Measuring labor inefficiency in post offices. In M. Marchand, P. Pestieau, & H. Tulkens (Eds.) *The performance of public enterprises: concepts and measurements* (pp. 243–267). Amsterdam: North-Holland.

Farrell, M. J. (1957). The measurement of productive efficiency. *Journal of the Royal Statistical Society, Series A*, *120*, 253–281.

Gijbels, I., Mammen, E., Park, B. U., & Simar, L. (1999). On estimation of monotone and concave frontier functions. *Journal of the American Statistical Association*, *94*(445), 220–228.

Koopmans, T. C. (1951). An analysis of production as an efficient combination of activities. In T.C. Koopmans (Ed.) *Activitity analysis of production and allocation*, Cowles Commission for Research in Economics, Monograph 13. New York: Wiley.

Park, B., Simar, L., & Weiner, Ch. (2000). The FDH estimator for productivity efficiency scores: asymptotic properties. *Econometric Theory*, *16*, 855–877.

Serfling, R. J. (1980). *Approximation theorems of mathematical statistics*. Wiley Series in Probability and Mathematical Statistics. New York: Wiley.

Shephard, R. W. (1970). *Theory of cost and production function*. Princeton, New-Jersey: Princeton University Press.

Shorack, G. R. (2000). *Probability for statisticians*. New York: Springer.

Simar, L., & Wilson, P. (2008). Statistical inference in nonparametric frontier models: recent developments and perspectives. In Harold O. Fried, C.A. Knox Lovell, & Shelton S. Schmidt (Eds.) *The measurement of productive efficiency*, 2nd edn. Oxford: Oxford University Press.

Chapter 5
Panel Data, Factor Models, and the Solow Residual

Alois Kneip and Robin C. Sickles

Abstract In this paper we discuss the Solow residual (Solow, Rev. Econ. Stat. 39:312–320, 1957) and how it has been interpreted and measured in the neoclassical production literature and in the complementary literature on productive efficiency. We point out why panel data are needed to measure productive efficiency and innovation and thus link the two strands of literatures. We provide a discussion on the various estimators used in the two literatures, focusing on one class of estimators in particular, the factor model. We evaluate in finite samples the performance of a particular factor model, the model of Kneip, Sickles, and Song (A New Panel Data Treatment for Heterogeneity in Time Trends, Econometric Theory, 2011), in identifying productive efficiencies. We also point out that the measurement of the two main sources of productivity growth, technical change and technical efficiency change, may be not be feasible in many empirical settings and that alternative survey based approaches offer advantages that have yet to be exploited in the productivity accounting literature.

5.1 Introduction

In this chapter we discuss the Solow residual (Solow 1957) and how it has been interpreted and measured in the neoclassical production literature and in the complementary literature on productive efficiency. We point out why panel data are needed to measure productive efficiency and innovation and thus link the two strands

---

A. Kneip
Department of Economics, University of Bonn, Adenauerallee 24-26, 53113 Bonn, Germany
e-mail: akneip@uni-bonn.de

R.C. Sickles (✉)
Department of Economics - MS 22, Rice University, 6100 S. Main Street, Houston, Texas 77005-1892, USA
e-mail: rsickles@rice.edu

I. van Keilegom and P.W. Wilson (eds.), *Exploring Research Frontiers in Contemporary Statistics and Econometrics*, DOI 10.1007/978-3-7908-2349-3_5,

of literatures. We provide a discussion on the various estimators used in the two literatures, focusing on one class of estimators in particular, the factor model. We evaluate in finite samples the performance of a particular factor model, the model of Kneip et al. (2011), in identifying productive efficiencies. We also point out that the measurement of the two main sources of productivity growth, technical change and technical efficiency change, may be not be feasible in many empirical settings and that alternative survey based approaches offer advantages that have yet to be exploited in the productivity accounting literature.

The plan of the chapter is as follows. In the next section we discuss how productivity growth has been measured and how certain aspects of its evolution have been disregarded by classical economic modeling that abstracted from the realities of inefficiency in the production process. We also point out how closely linked technical change and technical efficiency change can appear and how it is often difficult to discern their differences in productivity growth decompositions. Section 5.3 discusses alternative survey based methods that may be implemented to assess the contributions of technical innovation and technical efficiency change to productivity growth through the development of a series of *Blue-chip* consensus country surveys that could be collected over time and which could serve as a new measurement data source to evaluate governmental industrial and competition policies. Section 5.4 outlines methods that have been proposed to measure productivity, efficiency, and technical change as well as focusing on the class of factor models which may have an advantage over other methods proposed to identify productive efficiencies. Section 5.5 focuses on one such factor model developed by Kneip et al. (2011) for generic stochastic process panel models and which we reparametrize to estimated time-varying and firm-specific efficiency while allowing a common-stochastic trend to represent technical change. Concluding remarks are provided in Sect. 5.6.

5.2 Productivity Growth and Its Measurement

Productivity growth is the main determinant of changes in our standard of living. Although anecdotal evidence about particular levels of wealth creation is interesting it does not provide governments, sectors, or individual firms with an adequate picture of whether growth in living standards is economically significant and how the growth in living standards is distributed, both within countries and among countries. The linkages between productivity growth and living standards is clearly seen during different epochs for the U.S. economy in Fig. 5.1 (Koenig 2000). Growth in GDP per capita tends to rise and fall in conjunction with growth in labor productivity.

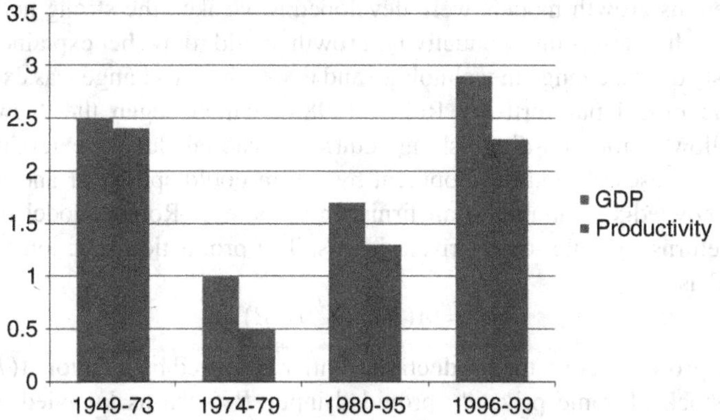

Fig. 5.1 Productivity growth and living standards (Percent per year)
Sources: Department of Commerce; Department of Labor; authors' calculations

5.2.1 *Classical Residual Based Partial and Total Factor Productivity Measurement*

Measurements of productivity usually rely on a ratio of some function of outputs (Y_i) to some function of inputs (X_i). To account for changing input mixes, modern index number analyses use some measure of total factor productivity (TFP). In its simplest form, this is a ratio of output to a weighted sum of inputs

$$TFP = \frac{Y}{\sum a_i X_i}. \tag{5.1}$$

Historically, there are two common ways of assigning weights for this index. They are to use either an arithmetic or geometric weighted average of inputs: *the arithmetic weighted average*, due to Kendrick (1961), uses input prices as the weights; *the geometric weighted average* of the inputs, attributable to Solow (1957), uses input expenditure shares as the weights. some reference point to be useful. *Solow's measure* is based on the Cobb-Douglas production function with constant returns to scale, $Y = AX_L^{\alpha}X_K^{1-\alpha}$ and leads to the TFP measure

$$TFP = \frac{Y}{X_L^{\alpha}X_K^{1-\alpha}}. \tag{5.2}$$

At cost minimizing levels of inputs, the α parameter describes the input expenditure share for labor. The TFP growth rate would be described by $T\dot{F}P = \frac{dY}{Y} - \left[\alpha\frac{dX_L}{X_L} + (1-\alpha)\frac{dX_K}{X_K}\right]$. In applied work, both sets of weights (Kendrick's and Solow's) are often inconsistent with the observed data.

Endogenous growth models were developed to weaken the strong neoclassical assumption that long-run productivity growth could only be explained by an exogenously driven change in technology and that technical change was exogenous. The classic model put forth by Romer (1986), which began the "new growth theory," allowed for non-diminishing returns to capital due to external effects. For example, research and development by a firm could spill over and affect the stock of knowledge available to all firms. In the simple Romer model firms face constant returns to scale to all private inputs. The production function frontier is formulated as

$$Y = A(R)f(K, L, R). \tag{5.3}$$

In the new growth theory, the production frontier is shifted by a factor $A(R)$ where R is the stock of some privately provided input R (such as knowledge) that is endogenously determined. What is its source? Arrow (1962) emphasized learning-by-doing. Recently, Blazek and Sickles (2010) have pursued this as an alternative to the stochastic frontier model. Romer (1986) modeled A as a function of the stock of research and development. Lucas (1988) modeled A as a function of stock of human capital.

Where multiple outputs exist, TFP can also be described as a ratio of an index number describing aggregate output levels(y_j) divided by an index number describing aggregate input levels(x_i). As such, they derive many of their properties based the assumptions of the underlying aggregator functions used. Fisher (1927) laid out a number of desirable properties for these index numbers. Many of these properties are easily achievable, while others are not. Following Jorgenson and Griliches (1972), a (logarithmic) total factor productivity index can be constructed as the difference between log output and log input indices, i.e.

$$\ln TFP = \ln y_t^1 - \ln x_t^1. \tag{5.4}$$

An implication of the endogenous growth model is that if a time trend is added to the standard neoclassical production function then the trend must be stochastic. This clearly has implications for stationarity (Reikard 2005). Recent work by Kneip et al. (2011) has addressed the estimation issues that are associated with estimating the endogenous technical change in the presence of technical efficiency change.

5.2.2 *Technical Efficiency in Production*

It is often quite difficult to separate the impacts of technical change from constraints in the use of the existing technology, or technical efficiency. An example of the overlay of technology (and its change) and efficiency (and its change) can be found in the classic story of the reason *behind* the specifications of the solid rocket boosters (SRB's) for the space shuttle (see, for example, one of the many URL's where it is documented at http://www.astrodigital.org/space/stshorse.html). The SRBs are made by Morton Thiokol at a factory in Utah. Originally, the engineers who designed the SRBs wanted to make them much fatter than they are. Unfortunately, the SRBs had to be shipped by train from the factory to the launch site in Florida and the railroad line runs through a tunnel in the mountains. The SRBs had to be made to fit through that tunnel. The width of that tunnel is just a little wider than the U.S. Standard Railroad Gauge (distance between the rails) of 4 feet, 8.5 inches. That is an odd number and begs the question of why that gauge was used? It was used because US railroads were designed and built by English expatriates who built them that way in England. The English engineers do so because the first rail lines of the 19th century were built by the same craftsmen who built the pre-railroad tramways, which used the gauge they used. The reason those craftsmen chose that gauge was because they used the same jigs and tools that were previously used for building wagons, and the wagons used that wheel spacing. The wagons used that odd wheel spacing since if the wagon makers and wheelwrights of the time tried to use any other spacing, the wheel ruts on some of the old, long distance roads would break the wagon axles. As a result, the wheel spacing of the wagons had to match the spacing of the wheel ruts worn into those ancient European roads. Those ancient roads were built by Imperial Rome for their legions and the roads have been used ever since. The initial ruts, which everyone else had to match for fear of destroying their wagons, were first made by Roman war chariots. Since the chariots were made by Imperial Roman chariot makers, they were all alike in the matter of wheel spacing. Why 4 feet, 8.5 inches? Because that was the width needed to accommodate the rear ends of two Imperial Roman war horses. Therefore, the railroad tunnel through which the late 20th century space shuttle SRBs must pass was excavated slightly wider than two 1st century horses' rear-ends and consequently, a major design feature of what is arguably the world's most advanced transportation system was specified by the width of the read-end of a horse.

The story is a bit of folk lore whimsy and has an oral and written tradition that is as old as the aging space shuttle fleet. Although this is just one of such anecdotes, it illustrates how constraints to adopting the most advanced technology may arise seemingly by a random process, in fact arise by historical precedent. We thus turn to an alternative to the Solow type neoclassical model of productivity and focus on a component neglected in the traditional neoclassical approach, technical inefficiency. Since the fundamental theoretical work by Debreu (1951), Farrell (1957), Shephard (1970) and Afriat (1972), researchers have established a method to measure the intrinsically unobservable phenomena of efficiency. Aigner et al. (1977), Battese and Cora (1977), and Meeusen and van den Broeck (1977) provided the econometric methods for the applications waiting to happen. The linear programming methodology, whose implementation was made transparent by Charnes et al. (1978), became available at about the same time. The U.S. and international emphasis on deregulation and the efficiencies accruing to increased international competition due to the movement to lower trade barriers provided a fertile research experiment for efficiency modelers and practitioners.

The efficiency score, as it is usually measured, is a residual. Parametric assumptions about the distribution of efficiency and its correlation structure often are made to sharpen the interpretation of the residual. However, that efficiency measurement should be highly leveraged by parametric assumptions is by no means a comforting resolution to this measurement problem. Productivity defined by the Solow residual is a reduced form concept, not one that can be given a structural interpretation. Different efficiency estimators differ on what identifying restrictions are imposed. Not surprisingly, different efficiency estimators often provide us with different cross-sectional and temporal decompositions of the Solow residual.[1] Kumbhakar and Lovell (2000) and Fried et al. (2008) have excellent treatments of this literature. It addresses the continuing debate on how the distributional assumptions made in Pitt and Lee (1981), Kumbhakar (1990), Battese and Coelli (1992), and others drive the estimates of efficiency. The robust and efficient estimators have been developed by Park et al. (1998, 2003, 2007), Adams et al. (1999), Adams and Sickles (2007). These share a number of generic properties with the estimators proposed by Schmidt and Sickles (1984) and Cornwell et al. (1990).

5.2.3 Difficulty in Measuring the Decomposition of Productivity Growth into Technical Change and Technical Efficiency Change

We point out below problems in decomposing productivity change into its innovation and its efficiency change components. One conclusion from this discussion is that it simply may not be possible from purely econometric models, no matter how sophisticated, to model structurally the role of innovation and the role of efficiency in determining TFP growth. We give two illustrations. The first is based on experience gleaned by Sickles as the Senior Research Coordinator for the Development Economic Policy Reform Analysis Project (DEPRA), USAID/Egyptian Ministry of Economy, Contract No. 263-0233-C-00-96-00001-00. A portion of this research was the basis for Getachew and Sickles (2007). The study analyzed the impact of regulatory and institutional distortions on the Egyptian private manufacturing sector from the mid 1980s to the mid 1990s. We focused on the impact of economic reforms undertaken since 1991. The second is based on work of Sickles and Streitwieser (1992, 1998) who addressed the impact of the Natural Gas Policy Act of 1978 on the U.S. interstate natural gas transmission industry.

5.2.3.1 How Can We Identify Specific Constraints at the Macro Level?

The Development Economic Policy Reform Analysis Project in Egypt was a USAID/World Bank project that began in the mid-1980s and lasted through the

[1] Since cross-sectional data are used, the efficiencies estimated are typically conditional expectations, as it is mentioned in Simar and Wilson (2010).

mid-1990s. The aim of the project was to transition from the planned economy left by the Soviet Union to a private sector market economy via a structural adjustment program. Initial efforts focused on macroeconomic stabilization which involved a reduction of the fiscal deficit through a variety of measures. These measures included:

1. Cuts in public investment and subsidization programs.
2. Tax reforms, particularly through the introduction of a general sales tax.
3. Improvements in collection.
4. Monetary policy tightening to fight inflation.

The structural adjustment program also involved extensive price liberalization that affected each sector of the Egyptian economy. This involved:

1. Adjustments of relative prices.
2. Removal of all export quotas, except for tanned hide, in the trade and financial sectors.
3. Lifting of tariffs on almost all imported capital goods.
4. Removal of constraints on nominal interest rate ceilings, administrative credit allocation, foreign exchange controls and prohibitions against international capital mobility.
5. Reform of labor laws, which gave employers the right to hire and lay off workers in accordance with economic conditions.

How do we develop a model that identifies such a plethora of structural changes in the Egyptian economy? One approach was undertaken by Getachew and Sickles (2007) who utilized a virtual cost system and were able to identify allocative distortions that existed before the reforms were undertaken and those that existed after the reforms had worked their way through the Egyptian private sector after the deregulatory reforms. Getachew and Sickles found substantial welfare benefits accruing to the Egyptian economy due to these reforms in total. Unfortunately, the specific determinants of the benefits of market reforms could not be ascertained since the specific constraints could not be modeled and thus incorporated into an estimable structural model.

5.2.3.2 How Can We Identify Specific Constraints at the Micro Level?

Another illustration is found in the regulatory change accompanying the U.S. Interstate Natural Gas Policy Act of 1978. The regulatory history of natural gas transmission industry is long and complicated. Figure 5.2 provides a schematic diagram that outlines the maximum ceiling price schedules from 1978 to 1985 and the 24 different price combinations over the period for different categories of natural gas (for details, see Sickles and Streitwieser 1992). As Fig. 5.2 points out, the regulations and their impact on the various firms involved in the deregulatory initiatives are enormously complex. A formal model of the constraints in an estimable structural econometric model is simply not feasible. One can clearly see

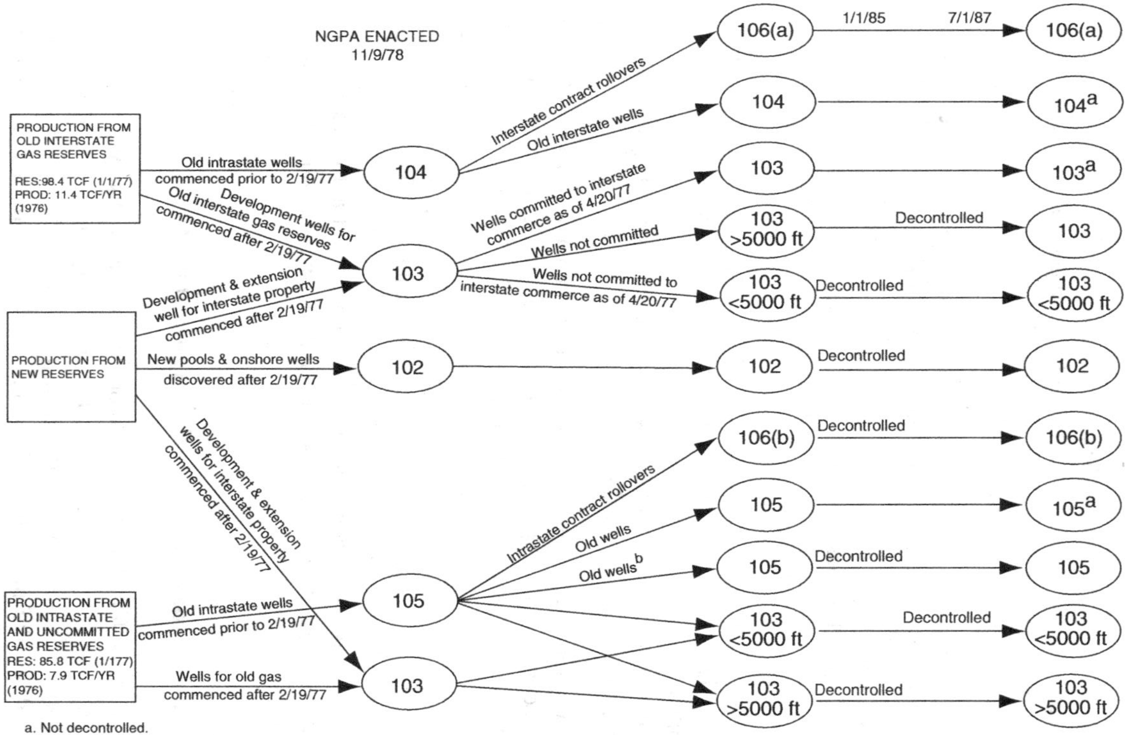

a. Not decontrolled.
b. Price greater than $ 1.00 per million Btu under a definite escalator clause.
Note: Stripper wells and high cost gas omitted.
Source: U.S. Energy Information Administration, Natural Gas Market (1981), p. 16.
Reprinter From: Teece, David J., "Structure and Organization of the Natural Gas Industry," The Energy Journal, 11 (1989), p. 17.

Fig. 5.2 Maximum ceiling price categories: NGPA Title I for onshore lower-48 natural gas above 15,000 feet

the difficulties inherent in any attempt to parsimoniously quantify the constraints, not to mention the difficulties one would have in ultimately interpreting how these constraints could impact optimal natural gas transmission decisions.

5.3 Alternatives to Measurement of Technical Change and Technical Efficiency Change

5.3.1 Survey-Based Methods for Decomposing Total Factor Productivity Growth into Technical Change and Technical Efficiency Change: A New Blue Chip Indicator

Various approaches to decomposing total factor productivity into sources that are due to efficiency change and due to technological change have been discussed. One popular index number approach based on the decomposing the Malmquist index (Caves et al. 1982) was introduced by Färe et al. (1992). Of course, regression based approaches using either traditional neoclassical growth models, growth models in which endogenous growth is allowed, or growth models in which inefficiency is explicitly introduced via a frontier technology offer potentially richer empirical specifications and a more structural determination of the sources of productivity growth. However, all approaches suffer due to poor empirical proxies for the measures of loosening constraints to business activity. One possibility to circumvent the paucity of reliable empirical measures of the determinants of productivity growth would be to conduct a structured survey of business leaders, political leaders World Bank, International Monetary Fund, and Non-governmental Organizations to identify what are the most important of an array of factors contributing to economic growth. The results of such a survey would allow us to parse out the contribution of efficiency change, in the form of loosening of binding constraints, to economic growth and its relative contribution vis-à-vis technical progress. The Blue Chip Economic Indicators each month survey America's top business economists and ask them to supply their forecasts of U.S. economic growth, inflation, and interest rates, among other business indicators. The survey began in 1976. The experts who make up the Blue-Chip panel are on the order of 50 or so economists and come from a cross section of manufacturing and financial services firms. The Blue Chip Economic Indicators are used by business journalists and by forecasting companies such as the Wall Street Journal, Forbes, and Reuters. The specific information contained in the survey contains forecasts for this year and next from each panel member as well as an average, or consensus, of their forecasts for the following measures of economic activity: Real GDP, GDP price index, Nominal GDP, Consumer price index, Industrial production, Real disposable personal income, Real personal consumption expenditures, Real non-residential fixed investment, Pre-tax corporate profits, 3-mo. Treasury bill rate, 10-yr. Treasury note yield, Unemployment rate, Total housing starts, Auto and light truck sales,

Real Net exports. Along with forecasts by each member of the panel is published the consensus forecast for each variable, as well as averages of the 10 highest and 10 lowest forecasts for each variable; a median forecast to eliminate the effects of extreme forecasts on the consensus; the number of forecasts raised, lowered, or left unchanged from a month ago; and a diffusion index that indicates shifts in sentiment that sometimes occur prior to changes in the consensus forecast.[2] One may question the accuracy of the Blue Chip indicators. However, in recent work by Chun (2009), the Blue Chip indicators were found to compare favorably with forecasts from the Diebold and Li (2006) model at short horizon forecasts of short to medium maturity interest rates. Development of a survey-based method to decompose total factor productivity growth in a technical change and a technical efficiency change component is motivated not only by an interest in sharper forecasts but also on the possibility that our econometrically based estimates may not be reliable or meaningful. A survey-based set of indicators of such a decomposition may well be all that we can hope for.

How might a questionnaire be constructed? What might be the best survey methods to use in order to solicit answers to such basic questions as:

Total factor productivity growth is the percentage change in production not attributable to changes in labor, capital, and other inputs. Historically total factor productivity growth in the U.S. has averaged 2% per year. Assuming that all contributions to total factor productivity growth much sum to 1 in percentage terms please answer the following 5 questions:

1. What portion of total factor productivity growth (regress) is due to the innovation provided by new technology?
2. What portion of total factor productivity growth (regress) is due to the better use of existing technology?
3. What portion of total factor productivity growth (regress) is due to changes in government regulations, business climate, or other institutional factors such as political stability and the democratic process?
4. What portion of total factor productivity growth (regress) is due to changes in the scale of operation?
5. What portion of total factor productivity growth (regress) is due to other factors? (Please list them and their relative importance.)

We expect that information of this sort, collected by a set of experts in countries of the world, will allow us to better understand the role of technology transfer, government regulation, institutional factors such as political stability and the democratic process, and market concentration on the engine for long term and sustainable economic growth: *Total Factor Productivity* growth.

---

[2]See Wolters Kluwer's Aspen Press website for the Blue Chip Economic Indicators publication by Randell E. Moore:

http://www.aspenpublishers.com/product.asp?catalog_name=Aspen&product_id=SS01934600 &cookie%5Ftest=1.

5.4 Measuring Technical Change and Efficiency Change Decompositions of Productivity Growth Decompositions

5.4.1 Index Number Procedures

Either index number or regression based approaches require panel data (at a minimum). The index number approach (Färe et al. 1992) begins by assuming a panel of firms (or countries, etc.) with $i = 1, \ldots, N$ firms, $t = 1, \ldots, T$ periods, $j = 1, \ldots, J$ inputs and $k = 1, \ldots, K$ outputs. Thus, x_{jit} is the level of input j used by firm i in period t and y_{kit} is the level of output k produced by firm i in period t. Assume an intertemporal production set where input and output observations from all time periods are used. The production technology, $\mathbf{S}$, is

$$\mathbf{S} = \{(x, y) \mid x \in \mathbb{R}_{+}^{J},\ y \in \mathbb{R}_{+}^{K},\ (x, y) \text{ is feasible}\}. \tag{5.5}$$

The efficiency scores are the distances from the frontier. An output-based distance function (Shephard 1970), OD, is defined as

$$\mathrm{OD}(x, y) = \min\{\lambda \mid (x, y/\lambda) \in \mathbf{S}\}. \tag{5.6}$$

Holding the input vector constant, this expression expands the output vector as much as possible without exceeding the boundaries of $\mathbf{S}$. An output efficient firm has a score of 1 and it is not possible for the firm to increase its output without increasing one or more of its inputs. Conversely, an output inefficient firm has $\mathrm{OD}(x, y) < 1$. The productivity index requires output distance functions calculated between periods. $\mathrm{OD}_t(x_{t+1}, y_{t+1}) = \min\{\lambda | (x_{t+1}, y_{t+1/\lambda}) \in \mathbf{S}_t\}$ has the technology of time t and scales outputs in time $t+1$ such that (x_{t+1}, y_{t+1}) is feasible in period t. The observed input-output combination may not have been possible in time t; the value of this expression can exceed one which would represent technical change. $\mathrm{OD}_{t+1}(x_t, y_t) = \min\{\lambda | (x_t, y_{t/\lambda}) \in \mathbf{S}_{t+1}\}$ has the technology of time $t+1$ and scales outputs in time t such that (x_t, y_t) is feasible in period $t+1$. The final equation can be expressed as

$$M(x_{t+1}, y_{t+1}, x_t, y_t) = \frac{\mathrm{OD}_{t+1}(x_{t+1}, y_{t+1})}{\mathrm{OD}_t(x_t, y_t)} \tag{5.7}$$

$$\times \left\{ \frac{\mathrm{OD}_t(x_{t+1}, y_{t+1})}{\mathrm{OD}_{t+1}(x_{t+1}, y_{t+1})} \frac{\mathrm{OD}_t(x_t, y_t)}{\mathrm{OD}_{t+1}(x_t, y_t)} \right\}^{1/2}. \tag{5.8}$$

$$= E_{t+1} \times A_{t+1}.$$

This index captures the dynamics of productivity change by incorporating data from two adjacent periods. E_{t+1} reflects changes in relative efficiency. A_{t+1} reflects changes in technology between t and $t+1$. For the index, a value below 1 indicates productivity decline while a value exceeding 1 indicates growth. For the index components, values below 1 signify a performance decline while values above 1 signify an improvement. There may be significant shortcomings of this approach, as noted by Førsund and Hjalmarsson (2008), due to potential vintage capital effects or its lack of any obvious inferential theory (Jeong and Sickles 2004).

5.4.2 Regression Based Approaches

Regression based approaches to decomposing productivity growth into technical change and efficiency change components can be explained using the following generic model. Assume that the multiple output/multiple input technology can be estimated parametrically using the output distance function. Since the output distance function, $OD(Y, X) \leq 1$, specifies the fraction of aggregated output (Y) produced by given aggregated inputs (X), it gives us a radial measure of technical efficiency. For an m-output, n-input production technology, the deterministic output distance function can be approximated by

$$\frac{\Pi_j^m Y_j^{\gamma_j}}{\Pi_k^n X_k^{\beta_k}} \leq 1, \tag{5.9}$$

where the γ_j's and the β_k's are weights representing the technology of the firm. If one simply multiplies through by the denominator, approximates the terms using a Young Index, a geometric mean with varying weights (Balk 2009), and adds a disturbance term v_{it} to take account of general statistical noise, and specify a nonnegative stochastic term u_{it} for the firm specific level of radial technical inefficiency, then a regression based approach to decomposing productivity growth into technical change and efficiency change can be specified. The Cobb-Douglas stochastic distance frontier model can be written as

$$0 = \sum_j \gamma_j \ln y_{j,it} - \sum_k \beta_k \ln x_{k,it} + v_{it} - u_{it}. \tag{5.10}$$

The output distance function is linearly homogeneous in outputs and if one imposes this restriction and then normalizes with respect to one y_i(the last) the following expression (Lovell et al. 1994) can be derived

$$-\ln(y_J) = \sum_j \gamma_j \ln \widehat{y}_{j,it} - \sum_k \beta_k \ln x_{k,it} + v_{it} - u_{it}, \tag{5.11}$$

where $y_J$is the normalized output and $\widehat{y}_j = y_j/y_J$, $j = 1, \ldots, J-1$. Let $X_{it}^* = -\ln(x_{k,it})$, $Y_{it}^* = \ln(\widehat{y}_{j,it})$, and $Y_{it} = -\ln(y_J)$. Then the stochastic distance frontier is

$$Y_{it} = X_{it}^{*\prime}\beta + Y_{it}^{*\prime}\gamma + v_{it} - u_{it},\ i = 1,...N, t = 1,...T. \tag{5.12}$$

Letting $\varepsilon_{it} = v_{it} - u_{it}$, $X_{it}' = [\,X_{it}^{*\prime}, Y_{it}^{*\prime}]$, $\xi = [\,\beta, \gamma\,]$, we obtain the familiar functional form for a stochastic frontier production model under a classical panel data setting

$$Y_{it} = X_{it}'\xi + \varepsilon_{it},\ i = 1,...N, t = 1,...T. \tag{5.13}$$

This is the generic model vehicle for estimating efficiency change using frontier methods. If we assume that innovations are available to all firms and that firms' specific idiosyncratic errors are due to relative inefficiencies then we can decompose sources of TFP growth by adding either an exogenous or a stochastic time trend (see also, Bai et al. 2007). The panel stochastic frontier model is quite flexible and robust. Technical efficiency of a particular firm (observation) can be consistently estimated. Estimation of the model and the separation of technical inefficiency from statistical noise and from a common technical change component does not require a set of specific assumptions about the parametric distribution of technical inefficiency (e.g., half-normal) and statistical noise (e.g., normal) and dependency structure. For example, it may be incorrect to assume that inefficiency is independent of the regressors since if a firm knows its level of technical inefficiency, this should affect its input choices. Pitt and Lee (1981) and Schmidt and Sickles (1984) have developed random and fixed effects as well as maximum likelihood based estimators for such panel frontier models. To allow for time varying and cross-sectional specific efficiency change one can use a parametrization chosen in Cornwell et al. (1990). They used a quadratic function of time $u_{it} = W_{it}'u_i = \theta_{i1} + \theta_{i2}t + \theta_{i3}t^2$. Other than a quadratic function of time, u_{it} has been modelled as $u_{it} = \gamma(t)\alpha_i = [1+\exp(bt+ct^2)]^{-1}\alpha_i$ (Kumbhakar 1990), and $u_{it} = \eta_{it}\alpha_i = \exp[-\eta(t-T)]\alpha_i$ (Battese and Coelli 1992). Both of these approaches used maximum likelihood estimation (MLE) to estimate efficiency. We now turn to other reduced form approaches for measuring the growth in the key components of TFP : efficiency and innovation.

5.4.3 *Bayesian Treatments for Time Varying Inefficiency*

Sickles and Tsionas (2008) consider a model similar to the KSS model with common factors whose number is unknown and whose effects are firm-specific. Bayesian inference techniques organized around MCMC are used to implement the models. The model is

$$y_{it} = x_{it}'\beta + \varphi_i\,(t) + v_{it}, \quad i = 1, \ldots, n, \quad t = 1, \ldots, T, \tag{5.14}$$

where x_{it} and β are $k \times 1$, and $\varphi_i\,(t)$ is a unit specific unknown function of time. They assume $v_{it} \overset{IID}{\sim} N\left(0, \sigma^2\right)$. The model can be written in the form $y_{it} = x_{it}'\beta +$

$\gamma_{it} + v_{it}$. For the ith individual we have $y_i = X_i\beta + \gamma_i + v_i$, $i = 1, \ldots, n$. Assuming $\gamma_{i1} \leq \cdots \leq \gamma_{iT}$, they assume a spline prior of the form

$$p(\beta, \sigma, \gamma) \propto \sigma^{-1} \prod_{i=1}^{n} \exp\left(-\frac{\gamma_i' Q \gamma_i}{2\omega^2}\right) = \sigma^{-1} \exp\left(-\frac{1}{2\omega^2}\gamma' (I_T \otimes Q)\gamma\right), \tag{5.15}$$

where $Q = D'D$, and D is the $(T-1) \times T$ matrix whose elements are $D_{tt} = 1$, $D_{t-1,t} = -1$ and zero otherwise. ω is a smoothness parameter which stands for the degree of smoothness. This prior says that $\gamma_{it} - \gamma_{i,t-1} \sim N(0, \omega^2)$ or $D\gamma_i \overset{IID}{\sim} N(0, \omega^2 I_{T-1})$, that is it assumes that the first derivative of functions $\varphi_i(t)$ is a smooth function of time. It is possible to allow for smooth second derivatives by using the formulation $\gamma_{it} - 2\gamma_{i,t-1} + \gamma_{i,t-2} \sim N(0, \omega^2)$, which can be written as $D^{(2)}\gamma_i \overset{IID}{\sim} N(0, \omega^2 I_{T-1})$. We can still define $Q = D^{(2)\prime}D^{(2)}$, and the analysis below goes through unmodified. Since γ_{i1} plays the role of an intercept, we can assume $\alpha_i \overset{IID}{\sim} N(0, \sigma_\alpha^2)$, $i = 1, \ldots, n$. The model generalizes Koop and Poirier (2004) in the case of panel data with individual-specific intercepts and time effects. Moreover, it does not rely on the conjugate prior formulation for the γ_{it}s which can be undesirable. The posterior kernel distribution is

$$p(\beta, \gamma, \sigma | Y, X, \omega) \propto \sigma^{-(nT+1)} \exp\left[-\frac{(Y - X\beta - \gamma)'(Y - X\beta - \gamma)}{2\sigma^2}\right] \times \exp\left[-\frac{1}{2\omega^2}\gamma'(I_T \otimes Q)\gamma\right], \tag{5.16}$$

where $\underset{(nT \times k)}{X} = [X_1', \ldots, X_n']'$, and $\underset{(nT \times 1)}{Y} = [y_1', \ldots, y_n']'$. Bayesian inference for this model can be implemented using Gibbs sampling.

5.4.4 The Latent Class Model

As discussed in Greene (2008), one way to extend the normal-half normal stochastic frontier model (or others) with respect to the distribution of v_i is the finite mixture approach suggested by Tsionas and Greene (2003). This is a class specific stochastic frontier model. The frontier model can be formulated in terms of J 'classes' so that within a particular class,

$$f_\varepsilon(\varepsilon_i | class = j) = \frac{2}{\sqrt{2\pi(\sigma_u^2 + \sigma_{vj}^2)}} \left[\Phi\left(\frac{-\varepsilon_i(\sigma_u/\sigma_{vj})}{\sqrt{\sigma_u^2 + \sigma_{vj}^2}}\right)\right] \exp\left(\frac{-\varepsilon_i^2}{2(\sigma_u^2 + \sigma_{vj}^2)}\right), \tag{5.17}$$
$$\varepsilon_i = y_i - \alpha - \beta^{\mathrm{T}} x_i.$$

Indexation is over classes and involves the variance of the symmetric component of ε_i, $\sigma_{v,j}$. The unconditional model is a probability weighted mixture over the J classes, $f_\varepsilon(\varepsilon_i) = \Sigma_j \pi_j \, \mathrm{f}_\varepsilon(\varepsilon_i | \text{ class} = j)$, $0 < \pi_j < 1$, $\Sigma_j \pi_j = 1$. Mixing probabilities are additional parameters to be estimated. The model preserves symmetry of the two-sided error component, but provides a degree of flexibility that is somewhat greater than the simpler half normal model. The mixture of normals is, with a finite number of classes, nonnormal. This model can be estimated by Bayesian (Tsionas and Greene 2003) or classical (Orea and Kumbhakar 2004; Tsionas and Greene 2003; Greene 2004, 2005) estimation methods. After estimation, a conditional (posterior) estimate of the class that applies to a particular observation can be deduced using Bayes theorem, i.e.

$$\mathrm{Prob}[class = j | y_i] = \frac{f(y_i | class = j)\mathrm{Prob}[class = j]}{\sum_{j=1}^{J} f(y_i | class = j)\mathrm{Prob}[class = j]} = \hat{\pi}_{j|i}. \quad (5.18)$$

Individual observations are assigned to the most likely class. Efficiency estimation is based on the respective class for each observation.

Orea and Kumbhakar (2004), Tsionas and Greene (2003) and Greene (2004, 2005) have extended this model in two directions. First, they allow the entire frontier model, not just the variance of the symmetric error term, to vary across classes. This represents a discrete change in the interpretation of the model. The mixture model is essentially a way to generalize the distribution of one of the two error components. For the fully mixed models, the formulation is interpreted as representing a latent regime classification. The second extension is to allow heterogeneity in the mixing probabilities;

$$\pi_{ij} = \frac{\exp(\theta_j^{\mathrm{T}} z_i)}{\Sigma_{j=1}^{J} \exp(\theta_j^{\mathrm{T}} z_i)}, \quad \theta_J = 0. \quad (5.19)$$

The rest of the model is a class specific stochastic frontier model

$$f_\varepsilon(\varepsilon_i | class = j) = \frac{2}{\sigma_j} \phi\left(\frac{\varepsilon_i | j}{\sigma_j}\right)\left[\Phi\left(\frac{-\lambda_j \varepsilon_i | j}{\sigma_j}\right)\right], \quad (5.20)$$

where $\varepsilon_i | j = y_i - \alpha_j - \beta_j^{\mathrm{T}} x_i$. This form of the model has all parameters varying by class. By suitable equality restrictions, subsets of the coefficients, such as the technology parameters, α and β, can be made generic.

5.4.5 The Semiparametric Model and Estimators of Technical Efficiency: The Park, Sickles, and Simar SPE Estimators

The models for which the SPE estimators have been derived vary on how the basic model assumptions have been modified to accommodate a particular issue of misspecification of the underlying efficiency model. A number of SPE estimators that differ on the basis of assumed orthogonality of effects and regressors, temporal variation in the efficiency effects, and correlation structure of the population disturbance have been considered and developed in a series of papers by Park and Park and Simar (1994) and Park et al. (1998, 2003, 2007). For example, when one believes that the effects and all of the regressors are dependent and are unwilling to specify a parametric distribution for the dependency structure then one can specify the joint distribution $h(\cdot,\cdot)$ using kernel smoothers. The Park, Sickles, and Simar (PSS) estimators are based on the theory of semiparametric efficient bounds estimators and utilize an orthogonalization of the scores of the likelihood function with respect to the parameters of interest and the nuisance parameters. The PSS estimators are also adaptive in the terminology of semi-nonparametric estimation theory.

5.4.6 Alternatives to the Semiparametric Efficient Estimators

There are a number of panel frontier estimators that have been used widely in the empirical efficiency literature. They differ from the SPE estimators based largely on assumptions made about the distribution of the unobserved efficiency effects and about the correlation of efficiency effects and regressors. In order to measure time variant heterogeneity, α_{it} can be specified as

$$\alpha_{it} = c_{i1}g_{1t} + c_{i2}g_{2t} + \cdots + c_{iL}g_{Lt}, \tag{5.21}$$

where c_{ir} are unknown parameters, and the *basis functions* g_{ir} are smooth, real-valued functions of x_{it}.This approach is more general than fitting polynomials and can be used to parsimoniously model virtually any temporal pattern of firm efficiency. The firm efficiencies are obtained from the structures of the g_{ir} and from the distribution of the effects α_i. The fixed and random effect models are nested in the mixed efficiency effects specification as are the CSS and SS estimators. Methods for estimating c_{ir}, g_{ir}, and L can be found in Kneip et al. (2011).

5.4.7 Using Factor Models to Estimate the Solow Residual

The literature on factor models and state-space representations of latent factors using the Kalman filter is quite lengthy and dense. First, we will give a very brief introduction to the factor models. Next we will try to provide some overview of the most recent papers. Then we will select a particular factor model introduced by Kneip, Sickles, and Song to decompose the Solow residual into a technical change and an efficiency change component. Breitung and Eickmeier (2005) provide a very review of factor models and we relay on it in what is discussed below.

5.4.7.1 Strict Factor Model

Strict factor models are the most simple of the factor model class and utilize the following basic assumptions

$$y_{it} = \lambda_{i1} f_{1t} + \cdots + \lambda_{ir} f_{rt} + u_{it} \tag{5.22}$$

$$= \lambda_i' f_t + u_{it} \tag{5.23}$$

or

$$y_t = \Lambda f_t + u_t \tag{5.24}$$

$$Y = F\Lambda' + U \tag{5.25}$$

where $\Lambda = [\lambda_1\, \lambda_2 \cdots \lambda_N]'$, $Y = [y_1\, y_2 \cdots y_T]'$, $F = [f_1\, f_2 \cdots f_T]'$, and $U = [u_1\, u_2 \cdots u_T]'$.

For the strict factor models it is usually assumed that u_t are mutually uncorrelated with $E[u_t] = 0$ and $E[u_t u_t'] = \Sigma = diag(\sigma_1^2, \sigma_2^2, \ldots, \sigma_N^2)$. Moreover, $E[f_t] = 0$. The principle components estimator, the most widely used of the various strict factor specifications, will be inconsistent for fixed N and $T \to \infty$ unless $\Sigma = \sigma^2 I$ as can be seen by considering the principle components estimator as an IV estimator.

5.4.7.2 Approximate Factor Models

When we allow for $N \to \infty$ we can avoid the restrictive assumptions of strict factor models (Chamberlain and Rothshield 1983; Stock and Watson 2002; Bai 2003) and in this case it is possible to allow for (weak) serial correlation for the idiosyncratic errors. However, persistent and non-ergodic processes are generally ruled out. Idiosyncratic errors can be allowed to be (weakly) cross-correlated and heteroscedastic and (weak) correlation among the factors and the idiosyncratic components are possible. With these and some other technical assumptions Bai (2003) establishes the consistency and asymptotic normality of the principle components estimator for Λ and f_t. However, as noted by Bai and Ng (2002), the small sample properties of this estimator may be severely affected whenever the data is cross-correlated.

5.4.7.3 Dynamic Factor Models

The dynamic model is given by

$$y_t = \Lambda_0 g_t + \Lambda_1 g_{t-1} + \cdots + \Lambda_m g_{t-m} + u_t, \tag{5.26}$$

where Λ_j are $N \times r$ matrices and g_t is a vector of q stationary factors. Idiosyncratic components of u_t are assumed to be independent (or weakly dependent) stationary processes. Forni et al. (2000) provide a method to estimate this model. Let $\eta_t = g_t - E[g_t | g_{t-1}, g_{t-2}, ...]$, $f_t = [g_t, g_{t-1}, \ldots, g_{t-m}]'$ (which is $r = (m+1) \times q$), and $\Lambda = [\Lambda_0, \Lambda_1, \ldots, \Lambda_m]$. In their first stage the usual principal components are estimated. Note that rather than f_t a rotated version of it, Qf_t, is estimated. The second step estimates a vector-autoregression (VAR) model given by

$$\hat{f}_t = A_1 \hat{f}_{t-1} + A_2 \hat{f}_{t-2} + \cdots + A_p \hat{f}_{t-p} + e_t. \tag{5.27}$$

Note that the rank of the covariance matrix for the e_t term is q since $\hat{f}_t$ includes estimation of lagged factors. If we let $\hat{W}_r$ be the matrix generated by the q largest eigen values of the covariance matrix of e_t, $\hat{\Sigma}_e = \frac{1}{T} \sum_{t=p+1}^{T} e_t e_t'$,then $\hat{\eta}_t = \hat{W}_r' \hat{e}_t$.

An important problem is to determine the number of factors. Forni et al. (2000) provide an informal criterion based on the proportion of explained variances. Bai and Ng (2007) and Stock and Watson (2005) suggest consistent selection procedures based on principal components. Also, information criteria and tests of the number of factors are suggested by Breitung and Kretschmer (2005). Pesaran (2006) is an interesting paper since it has potential for productivity analysis, in particular frontier production. His paper deals with estimation and inference in panel data models with a general multifactor error structure. The unobserved factors and the individual-specific errors are allowed to follow arbitrary stationary processes and the number of unobserved factors need not be estimated. Individual-specific regressors are filtered with cross-section averages and when the cross-section dimension (N) tends to infinity, the differential effects of unobserved common factors are eliminated.

Carriero et al. (2008) look at the forecasting performances of factor models, large scale Bayesian VARs, and multivariate boosting, while Marcellino and Schumacher (2007) focus on factor models that can handle unbalanced datasets in their analysis of the German economy. The approach followed by Doz et al. (2006) and Kapetanios and Marcellino (2009) casts the large factor model in state-space form. Kapetanios and Marcellino (2009) estimate the factors using subspace algorithms, while Doz et al. (2006) exploit the Kalman filter and kernel smoothers. We will focus below on a recent contribution by Kneip et al. (2011) who develop the asymptotic theory for general factor models using a combination of principal components and smoothing spines. In that model not only are methods developed to select the number of factors but also address the potential for nonstationarity. The nonstationarity applies here in regard to a stochastic trend in the standard production function. Below we use the Kneip, Sickles, and Song approach to provide a method to decompose total factor productivity change into a technical change and a technical efficiency change component.

5.5 The Kneip, Sickles, and Song Factor Model Estimator

The Kneip et al. (2011) model specifies the factors in the following fashion

$$Y_{it} = \beta_0(t) + \sum_{j=1}^{p} \beta_j X_{itj} + v_i(t) + \epsilon_{it}, \quad i = 1, \ldots, n, t = 1, \ldots, T, \tag{5.28}$$

where denotes a general average function, and $v_i(t)$ are non-constant individual effects. In the context of the production decomposition we consider here think of $\beta_0(t)$ as an exogenous or stochastic long term trend due to technical change in production (Y_{it}) and the $v_i(t)$ as the firm technical efficiency terms in a stochastic frontier production function. Details of the estimator are given in KSS. $\beta_0(t)$ can be eliminated by using centered variables $Y_{it} - \bar{Y}_t$, $X_{ijt} - \bar{X}_{tj}$, where $\bar{Y}_t = \frac{1}{n}\sum_i Y_{it}$ and $\bar{X}_{tj} = \frac{1}{n}\sum_i X_{itj}$ and can be viewed as a nuisance parameter, although in the context of production analysis we will use it to identify the common technical change factor, common to all firms. This is just the diffused technical change that is appropriated by each firm in the industry. With this normalization, we can write the model as

$$Y_{it} - \bar{Y}_t = \sum_{j=1}^{p} \beta_j (X_{itj} - \bar{X}_{tj}) + v_i(t) + \epsilon_{it} - \bar{\epsilon}_i, \quad i = 1, \ldots, n, t = 1, \ldots, T, \tag{5.29}$$

with $\bar{\epsilon}_t = \frac{1}{n}\sum_i \epsilon_{it}$. Identification requires that all variables X_{itj}, $j = 1, \ldots, p$ possess a considerable variation over t. Our focus lies on analyzing $v_i(t)$, $t = 1, \ldots, T$ which of course is motivated by our application in the field of stochastic frontier analysis wherein individual effects determine technical efficiencies and are the main quantity of interest. The coefficients β as well as the functions v_i can be estimated by semiparametric techniques using partial spline estimation where the basic underlying assumption is that $v_i(t)$ represent "smooth" time trends. KSS generalize the usual concept of smoothness by relying on second order differences which also allows them to deal with stochastic processes, for example, random walks. They assume the functions v_i can be represented as a weighted average of an unknown number $L \in \{1, 2, \ldots\}$ of basis functions (common factors) $g_1, \ldots, g_L$ given by

$$v_i(t) = \sum_{r=1}^{L} \theta_{ir} g_r(t), \tag{5.30}$$

with unknown factor loadings θ_{ir},in which case the centered model can be rewritten

$$Y_{it} - \bar{Y}_t = \sum_{j=1}^{p} \beta_j (X_{itj} - \bar{X}_{tj}) + \sum_{r=1}^{L} \theta_{ir} g_r(t) + \epsilon_{it} - \bar{\epsilon}_t, \quad i = 1, \ldots, n, t = 1, \ldots, T. \tag{5.31}$$

Parametric mixed effects models of this form are widely used in applications and assume that individual effects can be modeled by linear combinations of *pre-specified* basis function (e.g. polynomials). Cornwell et al. (1990) assume that the v_i can be modeled by quadratic polynomials which in our notation corresponds to an $L = 3$ and g_1, g_2, g_3 forming a polynomial basis. Battese and Coelli (1992) propose a model with $L = 1$ and $g_1(t) = \exp(-\eta(t - T))$ for some $\eta \in \mathbb{R}$. The underlying qualitative assumption is that there exist some common structure characterizing all $v_1, \ldots, v_n$ and that (5.30) is always fulfilled if the empirical covariance matrix $\Sigma_{n,T}$ of the vectors $(v_i(1), \ldots, v_i(T))'$, $i = 1, \ldots, n$, possesses rank L. This is the setup of factor models considered by Bai (2003, 2005) and Ahn et al. (2005) although the focus of KSS is to analyze non-stationary but smooth time trends. We will outline the basic steps in the estimation process. *Estimation* is based on the fact that under the above normalization $g_1, g_2, \ldots$ are to be obtained as (functional) principal components of the sample $v_1 = (v_1(1), \ldots, v_1(T))', \ldots, v_n = (v_n(1), \ldots, v_n(T))'$. If we let $\Sigma_{n,T} = \frac{1}{n} \sum_i v_i v_i{}'$denote the empirical covariance matrix of $v_1, \ldots, v_n$ (recall that $\sum_i v_i = 0$) and use $\lambda_1 \geq \lambda_2 \geq \cdots \geq \lambda_T$ as well as $\gamma_1, \gamma_2, \ldots, \gamma_T$ to denote the resulting eigenvalues and orthonormal eigenvectors of $\Sigma_{n,T}$,then some algebra reveals the following relationships

$$g_r(t) = \sqrt{T} \cdot \gamma_{rt} \text{ for all } r = 1, \ldots,\ t = 1, \ldots, T, \tag{5.32}$$

$$\theta_{ir} = \frac{1}{T} \sum_t v_i(t) g_r(t) \quad \text{for all } r = 1, 2, \ldots,\ i = 1, \ldots, n, \tag{5.33}$$

and

$$\lambda_r = \frac{T}{n} \sum_i \theta_{ir}^2 \quad \text{for all } r = 1, 2, \ldots. \tag{5.34}$$

Furthermore, for all $l = 1, 2, \ldots,$

$$\sum_{r=l+1}^{T} \lambda_r = \sum_{i,t} (v_i(t) - \sum_{r=1}^{l} \theta_{ir} g_r(t))^2 = \min_{\tilde{g}_1, \ldots, \tilde{g}_l} \sum_i \min_{\vartheta_{i1}, \ldots, \vartheta_{il}} \sum_t (v_i(t) - \sum_{r=1}^{l} \vartheta_{ir} \tilde{g}_r(t))^2 \tag{5.35}$$

The estimation algorithm can be represented in five basic steps.

Step 1: Determine estimates $\hat{\beta}_1, \ldots, \hat{\beta}_p$ and functional approximations $\hat{v}_1, \ldots, \hat{v}_n$ by minimizing

$$\sum_i \frac{1}{T} \sum_t \left(Y_{it} - \bar{Y}_t - \sum_{j=1}^{p} \beta_j (X_{itj} - \bar{X}_{tj}) - v_i(t) \right)^2 + \sum_i \kappa \frac{1}{T} \int_1^T (v_i^{(m)}(s))^2 ds \tag{5.36}$$

over all possible values of β and all m-times continuously differentiable functions $v_1, \ldots, v_n$ on $[1, T]$. Here $\kappa > 0$ is a preselected smoothing parameter and $v_i^{(m)}$ denotes the m-th derivative of v_i.

Step 2: Determine the empirical covariance matrix $\hat{\Sigma}_{n,T}$ of $\hat{v}_1 = (\hat{v}_1(1), \hat{v}_1(2), \ldots, \hat{v}_1(T))', \ldots, \hat{v}_n = (\hat{v}_n(1), \hat{v}_n(2), \ldots, \hat{v}_n(T))'$ by

$$\hat{\Sigma}_{n,T} = \frac{1}{n} \sum_i \hat{v}_i \hat{v}_i'$$

and calculate its eigenvalues $\hat{\lambda}_1 \geq \hat{\lambda}_2 \geq \ldots \hat{\lambda}_T$ and the corresponding eigenvectors $\hat{\gamma}_1, \hat{\gamma}_2, \ldots, \hat{\gamma}_T$.

Step 3: Set $\hat{g}_r(t) = \sqrt{T} \cdot \hat{\gamma}_{rt}$, $r = 1, 2, \ldots, L$, $t = 1, \ldots, T$, and for all $i = 1, \ldots, n$ determine $\hat{\theta}_{1i}, \ldots, \hat{\theta}_{Li}$ by minimizing

$$\sum_t \left(Y_{it} - \bar{Y}_t - \sum_{j=1}^{p} \hat{\beta}_j (X_{itj} - \bar{X}_{tj}) - \sum_{r=1}^{L} \vartheta_{ri} \hat{g}_r(t) \right)^2 \tag{5.37}$$

with respect to $\vartheta_{1i}, \ldots, \vartheta_{Li}$.

KSS develop the asymptotic theory underlying this particular factor model. Their main assumption is their **Assumption 5:** *The error terms* ϵ_{it} *are i.i.d. with* $\mathbf{E}(\epsilon_{it}) = 0$, $var(\epsilon_{it}) = \sigma^2 > 0$, *and* $\mathbf{E}(\epsilon_{it}^8) < \infty$. *Moreover,* ϵ_{it} *is independent from* $v_i(s)$ *and* $X_{is,j}$ *for all* t, s, j. They analyze the asymptotic behavior of the parameters of their factor model as $n, T \to \infty$. They do not impose any condition on the magnitude of the quotient T/n and they allow the smoothing parameter κ remain fixed or increase with n, T.

We consider below a range of stochastic frontier productivity models in a series of Monte Carlo experiments based on the panel data model (5.28)

$$Y_{it} = \beta_0(t) + \sum_{j=1}^{p} \beta_j X_{itj} + v_i(t) + \epsilon_{it}. \tag{5.38}$$

Two of the existing time-varying individual effects estimators are the random effects GLS (Cornwell et al. 1990) and MLE (Battese and Coelli 1992). We also compare the fixed and the random effects estimators (Schmidt and Sickles 1984). These estimators have been used extensively in the productivity literature that interprets time varying firm effects (time trends) as technical efficiencies. The CSS estimator allows for an arbitrary polynomial in time (usually truncated at powers larger than two) with different parameters for each firm. The BC estimator is a likelihood based estimator wherein the likelihood function is derived from a mixture of normal noise and an independent one-sided efficiency error, usually specified as a half-normal. In the BC estimator, efficiency levels are allowed to differ across firms but the temporal pattern of efficiency is the same for all firms. We simulate samples of size $n \in \{30, 100, 300\}$ with $T \in \{12, 30\}$ in a model with $p = 2$ regressors and with $\beta_0(t) = 0$ and compare the finite sample performance of four different stochastic frontier estimators. The error process ϵ_{it} is drawn randomly from i.i.d. $\mathbf{N}(0, 1)$. The values of true β are set equal to $(0.5, 0.5)$. In each Monte Carlo sample, the regressors are generated according to a bivariate VAR model as in Park et al. (2003, 2007)

$$X_{it} = RX_{i,t-1} + \eta_{it}, \text{ where } \eta_{it} \sim \mathbf{N}(0, I_2), \tag{5.39}$$

and

$$R = \begin{pmatrix} 0.4 & 0.05 \\ 0.05 & 0.4 \end{pmatrix}.$$

To initialize the simulation, we choose $X_{i1} \sim \mathbf{N}(0, (I_2 - R^2)^{-1})$ and generate the samples using (5.39) for $t \geq 2$. Then, the obtained values of X_{it} are shifted around three different means to obtain three balanced groups of firms from small to large. We fix each group at $\mu_1 = (5, 5)'$, $\mu_2 = (7.5, 7.5)'$, and $\mu_3 = (10, 10)'$. The idea is to generate a reasonable cloud of points for X.

We generate time-varying individual effects in the following ways:

$$\begin{aligned}
\text{DGP1} : v_{it} &= \theta_{i0} + \theta_{i1} t + \theta_{i2} t^2 \\
\text{DGP2} : v_{it} &= -\exp(-\eta(t - T)) u_i \\
\text{DGP3} : v_{it} &= \upsilon_{i1} g_{1t} + \upsilon_{i2} g_{2t} \\
\text{DGP4} : v_{it} &= -u_i
\end{aligned}$$

where θ_{ij} $(j = 0, 1, 2) \sim \mathbf{N}(0, 1)/10^2$, $\eta = 0.15$, $u_i \sim$ i.i.d. $|\mathbf{N}(0, 1)|$, υ_{ij} $(j = 1, 2) \sim \mathbf{N}(0, 1)$, $g_{1t} = \sin(\pi t/4)$ and $g_{2t} = \cos(\pi t/4)$. DGP1 is the GLS version but the fixed effects treatment is used in the experiments (CSS). We also consider a limited set of simulations in which the data generating process is a random walk. DGP2 is based on Battese and Coelli (1992). DGP3 is considered here to model effects with large temporal variations. DGP4 is the usual constant effects model. Thus, we may consider DGP3 and DGP4 as two extreme cases among the possible functional forms of time-varying individual effects.

For the KSS estimator, cubic smoothing splines were used to approximate v_{it} in Step 1, and the smoothing parameter κ was selected by using generalized cross-validation.[3] Most simulation experiments were repeated 1,000 times except the cases for $n = 300$ for which 500 replications were carried out. To measure the performances of the effect and efficiency estimators, we used normalized mean squared error (MSE)

$$R(\widehat{v}, v) = \frac{\sum_{i,t} (\widehat{v}_{it} - v_{it})^2}{\sum_{i,t} v_{it}^2}.$$

[3] We let $\kappa = (1 - p)/p$ and chose p among a selected grid of 9 equally spaced values between 0.1 and 0.9 so that generalized cross-validation rule is minimized.

For the estimates of technical efficiency, we also considered the Spearman rank order correlations of true average technical efficiency across the simulations and the estimates of technical efficiency based on the different estimators. Before we present the simulation results, we briefly introduce the other estimators. For the Within and generalized least squares (GLS) estimators which treat the effects as temporally varying, once individual effects v_{it} are estimated, technical efficiency is calculated as $TE = \exp\{v_i - \max(v_i)\}$ following Schmidt and Sickles (1984). Battese and Coelli (1992) employ the maximum likelihood estimation method to estimate the following equation

$$Y_{it} = \beta_0 + \sum_{j=1}^{p} \beta_j X_{itj} + \epsilon_{it} - u_{it}, \tag{5.40}$$

where the time-varying effects terms are defined as $u_{it} = \eta_{it} u_i = \{\exp[-\eta(t-T)]\} u_i$ for $i = 1, \ldots, n$. Technical efficiency is then calculated as $TE_{BC} = \exp(-u_{it})$. Cornwell et al. (1990) approximate time-varying effects by a quadratic function of time. The model can be written as

$$Y_{it} = X'_{it}\beta + W'_{it}\delta_i + \varepsilon_{it}, \tag{5.41}$$

where $W_{it} = [1, t, t^2]$. If W contains just a constant term then the model reduces to the standard panel data model with heterogeneity in the intercept. If we let $\delta_i = \delta_0 + u_i$ then the model can be rewritten as

$$Y_{it} = X'_{it}\beta + W'_{it}\delta_0 + \omega_{it}, \tag{5.42}$$

$$\omega_{it} = W'_{it}u_i + \varepsilon_{it} = v_i(t) + \epsilon_{it}, \tag{5.43}$$

or

$$Y = X\beta + W\delta_0 + \omega, \tag{5.44}$$

$$\omega = Qu + \varepsilon = v + \varepsilon. \tag{5.45}$$

The Within estimator for β is then

$$\widehat{\beta}_{cssw} = (X'M_Q X)^{-1} X'M_Q Y,$$

where $M_Q = I - Q(Q'Q)^{-1}Q'$, $Q = \text{diag}(W_i)$, $i = 1, \ldots, n$, . Technical efficiency is defined as $TE_{CSS} = \exp\{v_{it} - \max(v_{it})\}$. For the KSS estimator, technical efficiency is calculated similarly as for the CSS estimator.

We now we present the simulation results. Tables 5.1–5.4 present mean squared errors (MSE) of coefficients, effects, and efficiencies, and the Spearman rank order correlation coefficient of efficiencies for each DGP. Also, average optimal dimensions, L, chosen by $C(l)$ criterion are reported in the last column of second panel in each table. Note first that optimal dimension, L, is correctly chosen for the

Table 5.1 Monte carlo simulation results for DGP1

MSE of coefficients*						
N	T	Within	GLS	BC	CSS	KSS
30	12	0.9107	0.6039	0.4933	0.8863	0.4998
	30	4.5286	4.0001	1.1767	0.2329	0.1462
100	12	0.2635	0.1438	0.1454	0.2504	0.1170
	30	1.2219	1.0068	1.4172	0.0726	0.0410
300	12	0.0801	0.0402	0.0360	0.0790	0.0343
	30	0.3409	0.2848	0.1456	0.0258	0.0151
MSE of effects						
N	T	Within	GLS	CSS	KSS	L
30	12	0.6159	0.5692	0.4675	0.2278	1.1200
	30	0.4476	0.4455	0.0051	0.0037	1.0510
100	12	0.5940	0.5755	0.4438	0.1769	1.0620
	30	0.4539	0.4531	0.0050	0.0100	1.0590
300	12	0.6068	0.5990	0.5504	0.1964	1.0341
	30	0.4379	0.4376	0.0064	0.0025	1.0500
MSE of efficiencies						
N	T	Within	GLS	BC	CSS	KSS
30	12	0.3429	0.3255	0.1485	0.3329	0.0921
	30	0.6967	0.7005	0.8430	0.2069	0.0289
100	12	0.4415	0.4294	0.3817	0.3969	0.0529
	30	0.8305	0.8279	1.1184	0.2790	0.0236
300	12	0.5102	0.5070	0.4574	0.4575	0.0364
	30	0.9401	0.9400	1.6111	0.3470	0.0154
Spearman rank correlation of efficiencies						
N	T	Within	GLS	BC	CSS	KSS
30	12	0.5052	0.5004	0.8085	0.7692	0.9806
	30	0.4829	0.4834	0.7533	0.9841	0.9980
100	12	0.3886	0.3886	0.5656	0.7837	0.9923
	30	0.3885	0.3885	0.5900	0.9871	0.9993
300	12	0.3037	0.3037	0.6267	0.7771	0.9924
	30	0.2805	0.2805	0.5469	0.9878	0.9995

Note: * is multiplied by 10^2

KSS estimator in all DGPs[4] Thus, we can verify the validity of the dimension test $C(l)$ discussed in Sect. 5.2.

For DGP1, the performances of the KSS estimator are better than the other estimators by any standards. This is true even when the data is as small as $n = 30$ and $T = 12$. In particular, the KSS estimator outperforms the other estimators in terms of MSE of efficiency. Since the data are generated by DGP1, we may

[4]Although DGP1 consists of three different functions, $[1, t, t^2]$, t^2 term is dominating as T gets large. Thus a one dimensional model is sufficient to approximate the effects generated by DGP1.

Table 5.2 Monte carlo simulation results for DGP2

MSE of coefficients*						
N	T	Within	GLS	BC	CSS	KSS
30	12	2.2939	1.6274	0.3427	0.8901	0.4661
	30	161.0314	106.1230	9.6053	5.4253	0.1499
100	12	0.7709	0.6094	0.1149	0.2505	0.1206
	30	53.4336	39.4729	8.1635	1.9065	0.0403
300	12	0.2873	0.1760	0.0339	0.0800	0.0371
	30	18.4371	11.9706	1.3051	0.6689	0.0141
MSE of effects						
N	T	Within	GLS	CSS	KSS	*L*
30	12	0.3892	0.3753	0.0699	0.1401	1.0720
	30	0.7443	0.7351	0.0202	0.0705	1.0430
100	12	0.4678	0.4642	0.0701	0.2120	1.0350
	30	0.8029	0.8007	0.0217	0.1024	1.0050
300	12	0.4475	0.4452	0.0617	0.1966	1.0260
	30	0.7911	0.7902	0.0213	0.0986	1.0020
MSE of efficiencies						
N	T	Within	GLS	BC	CSS	KSS
30	12	0.2260	0.1951	0.0321	0.2586	0.0786
	30	0.7924	0.7321	0.0096	0.5236	0.0544
100	12	0.2598	0.2473	0.0400	0.2944	0.0787
	30	0.7361	0.7548	0.0091	0.5788	0.0116
300	12	0.2695	0.2618	0.0338	0.3607	0.0916
	30	0.7542	0.7342	0.0213	0.5568	0.0040
Spearman rank correlation of efficiencies						
N	T	Within	GLS	BC	CSS	KSS
30	12	0.8941	0.8914	0.9950	0.9716	0.9976
	30	0.6239	0.6293	0.9993	0.8871	0.9946
100	12	0.8283	0.8249	0.9981	0.9784	0.9966
	30	0.5349	0.5342	0.9997	0.8917	0.9999
300	12	0.8448	0.8446	0.9982	0.9726	0.9938
	30	0.5478	0.5479	0.9982	0.8820	1.0000

Note: * is multiplied by 10^2

expect that CSS estimator performs well. This is true for $T = 30$. However, if T is small ($T = 12$), the CSS estimator is no better than the other estimators. The performances of Within, GLS, and BC estimators generally get worse as T increases. Results in Table 5.1a, generated from a random walk data generating process, are comparable to those in Table 5.1. For DGP2, when data is generated using the model specification of the BC estimator the performances of the KSS estimator is comparable to or sometimes better than that of the BC estimator. The BC estimator seems to work fine for the estimation of effects and efficiencies. In terms of MSE of coefficients, however, it appears that the BC estimator is not

Table 5.3 Monte carlo simulation results for DGP3

MSE of coefficients*						
N	T	Within	GLS	BC	CSS	KSS
30	12	1.6631	0.6852	0.6986	2.7261	0.7099
	30	0.5340	0.2621	0.2779	0.6766	0.1821
100	12	0.4224	0.1597	0.1649	0.6866	0.1290
	30	0.1468	0.0667	0.0715	0.1853	0.0396
300	12	0.1549	0.0606	0.0638	0.2429	0.0378
	30	0.0516	0.0250	0.0281	0.0649	0.0138
MSE of effects						
N	T	Within	GLS	CSS	KSS	*L*
30	12	1.0897	1.0259	1.1143	0.2710	2.1609
	30	1.0432	1.0240	1.0840	0.1140	2.0483
100	12	1.0602	1.0393	1.0672	0.2351	2.0585
	30	1.0364	1.0294	1.0829	0.0929	2.0102
300	12	1.0424	1.0353	1.0197	0.2081	2.0061
	30	1.0307	1.0285	1.0734	0.0822	2.0021
MSE of efficiencies						
N	T	Within	GLS	BC	CSS	KSS
30	12	2.1298	2.4086	7.9252	1.4860	0.2583
	30	2.2636	2.5640	5.0451	1.6066	0.1031
100	12	2.4655	2.6934	12.8728	1.4582	0.2175
	30	7.1729	7.6171	18.6293	4.2421	0.1109
300	12	3.8455	3.9679	25.7966	1.9365	0.2085
	30	8.9848	9.2055	26.4074	4.8352	0.1122
Spearman rank correlation of efficiencies						
N	T	Within	GLS	BC	CSS	KSS
30	12	0.1754	0.1729	0.0408	0.2535	0.9298
	30	0.0597	0.0600	−0.0181	0.0019	0.9842
100	12	0.2050	0.2051	0.1513	0.2674	0.9277
	30	0.0499	0.0498	0.0477	0.0325	0.9731
300	12	0.2131	0.2130	0.0754	0.2615	0.9236
	30	0.0575	0.0574	0.0136	−0.0248	0.9691

Note: * is multiplied by 10^2

reliable when T is large ($T = 30$). The Within and GLS estimators also suffer from substantial distortions when T is large. DGP3 generates effects with large temporal variations. Hence, simple functions of time such as used in the CSS or BC estimators are not sufficient for this type of DGP. However, the KSS estimator does not impose any specific forms on the temporal pattern of effects, and thus it can approximate any shape of time varying effects. We may then expect good performances of the KSS estimator even in this situation, and results in Table 5.3 confirm such belief. On the other hand, the other estimators suffer from severe distortions in the estimates

Table 5.4 Monte carlo simulation results for DGP4

MSE of coefficients*						
N	T	Within	GLS	BC	CSS	KSS
30	12	0.5732	0.3586	0.3734	0.8634	0.6515
	30	0.2023	0.1513	0.1504	0.2319	0.2292
100	12	0.1741	0.1346	0.1260	0.2529	0.1816
	30	0.0571	0.0537	0.0510	0.0695	0.0596
300	12	0.0609	0.0360	0.0364	0.0910	0.0617
	30	0.0218	0.0164	0.0142	0.0258	0.0221
MSE of effects						
N	T	Within	GLS	CSS	KSS	*L*
30	12	0.4390	0.3500	1.2061	0.5407	1.0250
	30	0.1681	0.1465	0.4526	0.2217	1.0130
100	12	0.2769	0.2631	0.8046	0.2988	1.0300
	30	0.1082	0.1065	0.3145	0.1186	1.0200
300	12	0.2689	0.2614	0.7959	0.2799	1.0250
	30	0.0969	0.0954	0.2871	0.1015	1.0220
MSE of efficiencies						
N	T	Within	GLS	BC	CSS	KSS
30	12	0.1211	0.0993	0.1178	0.2600	0.1344
	30	0.0488	0.0421	0.0416	0.1205	0.0595
100	12	0.1719	0.1622	0.0478	0.3488	0.1778
	30	0.0798	0.0763	0.0252	0.1857	0.0829
300	12	0.2124	0.2075	0.0449	0.4120	0.2157
	30	0.0914	0.0907	0.0231	0.2168	0.0938
Spearman rank correlation of efficiencies						
N	T	Within	GLS	BC	CSS	KSS
30	12	0.9964	0.9742	0.9738	0.9481	0.9955
	30	0.9982	0.9804	0.9787	0.9757	0.9977
100	12	0.9989	0.9883	0.9896	0.9106	0.9987
	30	0.9997	0.9946	0.9949	0.9528	0.9996
300	12	0.9997	0.9997	0.9995	0.8946	0.9996
	30	0.9997	0.9995	0.9997	0.9588	0.9997

Note: * is multiplied by 10^2

of effects and efficiencies, although coefficient estimates look reasonably good. In particular, rank correlations of efficiencies are almost zero when T is large.

DGP4 represents the reverse situation so that there is no temporal variation in the effects. Hence, the Within and GLS estimators work very well. Now, our primary question is what are the performances of KSS estimator in this situation. As seen in Table 5.4, its performances are fairly well and comparable to those of the Within and GLS estimators. Therefore, the KSS estimator may be safely used even when temporal variation is not noticeable.

In summary, simulation experiments show that either if constant effects are assumed when the effects are actually time-variant, or if the temporal patterns of effects are misspecified, parameters as well as effect and efficiency estimates become severely biased. In these cases, large T increases the bias, and large n does not help solve the problem. On the other hand, our new estimator performs very well regardless of the assumption on the temporal pattern of effects, and may therefore be preferred to other existing estimators in these types of empirical settings, among potentially many others.

5.6 Conclusion

We have discussed the Solow residual and how it has been interpreted and measured in neoclassical production literature and in the complementary literature on productive efficiency. We have also pointed out why panel data are needed to measure productive efficiency and innovation and thus link the two strands of literature. We provided a discussion on the various estimators used in the two literatures, focusing on one in particular, the factor model and evaluated in finite samples the performance of a particular factor model, the KSS model.

Acknowledgements This paper is based in part on keynote lectures given by Sickles at the Pre-conference Workshop of the 2008 Asia-Pacific Productivity Conference, July 17–19, Department of Economics, National Taiwan University, Taipei, Taiwan, 2008; Anadolu University International Conference in Economics: Developments in Economic Theory, Modeling and Policy, Eskişehir, Turkey, June 17–19, 2009; and the 15th Conference on Panel Data, Bonn, July 3–5, 2009. The authors would like to thank Paul Wilson, Co-Editor of this festschrift honoring Leopold Simar, for his insightful comments and editorial oversight on our paper. The usual caveat applies.

References

Adams, R.M., Berger, A.N. & Sickles, R.C. (1999). Semiparametric approaches to stochastic panel frontiers with applications in the banking industry. *Journal of Business and Economic Statistics*, *17*, 349–358.

Adams, R.M., & Sickles, R.C. (2007). Semi-parametric efficient distribution free estimation of panel models. *Communication in Statistics: Theory and Methods*, *36*, 2425–2442.

Afriat, S. (1972). Efficiency estimation of a production function. *International Economic Review*, *13*, 568–598.

Ahn, S.C., Lee, Y., & Schmidt, P.J. (2005). Panel data models with multiple time-varying individual effects: application to a stochastic frontier production model. mimeo, Michigan State University.

Aigner, D.J., Lovell, C.A.K., & Schmidt, P. (1977) Formulation and estimation of stochastic frontier models. *Journal of Econometrics*, *6*, 21–37.

Arrow K.J. (1962). The economic implications of learning by doing. *Review of Economic Studies*, *29*, 155–173.

Battese, G.E., & Cora, G.S. (1977). Estimation of a production frontier model: with application to the pastoral zone of eastern Australia. *Australian Journal of Agricultural Economics, 21*, 169–179.

Bai, J. (2003). Inferential theory for factor models of large dimensions. *Econometrica, 71*, 135–171.

Bai, J. (2005). Panel data models with interactive fixed effects. April 2005, mimeo, Department of Economics, New York University.

Bai, J., & Ng, S. (2002). Determining the number of factors in approximate factor models. *Econometrica, 70*, 191–221.

Bai, J., & Ng, S. (2007). Determining the number of primitive shocks in factor models. *Journal of Business and Economic Statistics, 25*, 52–60.

Bai, J., Kao, C., & Ng, S. (2007). Panel cointegration with global stochastic trends. Center for Policy Research Working Papers 90, Center for Policy Research, Maxwell School, Syracuse University.

Balk, B. (2009). *Price and quantity index numbers: models for measuring aggregate change and difference*. New York: Cambridge University Press.

Baltagi, B., Egger, P., & Pfaffermayr, M. (2003). A generalized design for bilateral trade flow models. *Economics Letters, 80*, 391–397.

Baltagi, B. (2005). *Econometric Analysis of Panel Data, 3rd edition*, New Jersey: Wiley.

Battese, G.E. & Coelli, T.J. (1992). Frontier production functions, technical efficiency and panel data: with application to paddy farmers in India. *Journal of Productivity Analysis, 3*, 153–169.

Berger, A.N. (1993). "Distribution-Free" estimates of efficiency in U.S. banking industry and tests of the standard distributional assumption. *Journal of Productivity Analysis, 4*, 261–292.

Bernanke, B.S., & Boivin, J. (2003). Monetary policy in a data-rich environment. *Journal of Monetary Economics, 50*, 525–546.

Blazek, D., & Sickles, R.C. (2010). The impact of knowledge accumulation and geographical spillovers on productivity and efficiency: the case of U.S. shipbuilding during WWII. In Hall, S.G., Klein, L.R., Tavlas, G.S. & Zellner, A. (eds.), *Economic Modelling, 27*, 1484–1497.

Breitung, J., & Eickmeier, S. (2005). *Dynamic factor models. Discussion Paper Series 1: Economic Studies, No 38/2005*. Frankfurt: Deutsche Bundesbank.

Breitung, J., & Kretschmer, U. (2005). Identification and estimation of dynamic factors from large macroeconomic panels. Mimeo: Universitat Bonn.

Brumback, B.A., & Rice, J.A. (1998). Smoothing spline models for the analysis of nested and crossed samples of curves (with discussion). *Journal of the American Statistical Association, 93*, 961–94.

de Boor, C. (1978). *A Practical Guide to Splines*. New York: Springer.

Caves, D., Christensen, L.R, & Diewert, W.E. (1982). Multilateral comparisons of output, input, and productivity using superlative index numbers. *Economic Journal, 92*, 73–86.

Carriero, A., Kapetanios, G., & Marcellino, M. (2008). Forecasting large datasets with reduced rank multivariate models. Working Papers 617, Queen Mary, University of London, School of Economics and Finance.

Chang, Y. (2004). Bootstrap unit root tests in panels with cross-sectional dependency. *Journal of Econometrics, 120*, 263–293.

Chamberlain, G., & Rothschild, M. (1983). Arbitrage, factor structure and mean-variance analysis in large asset markets. *Econometrica, 51*, 1305–1324.

Charnes, A., Cooper, W.W., & Rhodes, E.L. (1978). Measuring the efficiency of decision making units. *European Journal of Operational Research, 2*, 429–444.

Chun, A. (2009). Forecasting interest rates and inflation: blue chip clairvoyants or econometrics? EFA 2009 Bergen Meetings Paper, Bergen, Norway.

Cornwell, C., Schmidt, P., & Sickles, R.C. (1990). Production frontiers with cross-sectional and time-series variation in efficiency levels. *Journal of Econometrics, 46*, 185–200.

Debreu, G. (1951). The coefficient of resource utilization. *Econometrica, 19*, 273–292.

Diebold, F.X., & Li, C. (2006). Forecasting the term structure of government bond yields. *Journal of Econometrics, 130*, 337–364.

Doz, C., Giannone, D., & Reichlin, L. (2006). A quasi maximum likelihood approach for large approximate dynamic factor models. ECB Working Paper 674.

Engle, R., Granger, C., Rice, J., & Weiss, A. (1986). Nonparametric estimates of the relation between weather and electricity sales. *Journal of American Statistical Association, 81*, 310–320.

Eubank, R.L. (1988). *Nonparametric regression and spline smoothing*. New York: Marcel Dekker.

Färe, R., Grosskopf, S., Lindgren, B., & Roos, P. (1992). Productivity changes in Swedish pharamacies 1980–1989: a non-parametric Malmquist approach. *Journal of Productivity Anlaysis, 3*, 85–101.

Färe R., Grosskopf, S., Norris, M., & Zhang, Z. (1994). Productivity growth, technical progress and efficiency change in industrialized countries. *American Economic Review, 84*, 66–83.

Farrell, M. (1957). The measurement of productive efficiency. *Journal of the Royal Statistical Society, Series A, 120*, 253–282.

Ferré, L. (1995). Improvement of some multivariate estimates by reduction of dimensionality. *Journal of Multivariate Analysis, 54*, 147–162.

Fisher, I. (1927). *The Making of Index Numbers*. Boston: Houghton-Mifflin.

Forni, M., & Lippi, M. (1997). *Aggregation and the microfoundations of dynamic macroeconomics*. Oxford: Oxford University Press.

Forni, M., & Reichlin, L. (1998). Let's get real: a factor analytic approach to disaggregated business cycle dynamics. *Review of Economic Studies, 65*, 653–473.

Forni, M., Hallin, M., Lippi, M., & Reichlin, L. (2000). The generalized dynamic factor model: identification and estimation. *Review of Economics and Statistics, 82*, 540–554.

Førsund, F., & Hjalmarsson, L.(2008). Dynamic Analysis of Structural Change and Productivity Measurement. Unpublished Working Paper, Mimeo.

Fried, H.O., Lovell, C.A.K., & Schmidt, S.S. (2008). The measurement of productive efficiency and productivity growth. Oxford University Press, Oxford.

Getachew, L., & Sickles, R.C. (2007). Allocative distortions and technical efficiency change in Egypt's private sector manufacturing industries: 1987–1996. *Journal of the Applied Econometrics, 22*, 703–728.

Greene, W. (2004). Fixed and random effects in stochastic frontier models. *Journal of Productivity Analysis, 23*, 7–32.

Greene, W. (2005). Reconsidering heterogeneity in panel data estimators of the stochastic frontier model. *Journal of Econometrics, 126*, 269–303.

Greene, W. (2008). In Fried, H., Lovell, C.A.K., & Schmidt, S. (eds.) *The Measurement of Productive Efficiency and Productivity Change, Chap. 2*. Oxford: Oxford University Press.

Härdle, W., Liang, H., & Gao, J. (2000). *Partially linear models*. Heidelberg: Physica-Verlag.

Jeon, B.M., & Sickles, R.C. (2004). The Role of environmental factors in growth accounting: a nonparametric analysis. *Journal of the Applied Economics, 19*, 567–591.

Jorgenson, D.W., & Griliches, Z. (1972). Issues in growth accounting: a reply to Edward F. Denison. *Survey of Current Business, 55* (part 2), 65–94.

Kao, C., & Chiang, M.H. (2000). On the estimation and inference of a cointegrated regression in panel data. *Advances in Econometrics, 15*, 179–222.

Kapetanios, G., & Marcellino, M., (2009). A parametric estimation method for dynamic factor models of large dimensions, *Journal of Time Series Analysis, 30*, 208–238.

Klee, E.C., & Natalucci, F.M. (2005). Profits and balance sheet developments at U.S. commercial banks in 2004. *Federal Reserve Bulletin*, Spring.

Kendrick, J. (1961). *Productivity trends in the United States*. Princeton: Princeton University Press for the National Bureau of Economic Research.

Koenig, E. (2000). Productivity growth. Federal Reserve Bank of Dallas, Expand Your Insight, March 1, http://www.dallasfed.org/eyi/usecon/0003growth.html.

Koop, G.M., & Poirier, D. (2004). Bayesian variants of some classical semiparametric regression techniques. *Journal of Econometrics, 123*(2), 259–282.

Kneip, A. (1994). Nonparametric estimation of common regressors for similar curve data. *Annals of Statistics, 22*, 1386–1427.

Kneip, A., & Utikal, K.J. (2001). Inference for density families using functional principal component analysis. *Journal of American Statistical Association, 96*, 519–532.
Kneip, A., Sickles, R.C., & Song, W. (2011). *A new panel data treatment for heterogeneity in time trends*. Econometric Theory, to appear.
Kumbhakar, S.C. (1990). Production Frontiers, panel data and time-varying technical inefficiency. *Journal of Econometrics, 46*, 201–211.
Kumbhakar, S., & Lovell, C.A.K. (2000). *Stochastic Frontier Analysis*. Cambridge: Cambridge University Press.
Lovell, C.A.K., Richardson, S., Travers, P., & Wood, L.L. (1994). Resources and functionings: a new view of inequality in Australia. In Eichorn, W. (Ed.) *Models and Measurement of Welfare and Inequality*, pp. 787–807. Berlin, Heidelberg, New York: Springer.
Lucas, R.E. (1988). On the Mechanics of Economic Development. *Journal of Monetary Economics, 22*, 3–42.
Maddala, G.S., & Kim, I.M. (1998). *Unit Roots, cointegration and structural change*. Cambridge: Cambridge University Press.
Mark, N.C., & Sul, D. (2003). Cointegration vector estimation by panel dlos and long-run money demand. *Oxford Bulletin of Economics and Statistics, 65*, 655–680.
Marcellino, M., & Schumacher, C. (2007). Factor-midas for now- and forecasting with ragged-edge data: a model comparison for German gdp. Discussion Paper Series 1: Economic Studies,34, Deutsche Bundesbank, Research Centre.
Meeusen, W., & van den Broeck, J. (1977). Efficiency estimation from Cobb-Douglas production functions with composed error. *International Economic Review, 18*, 435–444.
Nelson, C.R., & Plosser, C.I. (1982). Trends and random walks in macroeconomics time series: some evidence and implications. *Journal of Monetary Economics, 10*, 139–162.
Orea, C., & Kumbhakar, S. (2004). Efficiency measurement using a latent class stochastic frontier model. *Empirical Economics, 29*, 169–184.
Park, B.U., & Simar, L. (1994). Efficient semiparametric estimation in stochastic frontier models. *Journal of the American Statistical Association, 89*, 929–936.
Park, B.U., Sickles, R.C., & Simar, L. (1998). Stochastic frontiers: a semiparametric approach. *Journal of Econometrics, 84*, 273–301.
Park, B.U., Sickles, R.C., & Simar, L. (2003). Semiparametric efficient estimation of AR(1) panel data models. *Journal of Econometrics, 117*, 279–309.
Park, B.U., Sickles, R.C., & Simar, L. (2007). Semiparametric efficient estimation of dynamic panel data models. *Journal of Econometrics, 136*, 281–301.
Pesaran, M.H. (2006). Estimation and inference in large heterogeneous panels with a multifactor error structure. *Econometrica, 74*, 967–1012.
Pitt, M., & Lee, L.-F. (1981). The measurement and sources of technical inefficiency in the Indonesian weaving industry. *Journal of Development Economics, 9*, 43–64.
Ramsay, J., & Silverman, B. (1997). *Functional data analysis*. Heidelberg: Springer.
Rao, C.R. (1958). Some statistical methods for the comparison of growth curves. *Biometrics, 14*, 1–17.
Reikard, G. (2005). Endogenous technical advance and the stochastic trend in output: A neoclassical approach. *Research Policy, 34*, 1476–1490.
Romer, P.M. (1986). Increasing returns and long-run growth. *Journal of Political Economy, 94*, 1002–1037.
Schmidt, P., & Sickles, R.C. (1984). Production frontiers and panel data. *Journal of Business and Economic Statistics, 2*, 367–374.
Sickles, R.C., & Streitwieser, M. (1992). Technical inefficiency and productive decline in the U.S. interstate natural gas pipeline industry under the U.S. interstate natural gas policy act. *Journal of Productivity Analysis* (Lewin, A., & Lovell, C.A.K. Eds.), *3*, 115–130. Reprinted in *International Applications for Productivity and Efficiency Analysis*, (Thomas R. Gulledge, Jr., & Knox Lovell, C.A. Eds.). Boston: Kluwer.

Sickles, R.C., & Streitwieser, M. (1998). The structure of technology, substitution and productivity in the interstate natural gas transmission industry under the natural gas policy act of 1978. *Journal of Applied Econometrics, 13*, 377–395.

Sickles, R.C. (2005). Panel estimators and the identification of firm-specific efficiency levels in parametric, semiparametric and nonparametric settings. *Journal of Econometrics, 50, 126*, 305–334.

Sickles, R.C., & Tsionas, E.G. (2008). *A panel data model with nonparametric time effects*. Mimeo: Rice University.

Simar, L., & Wilson, P.W. (2010). Inference from cross-sectional stochastic frontier models. *Econometric Reviews, 29*, 62–98.

Shephard, R.W. (1970). *Theory of cost and production functions*. Princeton: Princeton University Press.

Solow, Robert M. (1957). Technical change and the aggregate production function. *Review of Economics and Statistics, 39*, 312–320.

Speckman, P. (1988). Kernel smoothing in partial linear models. *Journal of the Royal Statistical Society, Series B, 50*, 413–436.

Stock, J.H., & Watson, M.W. (2002). Forecasting using principal components from a large number of predictors. *Journal of the American Statistical Association, 97*, 1167–1179.

Stock, J.H., & Watson, M.W. (2005). *Implications of dynamic factor models for VAR analysis*. Mimeo: Princeton University.

Tsionas, E.G., & Greene, W. (2003). A panel data model with nonparametric time effects, Athens University of Business and Economics.

Utreras, F. (1983). Natural spline functions, their associated eigenvalue problem. *Numerical Mathematics, 42*, 107–117.

Chapter 6
Asymptotic Properties of Some Non-Parametric Hyperbolic Efficiency Estimators

Paul W. Wilson

Abstract A hyperbolic measure of technical efficiency was proposed by Fare et al. (The Measurement of Efficiency of Production, Kluwer-Nijhoff Publishing, Boston, 1985) wherein efficiency is measured by the simultaneous maximum, feasible reduction in input quantities and increase in output quantities. In cases where returns to scale are not constant, the non-parametric data envelopment analysis (DEA) estimator of hyperbolic efficiency cannot be written as a linear program; consequently, the measure has not been used in empirical studies except where returns to scale are constant, allowing the estimator to be computed by linear programming methods. This paper develops an alternative estimator of the hyperbolic measure proposed by Fare et al. (The Measurement of Efficiency of Production, Kluwer-Nijhoff Publishing, Boston, 1985). Statistical consistency and rates of convergence are established for the new estimator. A numerical procedure allowing computation of the original estimator is provided, and this estimator is also shown to be consistent, with the same rate of convergence as the new estimator. In addition, an unconditional, hyperbolic order-*m* efficiency estimator is developed by extending the ideas of Cazals et al. (J. Econometric. 106:1–25, 2002). Asymptotic properties of this estimator are also given.

6.1 Introduction

The performance of firms and other decision-making units in terms of technical efficiency, as well as allocative, cost, and other efficiencies, has received widespread attention in the economics, statistics, management science, and related literatures.

P.W. Wilson (✉)
Department of Economics, 222 Sirrine Hall, Clemson University, Clemson, South Carolina 29634-1309, USA
e-mail: pww@clemson.edu

I. van Keilegom and P.W. Wilson (eds.), *Exploring Research Frontiers in Contemporary Statistics and Econometrics*, DOI 10.1007/978-3-7908-2349-3_6,

In the case of private firms, estimates of inefficiency have been used to explain insolvency rates and merger activities, the effects of changes in regulatory environments, and overall industry performance. In the case of public and non-profit entities, estimates of inefficiency are intrinsically interesting because these entities do not face a market test, and inefficiency estimates often provide the only objective criteria for gauging performance. Measuring the performance of public entities may be important for allocating scarce public resources, for deciding which to eliminate during periods of consolidation, etc. In particular, identifying inefficient entities is a critical first step in any attempt to improve performance.

Non-parametric approaches to estimation of technical efficiency have been widely applied. Non-parametric methods usually involve the estimation of a production set or some other set by either the free-disposal hull (FDH) of sample observations, or the convex hull of the FDH. Methods based on the convex hull of the FDH are collectively referred to as data envelopment analysis (DEA). DEA is well known and has been applied widely: as of early 2004, DEA had been used in more than 1,800 articles published in some 490 refereed journals (Gattoufi et al. 2004).

The overwhelming majority of these applications have involved either input- or output-oriented measures of efficiency. In the input-orientation, one measures the proportion by which input quantities can feasibly be reduced without reducing output quantities, while in the output-orientation, one measures the proportion by which output quantities can feasibly be increased without increasing input quantities. The statistical properties of input-oriented DEA estimators have been established, and methods are now available for making statistical inferences about efficiency based on DEA; these results extend to output-oriented estimators after changes in notation.[1]

Unfortunately, however, the choice between input- versus output-orientation can give rather different, and perhaps misleading, indicators of inefficiency. To illustrate, consider the situation shown in Fig. 6.1 where a single input is used to produce a single output. Feasible combinations of input and output quantities are bounded from above by the convex curve labeled $\mathcal{P}^{\delta}$. Now consider the (technically inefficient) firm operating at point A; this firm can attain technical efficiency by reducing its input usage a small amount by moving to point A' while avoiding a decrease in output production. Alternatively, in can increase its outputs by a large amount by moving to the efficient point A'' without increasing its input usage. In the input-orientation, the firm operating at point A is slightly inefficient, but in the output-orientation, the firm is perhaps *grossly* inefficient. Similar reasoning reveals that, in contrast to the firm operating at point A, a firm operating at point B is slightly

[1]See Simar and Wilson (2000b) for a survey, and Kneip et al. (2008) for more recent results, on the statistical properties of DEA estimators. See Simar and Wilson (1998, 2000a) and Kneip et al. (2008, 2011) for details about the use of bootstrap methods to make inferences based on DEA.

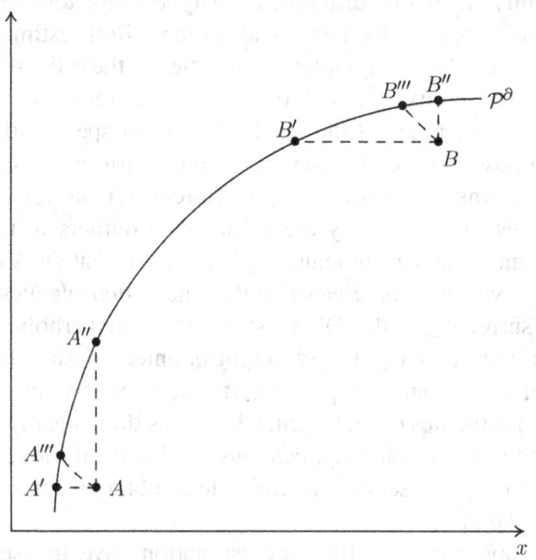

Fig. 6.1 Input, output, and hyperbolic efficiency

inefficient in the output-orientation, but very inefficient in the input orientation (i.e., the distance from B to B'' is much smaller than the distance from B to B').

Note that the choice between input- and output-orientation when measuring technical efficiency is largely arbitrary, in the sense that there is no notion of causality or endogeniety as in the case of regression models. As seen below in Sect. 6.2, assumptions about a joint distribution for input and output quantities with support over the production set bounded by $\mathcal{P}^\partial$ are needed, but from a statistical viewpoint, one is free to choose either an input- or output-oriented measure of technical efficiency.

One can also measure technical efficiency along hyperbolic paths, as proposed by Färe et al. (1985). The firm operating at point A fig. 6.1 can become technically inefficient by simultaneously reducing its output usage while increasing its output production by the same factor (this is explained in detail below in Sect. 6.2.1) and moving to point A''' on the production frontier. Similarly, the firm operating at point B can simultaneously reduce its input usage and expand its output usage by the same factor to arrive at point B''' on the production frontier. When technical efficiency is measured along the hyperbolic paths passing through points A and A''' and through points B an B''', the two firms operating at points A and B are seen to have similar levels of inefficiency. Moreover, their efficiency levels, when measured along these hyperbolic paths, are not dependent on the *slope* of the production frontier $\mathcal{P}^\delta$ in the neighborhood of the point at which either firm operates. This stands in contrast to the input- and output-oriented measures of technical efficiency.

I propose in this paper two different DEA-type estimators for the hyperbolic efficiency measure proposed by Färe et al. (1985). Both estimators are easy to compute. In addition, I derive asymptotic properties of the estimators.

Despite their popularity, DEA estimators have some obvious drawbacks. Although these estimators avoid the need for a priori specification of functional forms, they do impose (eventually) non-increasing returns to scale; i.e., they do not allow increasing returns to scale everywhere. Moreover, it has long been recognized that DEA estimates of inefficiency are sensitive to outliers in the data. Perhaps even more problematic, at least in many applications, is that DEA estimators suffer from the well-known *curse of dimensionality* that often plagues non-parametric estimators. Not surprisingly, the DEA estimators of hyperbolic efficiency share these limitations with their input- and output-oriented cousins. Regardless of the orientation (i.e., input, output, or hyperbolic), the number of observations required to obtain meaningful estimates of inefficiency increases dramatically with the number of inputs and outputs. In many applications, including the one in this chapter, there are simply too few observations available to obtain meaningful estimates of inefficiency using DEA.[2]

Some recent approaches to efficiency estimation have focused on estimating distances to some feature *near* the production frontier, as opposed to distance to the frontier itself. Specifically, efficiency can be measured relative to some notion of a *partial* frontier. This has several advantages; the estimators are robust with respect to outliers, achieve root-n rates of convergence, and are asymptotically normal. Daouia (2003), Aragon et al. (2005) and Daouia and Simar (2007) developed input- and output-oriented conditional order-α estimators and derived their asymptotic properties. These results were subsequently extended to an unconditional, hyperbolic order-α estimator by Wheelock and Wilson (2008).[3] Similarly, Cazals et al. (2002) proposed input- and output-oriented conditional order-m estimators and derived their asymptotic properties. In addition to the hyperbolic DEA estimators developed below, I also describe an unconditional, hyperbolic order-m estimator and derive its asymptotic properties. The properties are analogous to those obtained by Cazals et al. for the input- and output-oriented conditional order-m estimators.

The hyperbolic order-m estimator avoids a problem noted by Wheelock and Wilson (2008). Both the conditional order-m efficiency measures as well as the conditional order-α efficiency measures (either input- or output-oriented) measure distance to *different* partial frontiers with different slopes, depending on the orientation. This creates additional ambiguity for interpretation of inefficiency estimates, in addition to the ambiguity surrounding the choice between input- and

[2]One can find numerous published applications of DEA to datasets with 50–150 observations and 5 or more dimensions in the input-output space. DEA-based inefficiency estimates from such studies are likely meaningless in a statistical sense due to the curse of dimensionality problem (see Simar and Wilson (2000b) for discussion).

[3]Wheelock and Wilson (2008) also derived asymptotic properties for a hyperbolic FDH efficiency estimator.

output-orientation when DEA estimators are used as described above. As will be seen below in Sects. 6.2.2 and 6.5, and as with the unconditional hyperbolic order-α estimator, the unconditional hyperbolic order-m estimator developed below avoids this problem by estimating distance to a partial frontier that does not converge to the full frontier at some corner of the production set.

In the next section, I describe notation and a statistical model for efficiency estimation, as well as various measures of technical efficiency. In Sect. 6.3, I introduce the hyperbolic DEA estimators and their asymptotic properties. Followed by a brief discussion of the hyperbolic FDH estimator in Sect. 6.4, I present the unconditional, hyperbolic order-m estimator and its asymptotic properties in Sect. 6.5. Results of some Monte Carlo experiments designed to assess bias, variance, and mean-square error of the estimators are presented in Sect. 6.6, and conclusions are discussed in Sect. 6.7.

6.2 A Statistical Model of Production

6.2.1 The Basic Framework

In situations where non-parametric efficiency estimators are used, researchers typically have in mind a model of the production process, whether it is stated explicitly or not. Let $\boldsymbol{x} \in \mathbb{R}_+^p$ and $\boldsymbol{y} \in \mathbb{R}_+^q$ denote vectors of inputs and outputs, respectively, and denote the production possibilities set by

$$\mathcal{P} = \{(\boldsymbol{x}, \boldsymbol{y}) \mid \boldsymbol{x} \text{ can produce } \boldsymbol{y}\} \subset \mathbb{R}_+^{p+q}. \tag{6.1}$$

The production possibilities set can be described in terms of its sections

$$\mathcal{X}(\boldsymbol{y}) = \{\boldsymbol{x} \in \mathbb{R}_+^p \mid (\boldsymbol{x}, \boldsymbol{y}) \in \mathcal{P}\}, \tag{6.2}$$

and

$$\mathcal{Y}(\boldsymbol{x}) = \{\boldsymbol{y} \in \mathbb{R}_+^q \mid (\boldsymbol{x}, \boldsymbol{y}) \in \mathcal{P}\}, \tag{6.3}$$

or input requirement sets and output correspondence sets, respectively.

The upper boundary of $\mathcal{P}$, denoted $\mathcal{P}^\partial$, is sometimes referred to as the *technology* or the *production frontier*, and is given by the intersection of $\mathcal{P}$ and the closure of its compliment; i.e.,

$$\mathcal{P}^\partial = \{(\boldsymbol{x}, \boldsymbol{y}) \mid (\delta^{-1}\boldsymbol{x}, \delta\boldsymbol{y}) \notin \mathcal{P} \ \forall \ \delta > 1\}. \tag{6.4}$$

Similarly, the efficient subset of $\mathcal{X}(\boldsymbol{y})$ is given by

$$\mathcal{X}^\partial(\boldsymbol{y}) = \{\boldsymbol{x} \in \mathbb{R}_+^p \mid \boldsymbol{x} \in \mathcal{X}(\boldsymbol{y}),\ \theta\boldsymbol{x} \notin \mathcal{X}(\boldsymbol{y}) \ \forall \ \theta \in (0, 1)\}. \tag{6.5}$$

The efficient subset of $\mathcal{Y}(\boldsymbol{x})$ is

$$\mathcal{Y}^{\partial}(\boldsymbol{x}) = \{\boldsymbol{y} \in \mathbb{R}_{+}^{q} \mid \boldsymbol{y} \in \mathcal{Y}(\boldsymbol{x}),\ \lambda \boldsymbol{y} \notin \mathcal{Y}(\boldsymbol{x})\ \forall\ \lambda > 1\}. \tag{6.6}$$

Typical assumptions (e.g., see Färe 1988) include:

Assumption 6.2.1 *The production set $\mathcal{P}$ is convex and compact.*

Assumption 6.2.2 *The production set $\mathcal{P}$ allows free disposability; i.e., for $\widetilde{\boldsymbol{x}} \geq s$, $\widetilde{\boldsymbol{y}} \leq \boldsymbol{y}$, if $(\boldsymbol{x}, \boldsymbol{y}) \in \mathcal{P}$, then $(\widetilde{\boldsymbol{x}}, \boldsymbol{y}) \in \mathcal{P}$ and $(\boldsymbol{x}, \widetilde{\boldsymbol{y}}) \in \mathcal{P}$.*

Assumption 6.2.3 *All production requires the use of some inputs, i.e., $(\boldsymbol{x}, \boldsymbol{y}) \notin \mathcal{P}$ if $\boldsymbol{x} = 0$, $\boldsymbol{y} \geq 0$, $\boldsymbol{y} \neq 0$.*

Assumption 6.2.3 merely says that lunch is not free, while Assumption 6.2.2 is equivalent to an assumption of (weak) monotonicity of the frontier $\mathcal{P}^{\partial}$. Assumption 6.2.1 is needed to establish statistical consistency of the DEA estimators discussed below in Sect. 6.3. FDH and order-m estimators, however, do not require convexity, and remain consistent when $\mathcal{P}$ is not convex.

The Farrell (1957) input measure of efficiency

$$\theta(\boldsymbol{x}, \boldsymbol{y}) \equiv \inf\{\theta > 0 \mid (\theta \boldsymbol{x}, \boldsymbol{y}) \in \mathcal{P}\} \tag{6.7}$$

measures distance from an arbitrary point $(\boldsymbol{x}, \boldsymbol{y}) \in \mathbb{R}_{+}^{p+q}$ to the boundary $\mathcal{P}^{\partial}$ along the ray $(\theta \boldsymbol{x}, \boldsymbol{y})$, $\theta > 0$. For a firm operating at $(\boldsymbol{x}, \boldsymbol{y}) \in \mathcal{P}$, the corresponding efficient level of inputs, denoted $\boldsymbol{x}^{\partial}(\boldsymbol{y})$, is given by

$$\boldsymbol{x}^{\partial}(\boldsymbol{y}) \equiv \theta(\boldsymbol{x}, \boldsymbol{y})\boldsymbol{x}. \tag{6.8}$$

Similarly, the Farrell (1957) output efficiency measure

$$\lambda(\boldsymbol{x}, \boldsymbol{y}) \equiv \sup\{\lambda > 0 \mid (\boldsymbol{x}, \lambda \boldsymbol{y}) \in \mathcal{P}\} \tag{6.9}$$

measures distance from an arbitrary point $(\boldsymbol{x}, \boldsymbol{y}) \in \mathbb{R}_{+}^{p+q}$ to the boundary $\mathcal{P}^{\partial}$ along the ray $(\boldsymbol{x}, \lambda \boldsymbol{y})$, $\lambda > 0$. Again, for a firm operating at $(\boldsymbol{x}, \boldsymbol{y}) \in \mathcal{P}$, the corresponding efficient level of outputs is given by

$$\boldsymbol{y}^{\partial}(\boldsymbol{x}) \equiv \lambda(\boldsymbol{x}, \boldsymbol{y})\boldsymbol{y}. \tag{6.10}$$

The measures $\theta(\boldsymbol{x}, \boldsymbol{y})$ and $\lambda(\boldsymbol{x}, \boldsymbol{y})$ provide measures of the *technical efficiency* of a firm producing levels $\boldsymbol{y}$ of outputs from levels $\boldsymbol{x}$ of inputs.[4] A firm operating at a point $(\boldsymbol{x}, \boldsymbol{y})$ in the interior of $\mathcal{P}$ could become technically efficient by moving to either $(\boldsymbol{x}^{\partial}(\boldsymbol{y}), \boldsymbol{y})$, $(\boldsymbol{x}, \boldsymbol{y}^{\partial}(\boldsymbol{x}))$, or perhaps some other point along the frontier $\mathcal{P}^{\partial}$.

[4]The Farrell (1957) input and output distance functions defined in (6.7)–(6.9) are reciprocals of the corresponding Shephard (1970) measures.

Let $\mathcal{V}(\mathcal{P})$ denote the convex cone of the set $\mathcal{P}$. Then $\mathcal{P}^{\partial}$ is characterized by globally constant returns to scale (CRS) if and only if $\mathcal{P} = \mathcal{V}(\mathcal{P})$. Otherwise, $\mathcal{P} \subset \mathcal{V}(\mathcal{P})$, and $\mathcal{P}^{\partial}$ exhibits variable returns to scale (VRS).

Under CRS, $\theta(\boldsymbol{x}, \boldsymbol{y}) = \lambda(\boldsymbol{x}, \boldsymbol{y})^{-1}$. However, with VRS, the choice of orientation (either input or output) can have a large impact on measured efficiency. As discussed in Sect. 6.1, with VRS, a large firm could conceivably lie close to the frontier $\mathcal{P}^{\partial}$ in the output direction, but far from $\mathcal{P}^{\partial}$ in the input direction. Similarly, a small firm might lie close to $\mathcal{P}^{\partial}$ in the input direction, but far from $\mathcal{P}^{\partial}$ in the output direction. Such differences are related to the slope and curvature of $\mathcal{P}^{\partial}$. Moreover, there seems to be no criteria telling the applied researcher whether to use the input- or output-orientation. In the case of parametric, stochastic frontier models along the lines of Aigner et al. (1977), one specifies a production, cost, or other relationship, which determines how efficiency is to be measured; e.g., when a production function is specified, efficiency is measured in the output direction. By contrast, the model specified by Assumptions 6.2.1, 6.2.2, and 6.2.3 leaves open the question of the direction in which efficiency might be measured.

The hyperbolic graph efficiency measure

$$\gamma(\boldsymbol{x}, \boldsymbol{y}) \equiv \inf\{\gamma > 0 \mid (\gamma \boldsymbol{x}, \gamma^{-1}\boldsymbol{y}) \in \mathcal{P}\} \tag{6.11}$$

defined by Färe et al. (1985) provides an alternative measure of technical efficiency. This measure gives distance from the fixed point $(\boldsymbol{x}, \boldsymbol{y})$ to $\mathcal{P}^{\partial}$ along the hyperbolic path $(\gamma \boldsymbol{x}, \gamma^{-1}\boldsymbol{y})$, $\gamma \in \mathbb{R}^1_{++}$. Measuring efficiency along a hyperbolic path avoids some of the ambiguity cited above.[5]

By construction, $\gamma(\boldsymbol{x}, \boldsymbol{y}) \leq 1$ for $(\boldsymbol{x}, \boldsymbol{y}) \in \mathcal{P}$. While it is easy to show that under CRS,

$$\gamma(\boldsymbol{x}, \boldsymbol{y}) = \theta(\boldsymbol{x}, \boldsymbol{y})^{1/2} = \lambda(\boldsymbol{x}, \boldsymbol{y})^{-1/2}, \tag{6.12}$$

no such relationship exists under VRS. Just as the measures $\theta(\boldsymbol{x}, \boldsymbol{y})$ and $\lambda(\boldsymbol{x}, \boldsymbol{y})$ provide measures of the *technical efficiency* of a firm operating at a point $(\boldsymbol{x}, \boldsymbol{y}) \in \mathcal{P}$, so does $\gamma(\boldsymbol{x}, \boldsymbol{y})$, but along a hyperbolic path to the frontier $\mathcal{P}^{\partial}$. Using $\gamma(\boldsymbol{x}, \boldsymbol{y})$, any point in the interior of $\mathcal{P}$ can be projected onto $\mathcal{P}^{\partial}$.

Definition 6.2.1 *For any $(\boldsymbol{x}, \boldsymbol{y}) \in \mathcal{P}$, the corresponding hyperbolic-efficient input-output combination denoted by $(\boldsymbol{x}^{\partial}(\boldsymbol{x}, \boldsymbol{y}), \boldsymbol{y}^{\partial}(\boldsymbol{x}, \boldsymbol{y})) \in \mathcal{P}^{\partial}$ is given by*

[5] Alternatively, one might consider estimating the directional distance function

$$\varphi(\boldsymbol{x}, \boldsymbol{y}) = \sup\{\varphi \mid ((1 - \varphi)\boldsymbol{x}, (1 + \varphi)\boldsymbol{y}) \in \mathcal{P}\},$$

which is a special case of the general directional distance function proposed by Chambers et al. (1996). While this distance function can be estimated by linear programming methods, proofs of asymptotic properties such as consistency, rate of convergence, etc. remain elusive.

$$
\begin{aligned}
x^{\partial}(x, y) &= x\gamma(x, y), \\
y^{\partial}(x, y) &= y\gamma(x, y)^{-1}.
\end{aligned} \tag{6.13}
$$

Assumptions 6.2.1–6.2.3 define an *economic* model, but additional assumptions are needed to define a *statistical* model. First, assume independent sampling of input/output combinations:

Assumption 6.2.4 *The sample observations* $\mathcal{S}_n = \{(x_i, y_i)\}_{i=1}^n$ *are realizations of identically, independently distributed (iid) random variables with probability density function* $f(x, y)$ *with support over* $\mathcal{P}$.

Next, along the lines of Kneip et al. (1998), assume that the probability of observing firms in a neighborhood of the boundary of $\mathcal{P}$ approaches unity as the sample size increases:

Assumption 6.2.5 *At the frontier, the density* f *is strictly positive, i.e.,* $f_0 = f(x_0^{\partial}, y_0^{\partial}) > 0$*, and sequentially Lipschitz continuous, i.e., for all sequences* $(x_n, y_n) \in \mathcal{P}$ *converging to* $(x_0^{\partial}, y_0^{\partial})$, $|f(x_n, y_n) - f(x_0^{\partial}, y_0^{\partial})| \leq c_1 ||(x_n, y_n) - (x_0^{\partial}, y_0^{\partial})||$ *for some positive constant* c_1.

Finally, an assumption about the smoothness of the frontier is needed:

Assumption 6.2.6 *For all* (x, y) *in the interior of* $\mathcal{P}$, $\theta(x, y)$, $\lambda(x, y)$, *and* $\gamma(x, y)$ *are twice continuously differentiable in both of their arguments.*

The characterization of the smoothness condition in Assumption 6.2.6 is stronger than required; e.g., Kneip et al. (1998) require only Lipschitz continuity for the distance functions, which is implied by the simpler, but stronger requirement presented here. As noted earlier, Assumptions 6.2.1–6.2.3 are standard in microeconomic theory of the firm, while Assumptions 6.2.4–6.2.6 complete the definition of a statistical model.

6.2.2 Order-*m* Measures

Cazals et al. noted that the density $f(x, y)$ introduced in Assumption 6.2.4 implies a probability function

$$H(x, y) = \Pr(X \leq x, Y \geq y). \tag{6.14}$$

This is a non-standard probability distribution function, given the direction of the inequality for Y; nonetheless, it is well-defined. This function gives the probability of drawing an observation from $f(x, y)$ that weakly *dominates* the firm operating at $(x, y) \in \mathcal{P}$ in the sense that an observation $(\widetilde{x}, \widetilde{y})$ weakly dominates (x, y) if $\widetilde{x} \leq x$ and $\widetilde{y} \geq y$. Clearly, $H(x, y)$ is monotone, nondecreasing in x and monotone, non-increasing in y.

Using $H(\boldsymbol{x}, \boldsymbol{y})$, the efficiency measures defined in (6.7), (6.9), and (6.11) can be given a probabilistic interpretation by defining the measures in terms of $H(\boldsymbol{x}, \boldsymbol{y})$ as follows:

$$\theta(\boldsymbol{x}, \boldsymbol{y}) = \inf\{\theta > 0 \mid H(\theta\boldsymbol{x}, \boldsymbol{y}) > 0\}, \tag{6.15}$$

$$\lambda(\boldsymbol{x}, \boldsymbol{y}) = \sup\{\lambda > 0 \mid H(\boldsymbol{x}, \lambda\boldsymbol{y}) > 0\}, \tag{6.16}$$

and

$$\gamma(\boldsymbol{x}, \boldsymbol{y}) = \inf\{\gamma > 0 \mid H(\gamma\boldsymbol{x}, \gamma^{-1}\boldsymbol{y}) > 0\}. \tag{6.17}$$

Cazals et al. used the distribution function defined in (6.14) to define a *partial* frontier of order m. Using the definition of conditional probability and applying Bayes' theorem, (6.14) can be written as

$$\begin{aligned} H(\boldsymbol{x}, \boldsymbol{y}) &= \underbrace{\Pr(\boldsymbol{X} \le \boldsymbol{x} \mid \boldsymbol{Y} \ge \boldsymbol{y})}_{=F_{x|y}(x_0|y_0)} \times \underbrace{\Pr(\boldsymbol{Y} \ge \boldsymbol{y})}_{=S_y(y)} \\ &= \underbrace{\Pr(\boldsymbol{Y} \ge \boldsymbol{y} \mid \boldsymbol{X} \le \boldsymbol{x})}_{=S_{y|x}(y|x)} \times \underbrace{\Pr(\boldsymbol{X} \le \boldsymbol{x})}_{=F_x(x)} \end{aligned} \tag{6.18}$$

(the terms on the right-hand side of (6.18) also appear in Daraio and Simar 2005).

Now for a given level of outputs $\boldsymbol{y}$ in the interior of the support of $\boldsymbol{Y}$, consider the set of m iid random variables $\{\boldsymbol{X}_j\}_{j=1}^m$, $\boldsymbol{X}_j \in \mathbb{R}_+^p$, drawn from the conditional distribution $F_{x|y}(\cdot \mid \boldsymbol{y})$. Define the (random) set

$$\mathcal{P}_m(\boldsymbol{y}) = \left\{ (\widetilde{\boldsymbol{x}}, \widetilde{\boldsymbol{y}}) \in \mathbb{R}_+^{p+q} \mid \bigcup_{j=1}^{m} \widetilde{\boldsymbol{x}} \ge \boldsymbol{X}_j,\ \widetilde{\boldsymbol{y}} \ge \boldsymbol{y} \right\}. \tag{6.19}$$

Replacing $\mathcal{P}$ in (6.7) with $\mathcal{P}_m(\boldsymbol{y})$ yields the (random) efficiency measure

$$\theta_m(\boldsymbol{x}, \boldsymbol{y}) \equiv \inf\{\theta > 0 \mid (\theta\boldsymbol{x}, \boldsymbol{y}) \in \mathcal{P}_m(\boldsymbol{y})\} \tag{6.20}$$

The expectation of this random efficiency measure follows trivially from Cazals et al. (2002, Theorem 5.1); in particular, if $E(\theta_m(\boldsymbol{x}, \boldsymbol{y}))$ exists, then

$$\overline{\theta}_m(\boldsymbol{x}, \boldsymbol{y}) \equiv E(\theta_m(\boldsymbol{x}, \boldsymbol{y})) = \int_0^\infty \left[1 - F_{x|y}(u\boldsymbol{x} \mid \boldsymbol{y})\right]^m du. \tag{6.21}$$

Cazals et al. define, for any $\boldsymbol{x} \in \mathbb{R}_+^p$, the expected minimum input level of order m (defined for all $\boldsymbol{y}$ in the interior of the support of $\boldsymbol{Y}$) as

$$\boldsymbol{x}_m^\partial(\boldsymbol{y}) = \boldsymbol{x} E(\theta_m(\boldsymbol{x}, \boldsymbol{y})), \tag{6.22}$$

where the expectation is assumed to exist. Cazals et al. (2002, Theorem 5.2) establishes that for any $\boldsymbol{x} \in \mathbb{R}_+^p$ and for any $\boldsymbol{y}$ in the interior of the support of $\boldsymbol{Y}$,

$$\lim_{m\to\infty} \overline{\theta}_m(\boldsymbol{x}, \boldsymbol{y}) = \theta(\boldsymbol{x}, \boldsymbol{y}) \tag{6.23}$$

and

$$\lim_{m\to\infty} \boldsymbol{x}^{\partial}_m(\boldsymbol{y}_0) = \boldsymbol{x}^{\partial}(\boldsymbol{y}_0). \tag{6.24}$$

Cazals et al. (2002) define similar, output-oriented measures in Appendix A of their paper. The output-oriented measures have properties analogous to the input-oriented measures.

It is straightforward to extend the ideas of Cazals et al. (2002) to obtain an unconditional, hyperbolic measure of order-m efficiency. Consider a set of m iid random variables $\{(\boldsymbol{X}_j, \boldsymbol{Y}_j)\}_{j=1}^m$ drawn from the density $f(\boldsymbol{x}, \boldsymbol{y})$ defined in Assumption 6.2.5 with bounded support over $\mathcal{P}$. Define the random set

$$\mathcal{P}_m \equiv \bigcup_{j=1}^{m} \left\{(\boldsymbol{x}, \boldsymbol{y}) \in \mathbb{R}_+^{p+q} \mid \boldsymbol{x} \geq \boldsymbol{X}_j,\ \boldsymbol{y} \leq \boldsymbol{Y}_j\right\}. \tag{6.25}$$

Then for any $(\boldsymbol{x}, \boldsymbol{y}) \in \mathbb{R}_+^{p+q}$, define the random distance measure

$$\gamma_m(\boldsymbol{x}, \boldsymbol{y}) \equiv \inf\{\gamma \mid (\gamma\boldsymbol{x}, \gamma^{-1}\boldsymbol{y}) \in \mathcal{P}_m\}. \tag{6.26}$$

The next result is analogous to the result in (6.21) obtained by Cazals et al. (2002).

Theorem 6.2.1 *If $E\left(\gamma_m(\boldsymbol{x}, \boldsymbol{y})\right)$ exists, then*

$$\overline{\gamma}_m(\boldsymbol{x}, \boldsymbol{y}) \equiv E\left(\gamma_m(\boldsymbol{x}, \boldsymbol{y})\right) = \int_0^\infty \left[1 - H(u\boldsymbol{x}, u^{-1}\boldsymbol{y})\right]^m du. \tag{6.27}$$

Proof. The distribution function of $\gamma_m(\boldsymbol{x}, \boldsymbol{y})$ is given by

$$\begin{aligned}
\Pr(\gamma_m(\boldsymbol{x}, \boldsymbol{y}) \leq u) &= \Pr\left[\min_{i=1,\ldots,m}\left(\max_{\substack{j=1,\ldots,p\\ k=1,\ldots,q}}\left(\frac{\boldsymbol{X}_i^j}{\boldsymbol{x}^j}, \frac{\boldsymbol{y}^k}{\boldsymbol{Y}_i^k}\right)\right) \leq u\right] \\
&= 1 - \Pr\left[\min_{i=1,\ldots,m}\left(\max_{\substack{j=1,\ldots,p\\ k=1,\ldots,q}}\left(\frac{\boldsymbol{X}_i^j}{\boldsymbol{x}^j}, \frac{\boldsymbol{y}^k}{\boldsymbol{Y}_i^k}\right)\right) > u\right] \\
&= 1 - \left[\Pr\left(\max_{\substack{j=1,\ldots,p\\ k=1,\ldots,q}}\left(\frac{\boldsymbol{X}_i^j}{\boldsymbol{x}^j}, \frac{\boldsymbol{y}^k}{\boldsymbol{Y}_i^k}\right) > u\right)\right]^m \\
&= 1 - \left[1 - \Pr\left(\max_{\substack{j=1,\ldots,p\\ k=1,\ldots,q}}\left(\frac{\boldsymbol{X}_i^j}{\boldsymbol{x}^j}, \frac{\boldsymbol{y}^k}{\boldsymbol{Y}_i^k}\right) \leq u\right)\right]^m \\
&= 1 - \left[1 - \Pr(\boldsymbol{X} \leq u\boldsymbol{x},\ \boldsymbol{y} \leq u\boldsymbol{Y})\right]^m
\end{aligned}$$

$$= 1 - \left[1 - \Pr \boldsymbol{X} \leq u\boldsymbol{x},\ \boldsymbol{Y} \geq u^{-1}\boldsymbol{y})\right]^m$$
$$= 1 - \left[1 - H(u\boldsymbol{x}, u^{-1}\boldsymbol{y})\right]^m \tag{6.28}$$

where $\boldsymbol{x}^j$, $\boldsymbol{y}^k$ denote the jth and kth elements of $\boldsymbol{x}$ and $\boldsymbol{y}$. Denote this distribution function by $G(u)$. Then the probability density function of $\gamma_m(\boldsymbol{x}, \boldsymbol{y})$ is $dG(u) = g(u)\,du$, and

$$\begin{aligned} E\left(\gamma_m(\boldsymbol{x}, \boldsymbol{y})\right) &= \int_0^\infty t\, dG(t) \\ &= \int_0^\infty \int_0^t du\, dG(t) \\ &= \int_0^\infty \int_u^\infty dG(t)\, du \\ &= \int_0^\infty [1 - G(u)]\, du, \end{aligned} \tag{6.29}$$

establishing the result. In (6.29), the third line follows from the second line due to the fact that the integrand is non-negative; hence, the order of the integration can be reversed. ■

Definition 6.2.2 *The expected efficient frontier of order m, denoted $\mathcal{P}_m^\partial$, is defined for all $(\boldsymbol{x}, \boldsymbol{y})$ in the interior of $\mathcal{P}$ as*

$$\mathcal{P}_m^\partial \equiv \left\{(\widetilde{x}, \widetilde{y}) \mid \widetilde{x} = \boldsymbol{x}E(\gamma_m(\boldsymbol{x}, \boldsymbol{y})),\ \widetilde{y} = \boldsymbol{y}E(\gamma_m(\boldsymbol{x}, \boldsymbol{y}))^{-1},\ (\boldsymbol{x}, \boldsymbol{y}) \in \mathcal{P}\right\}. \tag{6.30}$$

Definition 6.2.3 *For any $(\boldsymbol{x}, \boldsymbol{y} \in \mathbb{R}_+^{p+q}$, the expected minimum input, maximum output combination of order m, denoted by $(\boldsymbol{x}_m^\partial(\boldsymbol{x}, \boldsymbol{y}), \boldsymbol{y}_m^\partial(\boldsymbol{x}, \boldsymbol{y})$, is given by*

$$\boldsymbol{x}_m^\partial(\boldsymbol{x}, \boldsymbol{y}) \equiv \boldsymbol{x}E(\gamma_m(\boldsymbol{x}, \boldsymbol{y})) \tag{6.31}$$

and

$$\boldsymbol{y}_m^\partial(\boldsymbol{x}, \boldsymbol{y}) \equiv \boldsymbol{y}E(\gamma_m(\boldsymbol{x}, \boldsymbol{y}))^{-1} \tag{6.32}$$

where the expectation exists.

The limiting behavior of the expected minimum input, maximum output combination of order m, as $m \to \infty$, is similar to that established by Cazals et al. (2002) for the expected minimum input of order m (conditional on $\boldsymbol{y}$) given in (6.23)–(6.24), as demonstrated by the following result.

Theorem 6.2.2 *Provided $E(\gamma_m(\boldsymbol{x}, \boldsymbol{y}))$ exists,*

$$(i) \qquad \lim_{m\to\infty} E(\gamma_m(\boldsymbol{x},\boldsymbol{y})) = \gamma(\boldsymbol{x},\boldsymbol{y});\ and \tag{6.33}$$

$$(ii) \qquad \lim_{m\to\infty} \left(\boldsymbol{x}_m^{\partial}(\boldsymbol{x},\boldsymbol{y}), \boldsymbol{y}_m^{\partial}(\boldsymbol{x},\boldsymbol{y})\right) = \left(\boldsymbol{x}^{\partial}(\boldsymbol{x},\boldsymbol{y}), \boldsymbol{y}^{\partial}(\boldsymbol{x},\boldsymbol{y})\right). \tag{6.34}$$

Proof. From (6.27),

$$E(\gamma_m(\boldsymbol{x},\boldsymbol{y}) = \int_0^{\gamma(x,y)} \left[1 - H(u\boldsymbol{x}, u^{-1}\boldsymbol{y}\right]^m du + \int_{\gamma(x,y)}^{\infty} \left[1 - H(u\boldsymbol{x}, u^{-1}\boldsymbol{y}\right]^m du. \tag{6.35}$$

For all $u \le \gamma(\boldsymbol{x},\boldsymbol{y})$, $H(u\boldsymbol{x}, u^{-1}\boldsymbol{y}) = 0$. For $u > \gamma(\boldsymbol{x},\boldsymbol{y})$, $[1 - H(u\boldsymbol{x}, u^{-1}\boldsymbol{y})] < 1$. Hence $\int_0^{\gamma(x,y)}\left[1 - H(u\boldsymbol{x}, u^{-1}\boldsymbol{y}\right]^m du = \gamma(\boldsymbol{x},\boldsymbol{y})$. In addition, by the Lebesgue dominated convergence theorem, $\lim_{m\to\infty}\int_{\gamma(x,y)}^{\infty}\left[1 - H(u\boldsymbol{x}, u^{-1}\boldsymbol{y}\right]^m du = 0$. Therefore (i) holds. The second part of the theorem follows after taking the limit in Definition 6.2.3 and applying Definition 6.2.1. □

Example 6.2.1 *As an illustration, let $p = q = 1$, and consider the DGP given by*

$$f(x,y) = \begin{cases} 4\pi^{-1}\ \forall\ x \in [0,1],\ y \in \left[0, \left(2x - x^2\right)^{1/2}\right] \\ 0 \qquad \textit{otherwise.} \end{cases} \tag{6.36}$$

The frontier $\mathcal{P}^{\partial}$ is the northwest quarter of a circle with radius 1 centered at $(1,0)$, shown as a solid curve in Fig. 6.2. It is straightforward to show that the marginal distributions are given by

$$F_x(x) = 4\pi^{-1}\left[\frac{x-1}{2}\left(2x - x^2\right)^{1/2} + \frac{1}{2}\sin^{-1}(x-1) + \frac{\pi}{4}\right], \tag{6.37}$$

and

$$F_y(y) = 4\pi^{-1}\left[\frac{1}{2}y\left(1 - y^2\right)^{1/2} + \frac{1}{2}\sin^{-1}(y)\right], \tag{6.38}$$

while the conditional distribution functions are given by

$$F_{x|y}(x \mid Y \ge y) = \begin{cases} 1 & \forall\ x \ge 1; \\ \left[1 - F_y(y)\right]^{-1} 4\pi^{-1} & \\ \quad \left[\frac{x-1}{2}(2x - x^2)^{1/2}\right. & \\ \quad +\frac{1}{2}\sin^{-1}(x-1) - yx & \\ \quad -\frac{1}{2}y(1-y^2)^{1/2} & \\ \quad \left. -\frac{1}{2}\sin^{-1}\left(-\sqrt{1-y^2}\right) + y\right] & \forall\ x \in \left[1 - \sqrt{1-y^2}, 1\right]; \\ 0 & \textit{otherwise,} \end{cases} \tag{6.39}$$

and

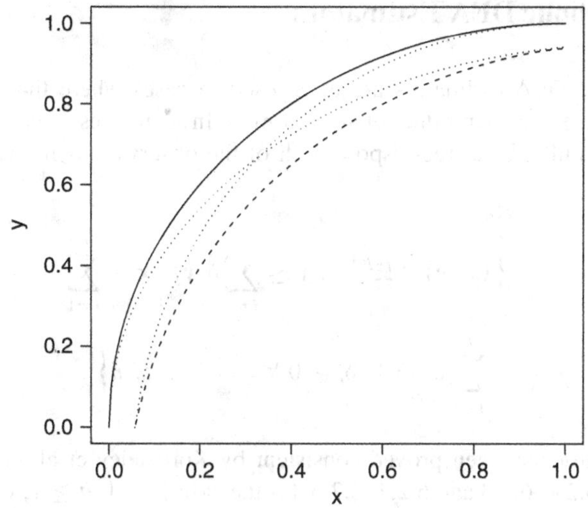

Fig. 6.2 Order-m frontiers ($m = 50$)

$$F_{y|x}(y \mid X \leq x) = \begin{cases} 1 & \forall\ y > (2x - x^2)^{1/2}; \\ F_x(x)^{-1} 4\pi^{-1}\Big\{ (x-1)\, y & \\ \quad + \frac{1}{2}\Big[y\,(1-y^2)^{1/2} + \sin^{-1}(y) \Big]\Big\} & \forall\ y \in [0, (2x - x^2)^{1/2}]; \\ 0 & \forall\ y \leq 0. \end{cases} \tag{6.40}$$

Moreover, it is clear that

$$H(x, y) = F_{x|y}(x \mid y)\left[1 - F_y(y)\right]. \tag{6.41}$$

The input-oriented and output-oriented conditional order-m frontiers (for $m = 50$) described by Cazals et al. (2002) are shown as dotted curves in Fig. 6.2; the steeper of the two curves is the input-oriented frontier. The unconditional, hyperbolic order-m frontier (again, for $m = 50$) can be found by computing $\overline{\gamma}_m(x, y)$ defined in (6.27) for each point on the frontier, and then computing points on the unconditional, hyperbolic order-m frontier using Definition 6.2.3. This is shown by the dashed curve in Fig. 6.2.

Example 6.2.1 illustrates the fact that the *conditional* order-m frontiers described by Cazals et al. (2002) have different slopes, depending on the orientation. By contrast, the unconditional, hyperbolic order-m frontier "parallels" the path of the full frontier $\mathcal{P}^{\partial}$. Since in the example $f(x, y)$ is uniform over $\mathcal{P}$, the unconditional hyperbolic order-m frontier is equidistant from the full frontier throughout its range.

6.3 Hyperbolic DEA Estimators

Non-parametric DEA estimators are often used in cases where the researcher is willing to assume that the production set is convex. In such cases, $\mathcal{P}$ can be estimated by the convex hull of the free-disposal hull of the observations in the sample $\mathcal{S}_n$, given by

$$\widehat{\mathcal{P}}_{\text{DEA},n} = \left\{ (\boldsymbol{x}, \boldsymbol{y}) \in \mathbb{R}_+^{p+q} \mid \boldsymbol{y} \leq \sum_{i=1}^{n} \delta_i \boldsymbol{y}_i,\ \boldsymbol{x} \geq \sum_{i=1}^{n} \delta_i \boldsymbol{x}_i, \right.$$
$$\left. \sum_{i=1}^{n} \delta_i = 1,\ \delta_i \geq 0\ \forall\ i = 1, \ldots, n \right\}. \tag{6.42}$$

This estimator has been proved consistent by Korostelev et al. (1995b) under Assumptions 6.2.1–6.2.3 and 6.2.4–6.2.6, for the case $p = 1, q \geq 1$, with

$$d_\Delta \left(\widehat{\mathcal{P}}_{\text{DEA},n}, \mathcal{P} \right) = O_p \left(n^{-2/(q+2)} \right) \tag{6.43}$$

where $d_\Delta(\cdot, \cdot)$ denotes the Lebesgue measure of the difference between two sets.

Replacing $\mathcal{P}$ on the right-hand sides of (6.7), (6.9), and (6.11) with $\widehat{\mathcal{P}}_{\text{DEA},n}$ gives (VRS) DEA estimators $\widehat{\theta}_{\text{DEA},n}(\boldsymbol{x}, \boldsymbol{y})$, $\widehat{\lambda}_{\text{DEA},n}(\boldsymbol{x}, \boldsymbol{y})$, and $\widehat{\gamma}_{\text{DEA},n}(\boldsymbol{x}, \boldsymbol{y})$, of $\theta(\boldsymbol{x}, \boldsymbol{y})$, $\lambda(\boldsymbol{x}, \boldsymbol{y})$, and $\gamma(\boldsymbol{x}, \boldsymbol{y})$, respectively. In the case of the input- or output-oriented measures, $\widehat{\theta}_{\text{DEA},n}(\boldsymbol{x}, \boldsymbol{y})$ and $\widehat{\lambda}_{\text{DEA},n}(\boldsymbol{x}, \boldsymbol{y})$ can be computed by solving familiar linear programs; for example, in the input-oriented case,

$$\widehat{\theta}_{\text{DEA},n}(\boldsymbol{x}, \boldsymbol{y}) = \min_{\substack{\theta \\ \delta_i,\, i=1,\ldots,n}} \left\{ \theta \mid \boldsymbol{y} \leq \sum_{i=1}^{n} \delta_i \boldsymbol{y}_i,\ \theta \boldsymbol{x} \geq \sum_{i=1}^{n} \delta_i \boldsymbol{x}_i, \right.$$
$$\left. \sum_{i=1}^{n} \delta_i = 1,\ \delta_i \geq 0\ \forall\ i = 1, \ldots, n \right\}. \tag{6.44}$$

See Simar and Wilson (2000b) for details using the output orientation. For the input-oriented case, Kneip et al. (1998) established consistency of $\widehat{\theta}_{\text{DEA},n}(\boldsymbol{x}, \boldsymbol{y})$ under Assumptions 6.2.1–6.2.3 and 6.2.4–6.2.6, with

$$\widehat{\theta}_{\text{DEA},n}(\boldsymbol{x}, \boldsymbol{y}) = \theta(\boldsymbol{x}, \boldsymbol{y}) + O_p \left(n^{-2/(p+q+1)} \right) \tag{6.45}$$

for $p, q \geq 1$. This result extends to the output oriented case after straightforward, if tedious, changes in the notation of the proof given by Kneip et al. (1998).

To date, no such results have been obtained for the hyperbolic DEA efficiency estimator $\widehat{\gamma}_{\text{DEA},n}(\boldsymbol{x}, \boldsymbol{y})$. In addition, until now, computation of the estimator when $\widehat{\mathcal{P}}_{\text{DEA},n}$ is used under VRS has been problematic since linear programming methods cannot be used. In fact, to my knowledge, the only cases where a hyperbolic DEA efficiency estimator has been used in empirical applications have imposed an assumption of CRS on the efficient frontier $\mathcal{P}^{\partial}$, allowing the hyperbolic DEA efficiency estimator to be computed as the square root of $\widehat{\theta}_{\mathrm{DEA},n}(\boldsymbol{x},\boldsymbol{y})$, analogous to (6.12).[6]

Using $\widehat{\mathcal{P}}_{\mathrm{DEA},n}$ under VRS, $\widehat{\gamma}_{\mathrm{DEA},n}(\boldsymbol{x},\boldsymbol{y})$ is the solution to the non-linear program

$$\widehat{\gamma}_{\mathrm{DEA},n}(\boldsymbol{x},\boldsymbol{y}) = \min_{\substack{\gamma \\ \delta_i,\, i=1,\ldots,n}} \Big\{ \gamma \mid \gamma^{-1}\boldsymbol{y} \le \sum_{i=1}^{n} \delta_i \boldsymbol{y}_i,\ \gamma\boldsymbol{x} \ge \sum_{i=1}^{n} \delta_i \boldsymbol{x}_i, \sum_{i=1}^{n} \delta_i = 1,\ \delta_i \ge 0\ \forall\ i = 1,\ \ldots,\ n \Big\} \quad (6.46)$$

(see Färe et al. 1985, p. 129). Rather than solving (6.46) directly, note that for $(\widetilde{\boldsymbol{x}},\widetilde{\boldsymbol{y}})$ on the boundary of $\widehat{\mathcal{P}}_{\mathrm{DEA},n}$, $\widehat{\gamma}_{\mathrm{DEA},n}(\widetilde{\boldsymbol{x}},\widetilde{\boldsymbol{y}}) = 1$ and $\Big(\widehat{\theta}_{\mathrm{DEA},n}(\widetilde{\boldsymbol{x}},\widetilde{\boldsymbol{y}}) \vee \widehat{\lambda}_{\mathrm{DEA},n}(\widetilde{\boldsymbol{x}},\widetilde{\boldsymbol{y}})^{-1}\Big) = 1$. This fact makes it easy to use the bisection method to solve for $\widehat{\gamma}_{\mathrm{DEA},n}(\boldsymbol{x},\boldsymbol{y})$ for an arbitrary point $(\boldsymbol{x},\boldsymbol{y}) \in \mathbb{R}_{+}^{p+q}$ using the following algorithm:

Algorithm #1

---

[1] Set $\gamma_a := 1$, $\gamma_b := 1$.
[2] Compute $\widehat{\theta}_{\mathrm{DEA},n}(\gamma_a\boldsymbol{x},\gamma_a^{-1}\boldsymbol{y})$ by solving (6.44); if a solution to the linear program does not exist, set $\widehat{\theta}_{\mathrm{DEA},n}(\gamma_a\boldsymbol{x},\gamma_a^{-1}\boldsymbol{y}) := \zeta$, where ζ is an arbitrary number greater than 1.
[3] If $\widehat{\theta}_{\mathrm{DEA},n}(\gamma_a\boldsymbol{x},\gamma_a^{-1}\boldsymbol{y}) > 1$, then set $\gamma_a := 0.5\gamma_a$.
[4] Repeat steps [2]–[3] until $\widehat{\theta}_{\mathrm{DEA},n}(\gamma_a\boldsymbol{x},\gamma_a^{-1}\boldsymbol{y}) < 1$.
[5] If $\widehat{\theta}_{\mathrm{DEA},n}(\gamma_b\boldsymbol{x},\gamma_b^{-1}\boldsymbol{y}) < 1$, then set $\gamma_a := 2\gamma_a$.
[6] Repeat step [5] until $\widehat{\theta}_{\mathrm{DEA},n}(\gamma_b\boldsymbol{x},\gamma_b^{-1}\boldsymbol{y}) > 1$.
[7] Set $\gamma_c := (\gamma_a + \gamma_b)/2$ and compute $\widehat{\theta}_{\mathrm{DEA},n}(\gamma_c\boldsymbol{x},\gamma_c^{-1}\boldsymbol{y})$.
[8] If $\widehat{\theta}_{\mathrm{DEA},n}(\gamma_c\boldsymbol{x},\gamma_c^{-1}\boldsymbol{y}) < 1$ then set $\gamma_a := \gamma_c$; otherwise, set $\gamma_b := \gamma_c$.
[9] If $(\gamma_b - \gamma_a) > \xi$, where ξ is a suitably small tolerance value, repeat steps [7]–[8].
[10] Set $\widehat{\gamma}_{\mathrm{DEA},n}(\boldsymbol{x},\boldsymbol{y}) := (\gamma_b + \gamma_a)/2$.

The procedure can be made accurate to an arbitrary degree by choosing (a priori) the convergence tolerance ξ to be suitably small. Setting ξ equal to 10^{-6} will

[6] In cases where CRS is assumed, the constraint $\sum_{i=1}^{n} \delta_i = 1$ is omitted from (6.42). In such cases, efficiency is estimated in terms of distance to the boundary of the convex cone of the sample observations, as opposed to the convex hull (of the free-disposal hull) of the sample observations.

result in precision to 5–6 decimal places, which is likely to exceed the number of significant digits in most data that researchers use.

Note that the speed of Algorithm #1 can be increased by forming, after the first computation of $\widehat{\theta}_{\mathrm{DEA},n}(\gamma_a \boldsymbol{x}, \gamma_a^{-1}\boldsymbol{y})$ in step [2], the set $\mathcal{S}_* = \{(\boldsymbol{x}_i, \boldsymbol{y}_i) \mid (\boldsymbol{x}_i, \boldsymbol{y}_i) \in \mathcal{S}_n,\ \widehat{\delta}_i > 0\}$ where the $\widehat{\delta}_i$ are the values of the δ_i in (6.44) when a minimum is attained. Then using this set of observations, rather than the set $\mathcal{S}_n$, in all future computations of $\widehat{\theta}_{\mathrm{DEA},n}(\gamma_a \boldsymbol{x}, \gamma_a^{-1}\boldsymbol{y})$ will result in a smaller constraint set when the linear program in (6.44) is solved. Nonetheless, the algorithm requires a number of iterations, and consequently its execution will require more time than a single solution of a linear program such as (6.44).

An alternative estimator of $\gamma(\boldsymbol{x}, \boldsymbol{y})$ offers computational burden similar to input- or output-oriented DEA estimators. Consider a mapping ϕ from $\mathbb{R}_+^p \times \mathbb{R}_+^q$ to $\mathbb{R}_+^p \times \mathbb{R}_+^q$:

$$\phi : (\boldsymbol{x}, \boldsymbol{y}) \mapsto (\boldsymbol{x}, \boldsymbol{y}^{-1}) \tag{6.47}$$

where $\boldsymbol{y}^{-1}$ is the vector whose elements are the inverses of the corresponding elements of $\boldsymbol{y}$. Denote $\boldsymbol{w} = \phi(\boldsymbol{x}, \boldsymbol{y})$. In addition, note that ϕ is a continuous, one-to-one transformation; hence $(\boldsymbol{x}, \boldsymbol{y}) = \phi^{-1}(\boldsymbol{w})$. Moreover, $\frac{\partial \phi(\boldsymbol{x},\boldsymbol{y})}{\partial \boldsymbol{y}} = -\boldsymbol{y}^{-2}$ and $\frac{\partial^2 \phi(\boldsymbol{x},\boldsymbol{y})}{\partial \boldsymbol{y} \partial \boldsymbol{y}'} = 2\mathrm{diag}(\boldsymbol{y}^{-3})$. Then (6.14) can be rewritten as

$$H(\boldsymbol{x}, \boldsymbol{y}) = H^*(\boldsymbol{w}) = \Pr(\boldsymbol{W} \leq \boldsymbol{w}). \tag{6.48}$$

Rewriting (6.17) as

$$\gamma(\boldsymbol{x}, \boldsymbol{y}) = \inf\{\gamma > 0 \mid H^*(\gamma \boldsymbol{w}) > 0\}, \tag{6.49}$$

it is clear that $\gamma(\boldsymbol{x}, \boldsymbol{y})$ is equivalent to an input-oriented efficiency score similar to that in (6.7) and (6.15) when the $(\boldsymbol{x}, \boldsymbol{y})$-space is transformed to the $\boldsymbol{w}$-space. Working in the $\boldsymbol{w}$-space, a new estimator of $\gamma(\boldsymbol{x}, \boldsymbol{y})$ is given by

$$\begin{aligned}
\widehat{\gamma}_{\mathrm{DEA}^*,n}(\boldsymbol{x}, \boldsymbol{y}) &= \min_{\substack{\gamma \\ \delta_i,\, i=1,\ldots,n}} \left\{ \gamma \mid \gamma \boldsymbol{w} \geq \sum_{i=1}^n \delta_i \boldsymbol{w}_i, \sum_{i=1}^n \delta_i = 1, \delta_i \geq 0\ \forall\ i = 1, \ldots, n \right\} \\
&= \min_{\substack{\gamma \\ \delta_i,\, i=1,\ldots,n}} \left\{ \gamma \mid \gamma \boldsymbol{y}^{-1} \geq \sum_{i=1}^n \delta_i \boldsymbol{y}_i^{-1},\ \gamma \boldsymbol{x} \geq \sum_{i=1}^n \delta_i \boldsymbol{x}_i, \right. \\
&\qquad \left. \sum_{i=1}^n \delta_i = 1,\ \delta_i \geq 0\ \forall\ i = 1,\ \ldots,\ n \right\}.
\end{aligned} \tag{6.50}$$

The following theorem establishes consistency and the convergence rate of the new estimator $\widehat{\gamma}_{\mathrm{DEA}^*,n}(\boldsymbol{x}, \boldsymbol{y})$, as well as its limiting distribution.

Theorem 6.3.1 *Under Assumptions 6.2.1–6.2.6, for all* $(\boldsymbol{x}, \boldsymbol{y})$ *in the interior of* $\mathcal{P}$,

(i) $\widehat{\gamma}_{DEA^*,n}(\boldsymbol{x},\boldsymbol{y}) = \gamma(\boldsymbol{x},\boldsymbol{y}) + O_p\left(n^{-2/(p+q+1)}\right)$; *and*

(ii) $$\lim_{n\to\infty} \Pr\left[n^{\frac{2}{p+q+1}} \left(\frac{\widehat{\gamma}_{DEA^*,n}(\omega\boldsymbol{x},\omega^{-1}\boldsymbol{y})}{\gamma(\omega\boldsymbol{x},\omega^{-1}\boldsymbol{y})} - 1 \right) \leq \delta \right] = F_{\boldsymbol{w}}(\delta) \;\forall\; \omega > 0,$$

where $F_{\boldsymbol{w}}(\delta)$ *is a well-defined, continuous probability distribution function.*

Proof. It is clear from (6.48) and (6.49) that in the $\boldsymbol{w}$-space, $\gamma(\boldsymbol{x},\boldsymbol{y})$ is an input-oriented efficiency measure along the lines of (6.7) and (6.15), where there are $(p+q)$ inputs and no outputs. Moreover, in the $\boldsymbol{w}$-space, $\widehat{\gamma}_{\mathrm{DEA}^*,n}(\boldsymbol{x},\boldsymbol{y})$ is an ordinary input-oriented (VRS) DEA efficiency estimator along the lines of (6.44), again with $(p+q)$ "inputs" and no outputs. Hence Theorem 1 of Kneip et al. (1998) establishes the result in (i), while Corollary 1 of Kneip et al. (2008) establishes the result in (ii). □

The result in Theorem 6.3.1 can now be used to establish consistency and the convergence rate of the estimator defined in (6.46).

Theorem 6.3.2 *Under Assumptions 6.2.1–6.2.6, for all* $(\boldsymbol{x},\boldsymbol{y})$ *in the interior of* $\mathcal{P}$, $\widehat{\gamma}_{DEA,n}(\boldsymbol{x},\boldsymbol{y})$ *is a consistent estimator of* $\gamma(\boldsymbol{x},\boldsymbol{y})$, *with*

$$\widehat{\gamma}_{DEA,n}(\boldsymbol{x},\boldsymbol{y}) = \gamma(\boldsymbol{x},\boldsymbol{y}) + O_p\left(n^{-2/(p+q+1)}\right).$$

Proof. $\widehat{\gamma}_{\mathrm{DEA},n}(\boldsymbol{x},\boldsymbol{y})$ is computed in Algorithm #1 in terms of distance to the convex hull of the free-disposal hull of the observations in $\mathcal{S}_n$. the set $\widehat{\mathcal{P}}_{\mathrm{DEA},n}$ in the $(\boldsymbol{x},\boldsymbol{y})$-space implies a set $\widehat{\mathcal{P}}^*_{\mathrm{DEA},n} = \{\boldsymbol{w} \mid \boldsymbol{w} = (\boldsymbol{x},\boldsymbol{y}^{-1}) \;\forall\; (\boldsymbol{x},\boldsymbol{y}) \in \widehat{\mathcal{P}}_{\mathrm{DEA},n}\}$ in the $\boldsymbol{w}$-space. In the $(\boldsymbol{x},\boldsymbol{y})$-space, the facets of $\widehat{\mathcal{P}}_{\mathrm{DEA},n}$ are subsets of $(p+q)$ dimensional hyperplanes, whereas in the $\boldsymbol{w}$-space, subsets of $\widehat{\mathcal{P}}^*_{\mathrm{DEA},n}$ corresponding to particular facets of $\widehat{\mathcal{P}}_{\mathrm{DEA},n}$ are strictly convex (from the origin in $\boldsymbol{w}$-space) due to the mapping from $(\boldsymbol{x},\boldsymbol{w})$ to $\boldsymbol{w}$. Consequently,

$$\widehat{\gamma}_{\mathrm{DEA}^*,n}(\boldsymbol{x},\boldsymbol{y}) \geq \widehat{\gamma}_{\mathrm{DEA},n}(\boldsymbol{x},\boldsymbol{y}) \geq \gamma(\boldsymbol{x},\boldsymbol{y}) \tag{6.51}$$

for any $(\boldsymbol{x},\boldsymbol{y}) \in \widehat{\mathcal{P}}_{\mathrm{DEA},n}$. Hence consistency holds due to Theorem 6.3.1 and (6.51). Moreover, $\widehat{\gamma}_{\mathrm{DEA},n}(\boldsymbol{x},\boldsymbol{y})$ is computed in Algorithm #1 using $\widehat{\theta}_{\mathrm{DEA},n}(\boldsymbol{x},\boldsymbol{y})$ to assess convergence of the bisection algorithm. Hence $\widehat{\gamma}_{\mathrm{DEA},n}(\boldsymbol{x},\boldsymbol{y})$ and $\widehat{\theta}_{\mathrm{DEA},n}(\boldsymbol{x},\boldsymbol{y})$ must converge at the same rate in n, and from Theorem 1 of Kneip et al. (1998), this rate is as given above. □

In finite samples, the two hyperbolic estimators will yield different estimates, but they are equivalent asymptotically, as the next result indicates:

Lemma 6.3.1 *Under the conditions in Theorem 6.3.1,*

$$\widehat{\gamma}_{DEA^*,n}(\boldsymbol{x},\boldsymbol{y}) - \widehat{\gamma}_{DEA,n}(\boldsymbol{x},\boldsymbol{y}) = O_p\left(n^{-2/(p+q+1)}\right). \tag{6.52}$$

Proof. The lemma follows directly from Theorems 6.3.1 and 6.3.2.

Since only asymptotic results exist for DEA estimators, the fact that the two estimators yield different values in finite samples should be of little consequence, apart from the fact that (6.51) in the proof of Theorem 6.3.2 indicates that the bias of $\widehat{\gamma}_{\text{DEA}^*,n}(\boldsymbol{x},\boldsymbol{y})$ is larger than that of $\widehat{\gamma}_{\text{DEA},n}(\boldsymbol{x},\boldsymbol{y})$. The limiting distribution in Theorem 6.3.1 is not directly useful for estimating confidence intervals in applications, but is useful for establishing consistency of either a double-smooth bootstrap or a sub-sampling bootstrap along the lines of Kneip et al. (2008). In the case of a sub-sampling bootstrap, the methods discussed by Simar and Wilson (2011) can be used to calibrate the sub-sample size.

As an illustration, the two hyperbolic DEA estimators were used to estimate $\gamma(\boldsymbol{x}_i,\boldsymbol{y}_i)$ for each of $n = 70$ observations on educational performance given by Charnes et al. (1981). The data contain observations on $p = 5$ input quantities and $q = 3$ output quantities. The results are shown in Table 6.1, where estimates obtained from the three hyperbolic DEA estimators are shown along with estimates obtained from the input- and output-oriented DEA estimators. Estimates obtained using Algorithm #1 are given in the second column of the table (observation numbers appear in the first column); estimates in the third column are from the estimator based on inverting the output data, while estimates in the fourth column are from the estimator based on the logarithmic transformation.

The results shown in Table 6.1 confirm that the hyperbolic estimators, though equivalent asymptotically, produce different estimates in finite samples, Moreover, the results can be seen to conform to the ordering given in (6.51). A more rigorous examination of the estimators' performance will be discussed later in Sect. 6.6.

6.4 The Hyperbolic FDH Estimator

For cases where the production set is not assumed to be convex, Deprins et al. (1984) proposed estimating $\mathcal{P}$ by the free-disposal hull (FDH) of the observations in $\mathcal{S}_n$, i.e.,

$$\widehat{\mathcal{P}}_{\text{FDH},n} = \bigcup_{(\boldsymbol{x}_i,\boldsymbol{y}_i)\in\mathcal{S}_n} \{(\boldsymbol{x},\boldsymbol{y}) \in \mathbb{R}_+^{p+q} \mid \boldsymbol{y} \le \boldsymbol{y}_i,\ \boldsymbol{x} \ge \boldsymbol{x}_i\}. \tag{6.53}$$

For the case with $p = 1, q \ge 1$, Korostelev et al. (1995a) proved that

$$d_\Delta\left(\widehat{\mathcal{P}}_{\text{FDH},n},\mathcal{P}\right) = O_p\left(n^{-1/(p+q)}\right). \tag{6.54}$$

Replacing $\mathcal{P}$ in (6.7), (6.9), or (6.11) gives FDH estimators $\widehat{\theta}_{\text{FDH},n}(\boldsymbol{x},\boldsymbol{y})$, $\widehat{\lambda}_{\text{FDH},n}(\boldsymbol{x},\boldsymbol{y})$, and $\widehat{\gamma}_{\text{FDH},n}(\boldsymbol{x},\boldsymbol{y})$ of the efficiency measures $\theta(\boldsymbol{x},\boldsymbol{y})$, $\lambda(\boldsymbol{x},\boldsymbol{y})$, and $\gamma(\boldsymbol{x},\boldsymbol{y})$, respectively. Park et al. (2000) proved consistency of the input-oriented estimator, with

Table 6.1 DEA efficiency estimates, Charnes et al. (1981) Data ($n = 70$)

Obs.	$\widehat{\gamma}_{\text{DEA},n}(\boldsymbol{x}, \boldsymbol{y})$	$\widehat{\gamma}_{\text{DEA}^*,n}(\boldsymbol{x}, \boldsymbol{y})$	$\widehat{\theta}_{\text{DEA},n}(\boldsymbol{x}, \boldsymbol{y})$	$\widehat{\lambda}_{\text{DEA},n}(\boldsymbol{x}, \boldsymbol{y})^{-1}$
1	0.9827	1.0000	0.9621	0.9687
2	0.9493	0.9705	0.9011	0.9015
3	0.9672	0.9922	0.9348	0.9360
4	0.9499	0.9622	0.9016	0.9030
5	1.0000	1.0000	1.0000	1.0000
6	0.9526	0.9805	0.9099	0.9049
7	0.9447	0.9688	0.8914	0.8936
8	0.9515	1.0000	0.9050	0.9056
9	0.9274	0.9402	0.8585	0.8615
10	0.9715	1.0000	0.9407	0.9483
11	1.0000	1.0000	1.0000	1.0000
12	1.0000	1.0000	1.0000	1.0000
13	0.9294	0.9366	0.8623	0.8650
14	0.9938	1.0000	0.9897	0.9847
15	1.0000	1.0000	1.0000	1.0000
16	0.9748	1.0000	0.9501	0.9506
17	1.0000	1.0000	1.0000	1.0000
18	1.0000	1.0000	1.0000	1.0000
19	0.9762	1.0000	0.9526	0.9532
20	1.0000	1.0000	1.0000	1.0000
21	1.0000	1.0000	1.0000	1.0000
22	1.0000	1.0000	1.0000	1.0000
23	0.9875	1.0000	0.9748	0.9754
24	1.0000	1.0000	1.0000	1.0000
25	0.9894	1.0000	0.9787	0.9791
26	0.9710	0.9905	0.9425	0.9432
27	1.0000	1.0000	1.0000	1.0000
28	0.9946	1.0000	0.9903	0.9877
29	0.9243	0.9586	0.8833	0.8478
30	0.9455	0.9540	0.8934	0.8950
31	0.9152	0.9244	0.8369	0.8383
32	1.0000	1.0000	1.0000	1.0000
33	0.9760	1.0000	0.9521	0.9531
34	0.9275	0.9518	0.8590	0.8615
35	1.0000	1.0000	1.0000	1.0000
36	0.8889	0.9186	0.7930	0.7883
37	0.9155	0.9431	0.8393	0.8391
38	1.0000	1.0000	1.0000	1.0000
39	0.9689	1.0000	0.9414	0.9371
40	0.9746	1.0000	0.9498	0.9498
41	0.9762	0.9971	0.9523	0.9534
42	0.9737	1.0000	0.9531	0.9476
43	0.9297	0.9725	0.8648	0.8648
44	1.0000	1.0000	1.0000	1.0000

(continued)

Table 6.1 (Continued)

Obs.	$\widehat{\gamma}_{\text{DEA},n}(\boldsymbol{x},\boldsymbol{y})$	$\widehat{\gamma}_{\text{DEA}^*,n}(\boldsymbol{x},\boldsymbol{y})$	$\widehat{\theta}_{\text{DEA},n}(\boldsymbol{x},\boldsymbol{y})$	$\widehat{\lambda}_{\text{DEA},n}(\boldsymbol{x},\boldsymbol{y})^{-1}$
45	1.0000	1.0000	1.0000	1.0000
46	0.9558	0.9765	0.9129	0.9143
47	1.0000	1.0000	1.0000	1.0000
48	1.0000	1.0000	1.0000	1.0000
49	1.0000	1.0000	1.0000	1.0000
50	0.9790	1.0000	0.9587	0.9583
51	0.9591	1.0000	0.9199	0.9199
52	1.0000	1.0000	1.0000	1.0000
53	0.9333	0.9543	0.8696	0.8722
54	1.0000	1.0000	1.0000	1.0000
55	0.9997	1.0000	0.9994	0.9994
56	1.0000	1.0000	1.0000	1.0000
57	0.9631	1.0000	0.9270	0.9281
58	1.0000	1.0000	1.0000	1.0000
59	1.0000	1.0000	1.0000	1.0000
60	0.9903	1.0000	0.9804	0.9809
61	0.9391	1.0000	0.8927	0.8815
62	1.0000	1.0000	1.0000	1.0000
63	0.9815	1.0000	0.9635	0.9632
64	0.9648	0.9929	0.9303	0.9314
65	0.9872	1.0000	0.9754	0.9737
66	0.9676	0.9866	0.9356	0.9368
67	0.9731	0.9957	0.9463	0.9476
68	1.0000	1.0000	1.0000	1.0000
69	1.0000	1.0000	1.0000	1.0000
70	0.9820	1.0000	0.9640	0.9647

$$\widehat{\theta}_{\text{FDH},n}(\boldsymbol{x},\boldsymbol{y}) = \theta(\boldsymbol{x},\boldsymbol{y}) + O_p\left(n^{-1/(p+q)}\right); \tag{6.55}$$

this result extends to the output oriented-estimator $\widehat{\lambda}_{\text{FDH},n}(\boldsymbol{x},\boldsymbol{y})$ after reversing the roles of the inputs and outputs in the proof given by Park et al. Park et al. also derived the limiting distribution of the input-oriented estimator, which is a Weibull distribution containing an unknown parameter that depends on features of the model.

In the input-oriented case, $\widehat{\theta}_{\text{FDH},n}(\boldsymbol{x},\boldsymbol{y})$ can be written as an integer programming problem:

$$\widehat{\theta}_{\text{FDH},n}(\boldsymbol{x},\boldsymbol{y}) = \min_{\substack{\theta \\ \delta_i,\, i=1,\,\dots,\,n}} \left\{ \gamma \mid \gamma^{-1}\boldsymbol{y} \leq \sum_{i=1}^{n} \delta_i \boldsymbol{y}_i, \right.$$
$$\left. \gamma\boldsymbol{x} \geq \sum_{i=1}^{n} \delta_i \boldsymbol{x}_i,\ \delta_i \in \{0,1\}\ \forall\ i = 1,\ \dots,\ n \right\} \tag{6.56}$$

Table 6.2 Hyperbolic order-m and FDH efficiency estimates, Charnes et al. (1981) Data ($n = 70$)

Obs.	$\widehat{\overline{\gamma}}_{m=35,n}(x, y)$	$\widehat{\overline{\gamma}}_{m=70,n}(x, y)$	$\widehat{\overline{\gamma}}_{m=140,n}(x, y)$	$\widehat{\gamma}_{\text{FDH},n}(x, y)$
1	1.1550	1.0729	1.0241	1.0000
2	1.0706	1.0269	1.0070	1.0000
3	1.1004	1.0517	1.0173	1.0000
4	1.0255	0.9962	0.9840	0.9800
5	1.2437	1.1074	1.0319	1.0000
6	1.1251	1.0449	1.0109	1.0000
7	1.0278	1.0076	1.0014	1.0000
8	1.1584	1.0801	1.0256	1.0000
9	1.0323	1.0137	1.0045	1.0000
10	1.1148	1.0549	1.0151	1.0000
11	1.1351	1.0773	1.0292	1.0000
12	1.1615	1.0829	1.0282	1.0000
13	1.0092	0.9945	0.9886	0.9866
14	1.1373	1.0481	1.0103	1.0000
15	1.4024	1.1802	1.0512	1.0000
16	1.2173	1.1151	1.0399	1.0000
17	1.3924	1.2054	1.0697	1.0000
18	1.1605	1.0820	1.0271	1.0000
19	1.1200	1.0625	1.0212	1.0000
20	1.2168	1.1198	1.0390	1.0000
21	1.1570	1.0732	1.0224	1.0000
22	1.1673	1.0833	1.0296	1.0000
23	1.1234	1.0648	1.0209	1.0000
24	1.1991	1.0960	1.0270	1.0000
25	1.1079	1.0548	1.0164	1.0000
26	1.1088	1.0544	1.0196	1.0000
27	1.1385	1.0554	1.0131	1.0000
28	1.1095	1.0372	1.0094	1.0000
29	1.1271	1.0646	1.0208	1.0000
30	1.0739	1.0260	1.0066	1.0000
31	0.9870	0.9646	0.9513	0.9454
32	1.3055	1.1372	1.0346	1.0000
33	1.1043	1.0487	1.0153	1.0000
34	1.0610	1.0283	1.0087	1.0000
35	1.1423	1.0729	1.0224	1.0000
36	1.0294	1.0062	1.0012	1.0000
37	1.0623	1.0149	1.0018	1.0000
38	1.2733	1.1220	1.0340	1.0000
39	1.1077	1.0560	1.0181	1.0000
40	1.1223	1.0604	1.0223	1.0000
41	1.0613	1.0278	1.0087	1.0000
42	1.1058	1.0356	1.0093	1.0000
43	1.0847	1.0404	1.0126	1.0000
44	1.4417	1.2562	1.0881	1.0000

(continued)

Table 6.2 (Continued)

Obs.	$\widehat{\gamma}_{m=35,n}(\boldsymbol{x},\boldsymbol{y})$	$\widehat{\gamma}_{m=70,n}(\boldsymbol{x},\boldsymbol{y})$	$\widehat{\gamma}_{m=140,n}(\boldsymbol{x},\boldsymbol{y})$	$\widehat{\gamma}_{\text{FDH},n}(\boldsymbol{x},\boldsymbol{y})$
45	1.2669	1.1148	1.0320	1.0000
46	1.0553	1.0317	1.0094	1.0000
47	1.1222	1.0608	1.0196	1.0000
48	1.4043	1.1678	1.0476	1.0000
49	1.1797	1.0656	1.0190	1.0000
50	1.0799	1.0412	1.0147	1.0000
51	1.1404	1.0699	1.0237	1.0000
52	1.1767	1.0895	1.0302	1.0000
53	1.0567	1.0086	0.9908	0.9872
54	1.1628	1.0701	1.0188	1.0000
55	1.1872	1.0991	1.0342	1.0000
56	1.3111	1.1093	1.0187	1.0000
57	1.1041	1.0488	1.0155	1.0000
58	1.4047	1.2078	1.0782	1.0000
59	1.8001	1.4280	1.1304	1.0000
60	1.0967	1.0430	1.0140	1.0000
61	1.1347	1.0625	1.0199	1.0000
62	1.7145	1.3917	1.1279	1.0000
63	1.1658	1.0827	1.0271	1.0000
64	1.0546	1.0179	1.0027	1.0000
65	1.2056	1.1161	1.0384	1.0000
66	1.0248	1.0051	0.9986	0.9979
67	1.0631	1.0269	1.0074	1.0000
68	1.3750	1.2009	1.0626	1.0000
69	1.6459	1.3328	1.1162	1.0000
70	1.1017	1.0408	1.0119	1.0000

However, recalling the decomposition of $H(\boldsymbol{x},\boldsymbol{y})$ in (6.18), it is clear that (6.15) can be written as

$$\theta(\boldsymbol{x},\boldsymbol{y}) = \inf\{\theta > 0 \mid F_{\boldsymbol{x}|\boldsymbol{y}}(\theta\boldsymbol{x} \mid \boldsymbol{y}) > 0\}. \tag{6.57}$$

The empirical analog of $F_{\boldsymbol{x}|\boldsymbol{y}}(\boldsymbol{x} \mid \boldsymbol{y})$ evaluated at $(\boldsymbol{x},\boldsymbol{y})$ is

$$\widehat{F}_{\boldsymbol{x}|\boldsymbol{y}}(\boldsymbol{x} \mid \boldsymbol{y}) = \frac{\sum_{i=1}^{n} I(\boldsymbol{x}_i \leq \boldsymbol{x}, \boldsymbol{y}_i \geq \boldsymbol{y})}{\sum_{i=1}^{n} I(\boldsymbol{y}_i \geq \boldsymbol{y})}. \tag{6.58}$$

Substituting this for $F_{\boldsymbol{x}|\boldsymbol{y}}(\theta\boldsymbol{x} \mid \boldsymbol{y})$ in (6.57) yields

$$\widehat{\theta}_{\text{FDH},n}(\boldsymbol{x},\boldsymbol{y}) = \inf\{\theta > 0 \mid \widehat{F}_{\boldsymbol{x}|\boldsymbol{y}}(\theta\boldsymbol{x} \mid \boldsymbol{y}) > 0\}, \tag{6.59}$$

which can be computed (for a fixed point $(\boldsymbol{x},\boldsymbol{y})$) as

$$\widehat{\theta}_{\mathrm{FDH},n}(\boldsymbol{x},\boldsymbol{y}) = \min_{i\in\mathcal{I}(\boldsymbol{y})}\left(\max_{j=1,\dots,p}\left(\frac{\boldsymbol{x}_i^j}{\boldsymbol{x}^j}\right)\right), \tag{6.60}$$

where $\mathcal{I}(\boldsymbol{y}) = \{i \mid \boldsymbol{y}_i \geq \boldsymbol{y},\ i = 1, \dots, n\}$. As a practical matter, it is easier to compute $\widehat{\theta}_{\mathrm{FDH},n}(\boldsymbol{x},\boldsymbol{y})$ using (6.60) instead of solving the integer program in (6.56).

Wheelock and Wilson (2008) extended the results of Park et al. (2000) to the hyperbolic FDH efficiency estimator. In particular, Wheelock and Wilson proved that the hyperbolic FDH estimator $\widehat{\gamma}_{\mathrm{FDH},n}(\boldsymbol{x},\boldsymbol{y})$ is a consistent estimator of $\gamma(\boldsymbol{x},\boldsymbol{y})$, converges at rate $n^{-1/(p+q)}$, and is asymptotically Weibull distributed (see Wheelock and Wilson 2008, Theorem 4.1 and Corollary 4.1, and corresponding proofs given in their appendix).

Computation of the hyperbolic FDH estimator is straightforward. The empirical analog of the distribution function defined in (6.14) based on the sample $\mathcal{S}_n$ is

$$\widehat{H}_n(\boldsymbol{x},\boldsymbol{y}) = n^{-1}\sum_{i=1}^{n} I(\boldsymbol{x}_i \leq \boldsymbol{x}_0, \boldsymbol{y}_i \geq \boldsymbol{y}_0), \tag{6.61}$$

where $I(\cdot)$ denotes the indicator function. Then $\widehat{\gamma}_{\mathrm{FDH},n}(\boldsymbol{x},\boldsymbol{y})$ can be written, analogously to (6.17), as

$$\widehat{\gamma}_{\mathrm{FDH},n}(\boldsymbol{x},\boldsymbol{y}) = \inf\left\{\widehat{H}_n(\gamma\boldsymbol{x},\gamma^{-1}\boldsymbol{y}) > 0\right\}. \tag{6.62}$$

For a fixed point $(\boldsymbol{x}_0, \boldsymbol{y}_0)$, the estimator in (6.62) can be computed by solving[7]

$$\widehat{\gamma}_{\mathrm{FDH},n}(\boldsymbol{x}_0,\boldsymbol{y}_0) = \min_{i=1,\dots,n}\left(\max_{\substack{j=1,\dots,p\\ k=1,\dots,q}}\left(\frac{\boldsymbol{x}_i^j}{\boldsymbol{x}_0^j}, \frac{\boldsymbol{y}_0^k}{\boldsymbol{y}_i^k}\right)\right). \tag{6.63}$$

As in Sect. 6.3, the data can be transformed from $(\boldsymbol{x},\boldsymbol{y})$-space to $\boldsymbol{w}$-space using the mapping ϕ. Then

$$\widehat{\gamma}_{\mathrm{FDH},n}(\boldsymbol{x},\boldsymbol{y}) = \inf\left\{\widehat{H}_n^*(\gamma\boldsymbol{w}^*) > 0\right\}, \tag{6.64}$$

which can be computed for a fixed point $\boldsymbol{w}$ by

$$\widehat{\gamma}_{\mathrm{FDH},n}(\boldsymbol{x},\boldsymbol{y}) = \min_{i=1,\dots,n}\left(\max_{j=1,\dots,(p+q)}\left(\frac{\boldsymbol{w}_i^j}{\boldsymbol{w}^j}\right)\right). \tag{6.65}$$

[7] Wheelock and Wilson (2008) gave a numerical algorithm for computing their unconditional, hyperbolic order-α quantile efficiency estimator. When $\alpha = 1$, their estimator is equivalent to (6.61) and (6.62). Typically, computing $\widehat{H}_n(\boldsymbol{x},\boldsymbol{y})$ using (6.61) will be faster than setting $\alpha = 1$ and applying the numerical algorithm given in Wheelock and Wilson.

This is identical in form to the input-oriented estimator $\widehat{\theta}(\boldsymbol{x},\boldsymbol{y})$ in (6.60) in $(\boldsymbol{x},\boldsymbol{y})$-space with no output dimensions.[8]

For the hyperbolic FDH efficiency estimator, it makes no difference whether the estimator is computed in the $(\boldsymbol{x},\boldsymbol{y})$-space or the $\boldsymbol{w}$-space; in either case, the resulting estimate is the same. This is due to the fact that the transformation ϕ preserves the ordering of observations in $\mathcal{S}_n$ each of the $(p+q)$ dimensions, and the estimator is computed in terms of distance to the free-disposal hull of the sample observations. This differs from the DEA case discussed above in Sect. 6.3, where the ϕ "bends" the facets of $\widehat{\mathcal{P}}_{\mathrm{DEA},n}$ in transforming from $(\boldsymbol{x},\boldsymbol{y})$-space to $\boldsymbol{w}$-space.

The last column of Table 6.2 gives hyperbolic FDH efficiency estimates for each of the 70 observations on educational performance appearing in Charnes et al. (1981). These estimates are analogous to the hyperbolic DEA estimates appearing in Table 6.1. Due to relaxation of the convexity assumption, many more estimates in the last column of Table 6.2 are equal to one than in Table 6.1.

6.5 The Hyperbolic Order-*m* Estimator

An estimator of the *expected* hyperbolic order-m efficiency defined in Theorem 6.2.1 is obtained by substituting the empirical distribution function given in (6.61) into (6.27) to obtain

$$\widehat{\overline{\gamma}}_{m,n}(\boldsymbol{x},\boldsymbol{y}) = \widehat{E}(\gamma_m(\boldsymbol{x},\boldsymbol{y})) = \int_0^{\infty}\left[1-\widehat{H}_n(u\boldsymbol{x},u^{-1}\boldsymbol{y})\right]^m du. \tag{6.66}$$

There is no simple expression for this estimator due to the multivariate nature of $\widehat{H}_n(\boldsymbol{x},\boldsymbol{y})$. Nonetheless, it can be computed using Monte Carlo methods along the lines of Cazals et al. (2002).

Consider a Monte Carlo sample $\mathcal{S}_n^{b,m} = \{(\boldsymbol{X}_{bi},\boldsymbol{Y}_{bi})\}_{i=1}^m$ obtained by drawing m times, independently and with replacement, from input-output pairs in $\mathcal{S}_n$. Then compute

$$\widehat{\gamma}_{m,n}^{b}(\boldsymbol{x},\boldsymbol{y}) = \min_{i=1,\ldots,m}\left(\max_{\substack{j=1,\ldots,p\\ k=1,\ldots,q}}\left(\frac{\boldsymbol{X}_{bi}^j}{\boldsymbol{x}^j},\frac{\boldsymbol{y}^k}{\boldsymbol{Y}_{bi}^k}\right)\right). \tag{6.67}$$

Repeat this exercise B times, for $b=1\ldots,\ B$. Then

$$\widehat{\overline{\gamma}}_{m,n}(\boldsymbol{x},\boldsymbol{y}) \approx B^{-1}\sum_{b=1}^{B}\widehat{\gamma}_{m,n}^{b}(\boldsymbol{x},\boldsymbol{y}), \tag{6.68}$$

[8] To see this, replace $\boldsymbol{x}$ in (6.60) with $\boldsymbol{w}$, set $q=1$, $\boldsymbol{y}=0$, and replace p with $(p+q)$. The resulting expression is equivalent to (6.65).

with the approximation improving as B increases.

The next theorem establishes asymptotic properties for the hyperbolic order-m efficiency estimator defined in (6.66).

Theorem 6.5.1 *Let Assumptions 6.2.2–6.2.6 hold, and assume $\mathcal{P}$ is compact. Then for any $(\boldsymbol{x}, \boldsymbol{y})$ for which there exists γ such that $(\gamma \boldsymbol{x}, \gamma^{-1}\boldsymbol{y}) \in \mathcal{P}$, and for any $m \geq 1$,*

$$(i) \quad \widehat{\overline{\gamma}}_{m,n}(\boldsymbol{x}, \boldsymbol{y}) \xrightarrow{a.s.} \overline{\gamma}_m(\boldsymbol{x}, \boldsymbol{y}); \text{ and}$$

$$(ii) \quad n^{1/2}\left(\widehat{\overline{\gamma}}_{m,n}(\boldsymbol{x}, \boldsymbol{y}) - \overline{\gamma}_m\right) \xrightarrow{d} N\left(0, \sigma_m^2(\boldsymbol{x}, \boldsymbol{y})\right)$$

where

$$\sigma_m^2(\boldsymbol{x}, \boldsymbol{y}) = E\left[m \int_0^\infty \left[1 - H(u\boldsymbol{x}, u^{-1}\boldsymbol{y})\right]^{m-1} I(\boldsymbol{w}_i \geq u)\, du - m\overline{\gamma}_{m,n}(\boldsymbol{x}, \boldsymbol{y})\right]^2$$

and $\boldsymbol{w}_i \geq u$ if $\boldsymbol{w}_i^j \geq u$ for each element $\boldsymbol{w}_i^j$ of $\boldsymbol{w}$, $j = 1, \ldots, (p+q)$.

Proof. By the strong law of large numbers, $\widehat{H}_n(\boldsymbol{x}\,\boldsymbol{y}) \xrightarrow{a.s.} H(\boldsymbol{x}, \boldsymbol{y})$. Part (i) of the result then follows from the Lebesgue dominated convergence theorem applied to (6.27) and (6.66).

To prove part (ii), use the transformation from $(\boldsymbol{x}, \boldsymbol{y})$ space to $\boldsymbol{w}$-space that was introduced in Sect. 6.3. After applying the transformation, the proof is similar to the proof of Theorem 3.1 in Cazals et al. (2002). In particular, after transforming to $\boldsymbol{w}$-space, (6.27) can be written as

$$\begin{aligned} \overline{\gamma}_{m,n}(\boldsymbol{x}, \boldsymbol{y}) &= \int_0^\infty [1 - H^*(u\boldsymbol{w})]^m\, du \\ &= \int_0^\infty [S_{c,\boldsymbol{w}}(u \mid z)]^m\, du, \end{aligned} \tag{6.69}$$

where $S_{c,\boldsymbol{w}}(u \mid z)$ is a conditional survivor function; the form of $S_{c,\boldsymbol{w}}(u \mid z)$ depends on the fixed point of interest $(\boldsymbol{x}, \boldsymbol{y})$, or equivalently, on $\boldsymbol{w}$. The conditioning is on an artificial output z=0; the conditioning is in effect redundant. The conditional survivor function has bounded support on $[0, \infty)$. In addition, $S_{c,\boldsymbol{w}}(u \mid z) = \frac{S(u,z)}{S_z(z)}$ where $S(u, z)$ is the joint survivor function and $S_z(z)$ is the marginal survivor function for z.

Now define the statistical functional

$$T(S) = \int_0^\infty [S_{c,\boldsymbol{w}}(u \mid z)]^m\, du \tag{6.70}$$

so that $T(S)$ is an operator that assigns a real value to any survivor function S. The operator is Fréchet differentiable with respect to the sup-norm; i.e, for any two survivor functions R and S,

$$T(R) - T(S) = DT_S(R - S) + \epsilon(R - S)||R - S||_\infty \tag{6.71}$$

where $||\cdot||_\infty$ denotes the sup-norm and where $\epsilon(V) \to 0$ when $|V||_\infty \to 0$.

A central limit theorem result then follows from arguments in Boos and Serfling (1980). The Fréchet derivative is

$$DT_S(V) = \frac{m}{S_z(0)^m}\int_0^\infty S_w(u)^{m-1}V(u)\,du - m\frac{\overline{\gamma}_{m,n}(\boldsymbol{x},\boldsymbol{y})}{S_z(0)}V(0), \tag{6.72}$$

noting that $S_w(u) = S_w(u,z)$ since $z = 0$. Noting that $DT_S(\widehat{S}_n - S) = DT_S(\widehat{S}_n)$ and using (6.72), (6.70) becomes

$$n^{1/2}\left[T(\widehat{S}_n) - T(S)\right] = \frac{n^{1/2}}{n}\sum_{i=1}^{n}\left[\frac{m}{S_z(0)^m}\int_0^\infty S_w(u)^{m-1}I(\boldsymbol{w}_i \geq u)\,du\right.$$
$$\left. -m\frac{\overline{\gamma}_{m,n}(\boldsymbol{x},\boldsymbol{y})}{S_z(0)}I(z_i \geq 0)\right] + \epsilon(\widehat{S}_n - S)\left[n^{1/2}||\widehat{S}_n - s||_\infty\right]. \tag{6.73}$$

By the Doveretzky, Kiefer, and Wolfowitz inequality (see Boos and Serfling 1980, p. 619), $n^{1/2}||\widehat{S}_n - S||_\infty = O_p(1)$. Moreover, $\widehat{S}_n$ converges uniformly, implying $\epsilon(\widehat{S}_n - S) \xrightarrow{p} 0$. Hence the second term on the right-hand side of (6.73) converges to 0 in probability. Standard central limit theorem arguments lead to the asymptotic normality result. In addition,

$$E\left[\frac{m}{S_z(0)^m}\int_0^\infty S_w(u)^{m-1}I(\boldsymbol{w}_i \geq u)\,du - m\frac{\overline{\gamma}_{m,n}(\boldsymbol{x},\boldsymbol{y})}{S_z(0)}I(z_i \geq 0)\right]$$
$$= n\left[m\int_0^\infty S(u)_w^{m-1}S_w(u)\,du - m\overline{\gamma}_{m,n}(\boldsymbol{x},\boldsymbol{y})\right] = 0, \tag{6.74}$$

establishing the zero mean in part (ii). The variance expression in part (ii) of the theorem follows after noting that $S_w(u) = \left[1 - H(u\boldsymbol{x}, u^{-1}\boldsymbol{y})\right]$. □

The root-n rate of convergence established by Theorem 6.5.1 is atypical in non-parametric estimation, although the conditional input- and output-oriented order-m estimators examined by Cazals et al. (2002), as well as the order-α efficiency estimators of Daouia and Simar (2007) and Wheelock and Wilson (2008) achieve this rate. The variance expression can be used to estimate asymptotic confidence intervals for $\overline{\gamma}_m(\boldsymbol{x},\boldsymbol{y})$ by replacing the unknown quantities with consistent estimates.

The limiting form of the unconditional, hyperbolic order-m estimator (as $m\to\infty$) is the hyperbolic FDH estimator, as demonstrated by the next lemma.

Lemma 6.5.1 *Under Assumptions 6.2.2–6.2.6, for fixed n,*

$$\lim_{m\to\infty}\widehat{\overline{\gamma}}_{m,n}(\boldsymbol{x},\boldsymbol{y}) = \widehat{\gamma}_{FDH,n}(\boldsymbol{x},\boldsymbol{y}).$$

Proof. Re-write (6.66) as

$$\begin{aligned}\widehat{\widehat{\gamma}}_{m,n}(x, y) &= \int_0^{\widehat{\gamma}_{\mathrm{FDH,n}}(x,y)} \left[1 - \widehat{H}_n(ux, u^{-1}y)\right]^m du \\ &\quad + \int_{\widehat{\gamma}_{\mathrm{FDH,n}}(x,y)}^{\infty} \left[1 - \widehat{H}_n(ux, u^{-1}y)\right]^m du \\ &= \widehat{\gamma}_{\mathrm{FDH,n}}(x, y) + \int_{\widehat{\gamma}_{\mathrm{FDH,n}}(x,y)}^{\infty} \left[1 - \widehat{H}_n(ux, u^{-1}y)\right]^m du. \quad (6.75)\end{aligned}$$

The second term on the right-hand side of (6.75) converges to 0 as $m \to \infty$ since $\left[1 - \widehat{H}_n(ux, u^{-1}y)\right] < 1 \; \forall \; u > \widehat{\gamma}_{\mathrm{FDH,n}}(x, y)$. □

Cazals et al. (2002) discuss how m can be sensibly chosen in applications using their input- and output-oriented conditional order-m estimators; one can use similar criteria with the unconditional hyperbolic order-m estimator. In addition, as with the Cazals et al. estimators, by letting $m \to \infty$ at an appropriate rate as $n \to \infty$, $\widehat{\widehat{\gamma}}_{m,n}(x, y)$ can be used to estimate $\gamma(x, y)$. This is apparent when one considers the result in Lemma 6.5.1 in conjunction with Corollary 4.1 in Wheelock and Wilson (2008), which demonstrates that $\widehat{\gamma}_{\mathrm{FDH},n}(x, y)$ converges to $\gamma(x, y)$ as $n \to \infty$. As remarked by Cazals et al., m can be viewed as a "trimming" parameter. The next result formalizes these ideas.

Theorem 6.5.2 *Under Assumptions 6.2.2–6.2.6, for $m = m(n)$ a sequence in n such that $m(n) = O(\beta n \log(n) H(x, y))$ where $\beta > \frac{1}{p+q}$,*

$$n^{1/(p+q)} \left(\widehat{\widehat{\gamma}}_{m(n),n}(x, y) - \gamma(x, y)\right) \xrightarrow{d} \textit{Weibull}\left(\mu_{\mathcal{H},0}^{p+q}, p + q\right)$$

where $\mu_{\mathcal{H},0}$ is a constant given in Wheelock and Wilson (2008, (A.7)).

Proof. By Corollary 4.1 appearing in Wheelock and Wilson (2008),

$$n^{1/(p+q)} \left(\widehat{\gamma}_{\mathrm{FDH},n}(x, y) - \gamma(x, y)\right) \xrightarrow{d} \mathrm{Weibull}\left(\mu_{\mathcal{H},0}^{p+q}, p + q\right). \quad (6.76)$$

Using (6.75), it is clear that

$$\begin{aligned}n^{1/(p+q)} \left(\widehat{\widehat{\gamma}}_{m(n),n}(x, y) - \gamma(x, y)\right) &= n^{1/(p+q)} \left[\widehat{\gamma}_{\mathrm{FDH},n}(x, y) - \gamma(x, y)\right] \\ &+ \int_{\widehat{\gamma}_{\mathrm{FDH,n}}(x,y)}^{\infty} \left[1 - \widehat{H}_n(ux, u^{-1}y)\right]^m du. \quad (6.77)\end{aligned}$$

Hence $m = m(n)$ must be such that the last term on the right-hand side of (6.77) is $o_p(1)$ as $n \to \infty$. Let $\widehat{S}_{xy}(u) = [1 - \widehat{H}_n(ux, u^{-1}by]$ so that the integral in this term can be written as $\int_{\widehat{\gamma}_{\mathrm{FDH,n}}(x,y)}^{\infty} \left[\widehat{S}_{xy}(u)\right]^m du$. Then, using the integral mean value theorem, this integral can be written as $\left(v_{(n(x,y))},\ v_1\right)\left[\widehat{S}_{xy}(\widetilde{u})\right]^m$ where $\widetilde{u} \in \left[v_1,\ v_{(n(x,y))}\right]$ and $n(\boldsymbol{x},\boldsymbol{y}) = \sum_{i=1}^{n} I(\boldsymbol{x}_i \le \boldsymbol{x}, \boldsymbol{y}_i \ge \boldsymbol{y})$. Since the support of $(\boldsymbol{X},\boldsymbol{Y})$ is compact, the ranges of both $\boldsymbol{X}$ and $\boldsymbol{Y}$ are bounded. Also, for $\widetilde{u} > \widehat{\gamma}_{\text{FDH,n}}(\boldsymbol{x},\boldsymbol{y})$, $\widehat{S}_{xy}(\widetilde{u})$ is bounded by $\frac{n(x,y)-1}{n(x,y)}$. Hence, the integral in the last term on the right-hand side of (6.77) is $o_p(1)$ provided $\left[\frac{n(x,y)-1}{n(x,y)}\right]^{m(n)} = O_p\left(n^{-\beta}\right)$, where $\beta > (p+q)^{-1}$. Since $\log\left(\frac{n(x,y)-1}{n(x,y)}\right) \asymp -\frac{1}{n(x,y)}$ and $n(\boldsymbol{x},\boldsymbol{y}) \asymp nH(\boldsymbol{x},\boldsymbol{y})$, it follows that $m(n)\log\left(\frac{n(x,y)-1}{n(x,y)}\right) \asymp \frac{m(n)}{nH(x,y)} = O\left(-\beta\log n\right)$; hence $m(n) = O\left(\beta n\log(n)H(\boldsymbol{x},\boldsymbol{y})\right)$. □

Example 6.5.1 *Figure 6.3 shows 250 observations from the DGP introduced in Example 6.2.1 in each of the figure's three panels (the data are the same across panels). In panel (a), the hyperbolic order-m ($m = 50$) frontier is shows as a dashed curve, and its estimate is traced by the irregular solid curve. Similarly, in panel (b), the conditional input order-m ($m = 50$) frontier is shown with its estimate, while in panel (c), the conditional output order-m ($m = 50$) frontier is shown with its estimate. Comparing the three panels in Fig. 6.3, it is apparent that the hyperbolic order-m frontier estimate is smoother than either of the conditional order-m frontier estimates. Conditioning on either input or output levels causes large jumps in the frontier estimates in panels (b) and (c).*

Columns 2–4 of Table 6.2 give hyperbolic order-m efficiency estimates for the Charnes et al. (1981) data. As expected, increasing m from 35 to 70 and then to 170 reduces the value of the efficiency estimates.

6.6 Monte Carlo Results

In order to examine using Monte Carlo methods the bias, variance, and mean-square error properties of the estimators discussed in Sects. 6.3, 6.4, and 6.5, it is first necessary to simulate a DGP. For p input quantities and q output quantities, consider the unit $(p+q-1)$-sphere centered at the origin in $(p+q)$-space.[9] Let z be a draw from the uniform distribution on the surface of the unit $(p+q-1)$-sphere centered at the origin. The vector z has length $(p+q)$, and can be partitioned by writing $z = \left[z_x\ z_y\right]$ where z_x has length p and z_y has length q.

Now define

$$\widetilde{\boldsymbol{x}} = 1 - |z_x| \tag{6.78}$$

and

$$\widetilde{\boldsymbol{y}} = |z_y|. \tag{6.79}$$

[9] Recall that for any natural number d, a unit d-sphere is the set of points in $(d+1)$-dimensional Euclidean space lying at distance one from a central point; the set of points comprises a d-dimensional manifold in Euclidean $(d+1)$-space.

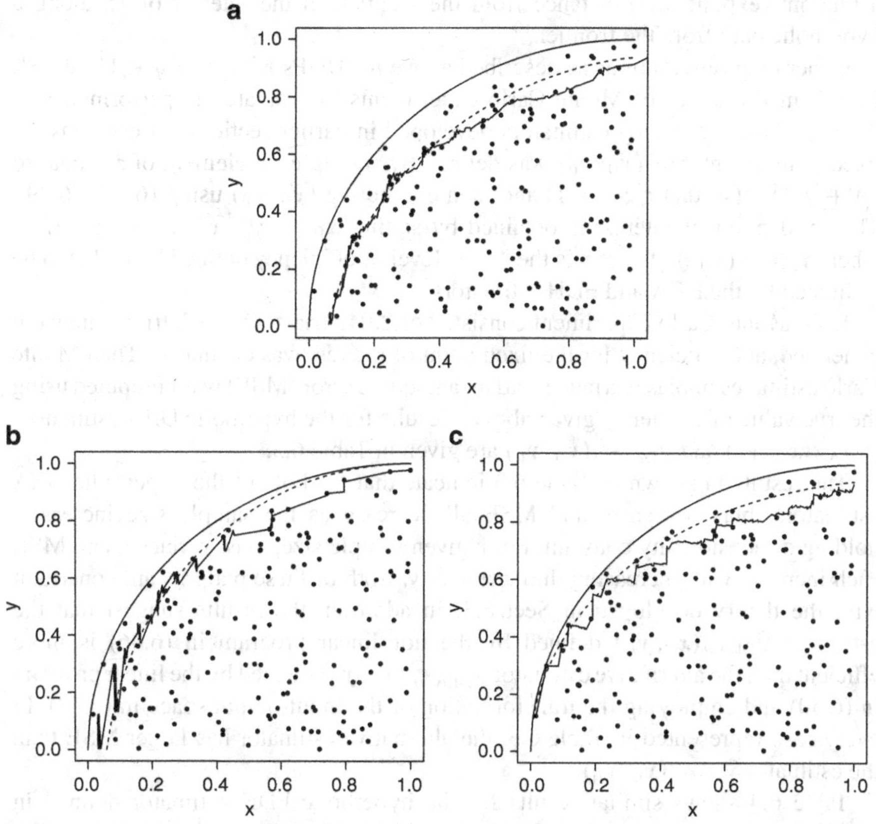

Fig. 6.3 Estimated order-m frontiers ($m = 50, n = 250$)

Then $(\widetilde{x} - 1, \widetilde{y})$ represents a draw from the uniform distribution on the surface defined by the intersection of the unit $(p + q - 1)$-sphere centered at the origin and the closed orthant in $\mathbb{R}^{p+q}$ defined by the Cartesian product $\mathbb{R}^p_- \times \mathbb{R}^q_+$. The addition of 1 to $-|z_x|$ on the right-hand side of (6.78) amounts to shifting this surface one unit in the positive direction along the axes in $\mathbb{R}^p$-space. In the case where $p = q = 1$, $(\widetilde{x}, \widetilde{y})$ represents a draw from the uniform distribution on the northwest quarter of a circle of radius 1 centered at $(1, 0)$ in $\mathbb{R}^2$.

A draw z from the uniform distribution on the surface of the unit $(p + q - 1)$-sphere can be simulated using the method of Muller (1959) and Marsaglia (1972). First, generate a pseudo-random vector $\boldsymbol{u}$ from the $(p + q)$-variate normal distribution with zero means and variance-covariance matrix equal to a $(p + q)$ by $(p + q)$ identity matrix. In order to introduce inefficiency, I first generate a point $(\widetilde{x}, \widetilde{y})$ lying on the frontier described above, and then project this point away from the frontier randomly by drawing a pseudo-random value $\nu \sim \text{Exp}(3)$ and computing $\boldsymbol{x} = \widetilde{x}(1 + \nu)$ and $\boldsymbol{y} = \widetilde{y}(1 + \nu)^{-1}$ to generate a point $(\boldsymbol{x}, \boldsymbol{y})$ lying a random (exponential) distance from the frontier, in the interior of $\mathcal{P}$, along a hyperbolic path from the frontier.[10]

Generating observations as described above for DGPs with $p = q \in \{1, 2, 3\}$, I performed a series of Monte Carlo experiments to evaluate the performance of the hyperbolic efficiency estimators developed in earlier sections. In each case, a fixed point of interest $(\boldsymbol{x}_0, \boldsymbol{y}_0)$ was defined by setting each element of z_0 equal to $(p+q)^{-1/2}$ (so that $z_0' z_0 = 1$) and then computing $(\widetilde{\boldsymbol{x}}_0, \widetilde{\boldsymbol{y}}_0)$ using (6.78)–(6.79). The fixed point of interest is obtained by setting $\boldsymbol{x}_0 = \gamma_0^{-1}\widetilde{\boldsymbol{x}}$ and $\boldsymbol{y}_0 = \gamma_0 \widetilde{\boldsymbol{y}}_0$, where $\gamma_0 = \gamma(\boldsymbol{x}_0, \boldsymbol{y}_0) = 0.6$ is the "true" level of efficiency defined by (6.11) to be estimated by the DEA and FDH estimators.

Each Monte Carlo experiment consisted of 2,048 trials. On each trial, data were generated, and efficiency for the fixed point of interest was estimated. Then Monte Carlo estimates of bias, variance, and mean-square error (MSE) was computed using the true value of efficiency given above. Results for the hyperbolic DEA estimators $\widehat{\gamma}_{\mathrm{DEA},n}(\boldsymbol{x}_0, \boldsymbol{y}_0)$ and $\widehat{\gamma}_{\mathrm{DEA}^*,n}(\boldsymbol{x}_0, \boldsymbol{y}_0)$ are given in Table 6.3.

The results in shown in Table 6.3 indicate that for both of the hyperbolic DEA estimators, bias, variance, and MSE all decrease as the sample size increases, holding dimensionality constant. For a given sample size, bias, variance, and MSE each increase with increasing dimensionality. Both of these patterns are consistent with the theory developed in Sect. 6.3. In addition, the results suggest that the estimator $\widehat{\gamma}_{\mathrm{DEA},n}(\boldsymbol{x}_0, \boldsymbol{y}_0)$ defined by the non-linear program in (6.46) is more efficient than the alternative estimator $\widehat{\gamma}_{\mathrm{DEA}^*,n}(\boldsymbol{x}_0, \boldsymbol{y}_0)$ defined by the linear program in (6.50) and employing the transformation of the input-output space in (6.47). In every case represented in Table 6.3, the alternative estimator has larger MSE than the estimator $\widehat{\gamma}_{\mathrm{DEA},n}(\boldsymbol{x}_0, \boldsymbol{y}_0)$.

Table 6.4 shows similar results for the hyperbolic FDH estimator defined in (6.62). As with the hyperbolic DEA estimators, for given dimensionality, increasing the sample size reduces both bias and variance, and hence MSE. Holding sample size constant, increasing the dimensionality leads to increases in both bias and variance, and hence MSE. These results are consistent with the theoretical results in Wheelock and Wilson (2008) that were discussed in Sect. 6.4.

Comparing results for the DEA estimators in Table 6.3 with those for the FDH estimator in Table 6.4 reveals that in each of the 12 cases considered, both of the DEA estimators have smaller bias, variance, and MSE than the FDH estimator. In addition, for a given dimensionality, MSE of the FDH estimator declines less rapidly than MSE of the DEA estimators. For example, with $p = q = 3$, dividing the MSE corresponding to $n = 1{,}000$ by the MSE corresponding to $n = 10$ yields 0.24845 for the FDH estimator, but only 0.002396 for the DEA estimator $\widehat{\gamma}_{\mathrm{DEA},n}(\boldsymbol{x}_0, \boldsymbol{y}_0)$. In other words, when the sample size increases from 10 to 1,000, the MSE of the DEA estimator declines by about 99.76%, while the MSE of the FDH estimator declines by only about 75.15%. This is to be expected since the FDH estimator has a slower convergence rate than the DEA estimators.

[10] The notation Exp(3) denotes an exponential distribution with parameter 3; hence $E(v) = 1/3$.

Table 6.3 Performance of hyperbolic DEA estimators

$p = q$	n	$\widehat{\gamma}_{\text{DEA},n}(x, y)$ Bias	Var	MSE	$\widehat{\gamma}_{\text{DEA}^*,n}(x, y)$ Bias	Var	MSE
1	10	9.6081E−02	3.7788E−03	1.3010E−02	1.2436E−01	6.8685E−03	2.2335E−02
1	100	1.7669E−02	1.0993E−04	4.2212E−04	2.2142E−02	1.7551E−04	6.6580E−04
1	1,000	3.5087E−03	3.9804E−06	1.6292E−05	4.4012E−03	6.2650E−06	2.5636E−05
1	10,000	7.4635E−04	1.7712E−07	7.3416E−07	9.3375E−04	2.7255E−07	1.1444E−06
2	10	1.7990E−01	6.2316E−03	3.8594E−02	2.4334E−01	1.1645E−02	7.0858E−02
2	100	4.5355E−02	2.2891E−04	2.2860E−03	6.4793E−02	5.1246E−04	4.7106E−03
2	1,000	1.5185E−02	2.3157E−05	2.5373E−04	2.1399E−02	4.7401E−05	5.0534E−04
2	10,000	5.5419E−03	2.8517E−06	3.3564E−05	7.5300E−03	4.0698E−05	9.7399E−05
3	10	2.1695E−01	6.7214E−03	5.3789E−02	3.0830E−01	1.7897E−02	1.1295E−01
3	100	5.8437E−02	1.8702E−04	3.6019E−03	9.0671E−02	5.2195E−04	8.7432E−03
3	1,000	2.4505E−02	2.7140E−05	6.2766E−04	3.6754E−02	2.6985E−04	1.6207E−03
3	10,000	1.1136E−02	4.8670E−06	1.2888E−04	1.6036E−02	3.9707E−04	6.5423E−04

Table 6.4 Performance of hyperbolic FDH estimator

		$\widehat{\gamma}_{\text{FDH},n}(\boldsymbol{x}, \boldsymbol{y})$		
$p = q$	n	Bias	Var	MSE
1	10	2.0246E−01	1.9895E−02	6.0887E−02
1	100	5.0597E−02	7.9889E−04	3.3589E−03
1	1,000	1.4683E−02	6.3263E−05	2.7886E−04
1	10,000	4.5313E−03	5.2999E−06	2.5833E−05
2	10	3.8546E−01	2.8594E−02	1.7717E−01
2	100	1.6011E−01	3.6050E−03	2.9240E−02
2	1,000	7.5936E−02	6.3692E−04	6.4032E−03
2	10,000	3.9857E−02	1.5461E−04	1.7432E−03
3	10	4.6633E−01	4.3366E−02	2.6083E−01
3	100	2.2331E−01	4.2423E−03	5.4110E−02
3	1,000	1.3022E−01	1.1161E−03	1.8075E−02
3	10,000	7.8372E−02	3.3769E−04	6.4799E−03

Performance of the hyperbolic order-m estimators discussed in Sect. 6.5 was examined in the same experimental framework. However, for a particular order m, it is difficult to determine analytically the "true" value $\overline{\gamma}_m(\boldsymbol{x}_0, \boldsymbol{y}_0)$ defined in Theorem 6.2.1 that is to be estimated by $\widehat{\overline{\gamma}}_{m,n}(\boldsymbol{x}_0, \boldsymbol{y}_0)$ defined in (6.66). In order to determine $\overline{\gamma}_m(\boldsymbol{x}_0, \boldsymbol{y}_0)$ for each value of m considered in each of the three cases where $p = q = 1$, 2, or 3, initial Monte Carlo experiments were performed with 2,048 trials and a sample size of 33,600,000. On each trial, $\widehat{\overline{\gamma}}_{m,n}(\boldsymbol{x}_0, \boldsymbol{y}_0)$ was approximated using (6.68) with $B = 1{,}000{,}000$; the results were then averaged over Monte Carlo trials in order to approximate $\overline{\gamma}_m(\boldsymbol{x}_0, \boldsymbol{y}_0)$. Then, a second set of Monte Carlo experiments were performed, identical to those in the first set except that the sample size was set at 42,000,000 – 25% larger than in the first set of experiments. The approximated values of $\overline{\gamma}_m(\boldsymbol{x}_0, \boldsymbol{y}_0)$ were compared to the corresponding approximations from the first set of experiments. For every pair of corresponding approximations, the results were identical to at least 5 digits to the right of the decimal point.

Using the "true" values of $\overline{\gamma}_m(\boldsymbol{x}_0, \boldsymbol{y}_0)$ obtained from the initial Monte Carlo experiments, I then performed a second set of experiments to evaluate the performance of hyperbolic order-m estimator $\widehat{\overline{\gamma}}_m(\boldsymbol{x}_0, \boldsymbol{y}_0)$ defined in (6.68) with $B = 2{,}000$ to reflect the number of replications in (6.68) an applied researcher might reasonably use. I considered four values of the order $m \in \{10, 50, 100, 200\}$ for each of the sample sizes used in the previous experiments. Results for estimated bias, variance, and MSE are given in Table 6.5.

The results in Table 6.5 confirm that bias, variance, and consequently MSE decrease as n increases, in every case. The results also indicate that for the same sample size n and the same number of dimensions, the order-m estimators have smaller MSE than the FDH estimator for the values of m that were considered.

Table 6.5 Performance of hyperbolic order-m estimators

		$\widehat{\overline{\gamma}}_{m=10,n}(\boldsymbol{x},\boldsymbol{y})$			$\widehat{\overline{\gamma}}_{m=50,n}(\boldsymbol{x},\boldsymbol{y})$		
$p=q$	n	Bias	Var	MSE	Bias	Var	MSE
1	10	7.7606E−02	2.0752E−02	2.6775E−02	1.2935E−01	1.9873E−02	3.6604E−02
1	100	6.4032E−03	1.2433E−03	1.2843E−03	1.1320E−02	7.3687E−04	8.6501E−04
1	1,000	4.2068E−04	1.2905E−04	1.2923E−04	9.0548E−04	6.6166E−05	6.6986E−05
1	10,000	1.2689E−04	2.1714E−05	2.1730E−05	6.2508E−05	7.1082E−06	7.1121E−06
2	10	8.7968E−02	2.8499E−02	3.6238E−02	1.7935E−01	2.8537E−02	6.0703E−02
2	100	7.3228E−03	2.2017E−03	2.2553E−03	1.7294E−02	2.5599E−03	2.8590E−03
2	1,000	7.9486E−04	2.2149E−04	2.2212E−04	1.5531E−03	2.2996E−04	2.3238E−04
2	10,000	9.9964E−05	3.5816E−05	3.5826E−05	1.3342E−04	2.6148E−05	2.6166E−05
3	10	1.2185E−01	5.2664E−02	6.7510E−02	1.9600E−01	4.3374E−02	8.1791E−02
3	100	9.5535E−03	2.7066E−03	2.7978E−03	1.7173E−02	2.4972E−03	2.7921E−03
3	1,000	1.0987E−03	2.6802E−04	2.6923E−04	2.3343E−03	2.3771E−04	2.4316E−04
3	10,000	3.3949E−04	4.7989E−05	4.8104E−05	3.5202E−04	2.6977E−05	2.7101E−05

		$\widehat{\overline{\gamma}}_{m=100,n}(\boldsymbol{x},\boldsymbol{y})$			$\widehat{\overline{\gamma}}_{m=200,n}(\boldsymbol{x},\boldsymbol{y})$		
$p=q$	n	Bias	Var	MSE	Bias	Var	MSE
1	10	1.5256E−01	1.9895E−02	4.3171E−02	1.6830E−01	1.9895E−02	4.8220E−02
1	100	1.5135E−02	7.1502E−04	9.4407E−04	2.0574E−02	7.5631E−04	1.1796E−03
1	1,000	1.3315E−03	5.5655E−05	5.7428E−05	1.8760E−03	5.0544E−05	5.4063E−05
1	10,000	8.3433E−05	5.8493E−06	5.8563E−06	1.2794E−04	5.0313E−06	5.0477E−06
2	10	2.2328E−01	2.8594E−02	7.8448E−02	2.5691E−01	2.8594E−02	9.4596E−02
2	100	2.6302E−02	2.9129E−03	3.6047E−03	3.9971E−02	3.3244E−03	4.9221E−03
2	1,000	2.3132E−03	2.5810E−04	2.6345E−04	3.5201E−03	2.9707E−04	3.0946E−04
2	10,000	2.6347E−04	2.7031E−05	2.7100E−05	4.6091E−04	3.0087E−05	3.0300E−05
3	10	2.4045E−01	4.3366E−02	1.0118E−01	2.7690E−01	4.3366E−02	1.2004E−01
3	100	2.6942E−02	3.1405E−03	3.8663E−03	4.2696E−02	3.8207E−03	5.6437E−03
3	1,000	3.4127E−03	3.0414E−04	3.1578E−04	5.1444E−03	4.0115E−04	4.2761E−04
3	10,000	3.5246E−04	3.3634E−05	3.3758E−05	5.8831E−04	4.2300E−05	4.2647E−05

Of course, Lemma 6.5.1 establishes that as $m \to \infty$, the order-m estimator converges to the FDH estimator; but for the particular values of m considered here, the order-m estimators have smaller MSE than the FDH estimator.

Comparing the MSEs of the DEA and order-m estimators reveals additional phenomena. For $p = q = 1$, the DEA estimator $\widehat{\gamma}_{\text{DEA},n}(\boldsymbol{x}_0, \boldsymbol{y}_0)$ has smaller MSE than the order-m estimators for any of the values of m that were considered. With $p = q = 2$, however, the corresponding MSE estimates are more similar between the DEA and order-m estimators, and when $p = q = 3$, the order-m estimators outperform the DEA estimator in terms of MSE in many cases, particularly as sample size increases beyond 10. This is to be expected; the DEA estimator incurs the curse of dimensionality, while the order-m estimators do not.

6.7 Conclusions

In this paper I provide two easily-computable estimators of the hyperbolic measure of technical efficiency proposed by Färe et al. (1985). The estimators do not require an assumption of constant returns to scale. Asymptotic results, including consistency, rates of convergence, and the limiting distribution are derived. In addition, an unconditional, hyperbolic order-m estimator of technical efficiency and its asymptotic properties are derived by extending results from Cazals et al. (2002) for their conditional, input-oriented order-m estimator.

As discussed in the Introduction, measuring efficiency along hyperbolic paths rather than along paths parallel either to input or output axes avoids results that might depend on the slope of the production frontier in the neighborhood where an inefficient observation is projected onto the frontier using either input- or output-oriented efficiency scores. This problem has been discussed by Wheelock and Wilson (2008, 2009). Wheelock and Wilson (2009) used the unconditional, hyperbolic order-α measure of efficiency to construct Malmquist-type productivity indices. Similar indices can be constructed using either the hyperbolic measure of efficiency $\gamma(\boldsymbol{x}, \boldsymbol{y})$ defined in (6.11) or the expected unconditional, hyperbolic order-m measure $\overline{\gamma}_m(\boldsymbol{x}, \boldsymbol{y})$ defined in (6.27).

Avoiding the choice between input- and output-orientation is potentially even more important in dynamic settings where Malmquist indices and their components are estimated. With cross-period comparisons needed to define such indices, the sensitivity of results to the choice of input- or output orientation is more likely to arise than in cross-sectional settings. Whereas a firm might lie near the middle of the range of the data for one period, it might lie near the steeply-sloped or nearly flat portions of the frontier prevailing in the other period, in which case estimates of productivity or efficiency change may be highly sensitive to the choice of input- or output-orientation.

Moreover, from a practical viewpoint, the cross-period comparisons used to estimate changes in technology often result in infeasible solutions when input- or output-oriented DEA or FDH estimators are used. In particular, it is sometimes the case that a firm's position in one period is either above or to the left of the frontier estimate in another period; in the former case, cross-period, input-oriented DEA or FDH efficiency estimates cannot be computed, while in the latter case, cross-period output-oriented estimates cannot be computed. Similar problems exist for the input- and output-oriented conditional order-m and order-α estimators of Cazals et al. (2002) and Daouia and Simar (2007). Measuring efficiency along hyperbolic paths avoids these problems.

Acknowledgements This work was made possible by the Palmetto cluster operated and maintained by the Clemson Computing and Information Technology group at Clemson University. In addition, I have benefited from numerous discussions with Léopold Simar and other members of the Institut de Statistique, Université Catholique de Louvain in Louvain-la-Neuve over the years. Any errors are solely my responsibility.

References

Aigner, D., Lovell, C.A.K., & Schmidt, P. (1977). Formulation and estimation of stochastic frontier production function models. *Journal of Econometrics, 6*, 21–37.

Aragon, Y., Daouia, A., & Thomas-Agnan, C. (2005). Nonparametric frontier estimation: A conditional quantile-based approach. *Econometric Theory, 21*, 358–389.

Boos, D.D., & Serfling, R.J. (1980). A note on differentials and the CLT and LIL for statistical functions, with application to M-estimates. *The Annals of Statistics, 8*, 618–624.

Cazals, C., Florens, J.P., & Simar, L. (2002). Nonparametric frontier estimation: A robust approach. *Journal of Econometrics, 106*, 1–25.

Chambers, R.G., Chung, Y., & Färe, R. (1996). Benefit and distance functions. *Journal of Economic Theory, 70*, 407–419.

Charnes, A., Cooper, W.W., & Rhodes, E. (1981). Evaluating program and managerial efficiency: An application of data envelopment analysis to program follow through. *Management Science, 27*, 668–697.

Daouia, A. (2003). *Nonparametric Analysis of Frontier Production Functions and Efficiency Measurement using Nonstandard Conditional Quantiles.* PhD thesis, Groupe de Recherche en Economie Mathématique et Quantititative, Université des Sciences Sociales, Toulouse I, et Laboratoire de Statistique et Probabilités, Université Paul Sabatier, Toulouse III, 2003.

Daouia, A., & Simar, L. (2007). Nonparametric efficiency analysis: A multivariate conditional quantile approach. *Journal of Econometrics, 140*, 375–400.

Daraio, C., & Simar, L. (2005). Introducing environmental variables in nonparametric frontier models: A probabilistic approach. *Journal of Productivity Analysis, 24*, 93–121.

Deprins, D., Simar, L., & Tulkens, H. (1984). Measuring labor inefficiency in post offices. In Marchand, M., Pestieau, P., &Tulkens, H. (Eds.). *The Performance of Public Enterprises: Concepts and Measurements*, pp. 243–267. Amsterdam: North-Holland.

Färe, R. (1988). *Fundamentals of Production Theory.* Berlin: Springer.

Färe, R., Grosskopf, S., & Lovell, C.A.K. (1985). *The Measurement of Efficiency of Production.* Boston: Kluwer-Nijhoff Publishing.

Farrell, M.J. (1957). The measurement of productive efficiency. *Journal of the Royal Statistical Society A, 120*, 253–281.

Gattoufi, S., Oral, M., & Reisman, A. (2004). Data envelopment analysis literature: A bibliography update (1951–2001). *Socio-Economic Planning Sciences, 38*, 159–229.

Kneip, A., Park, B., & Simar, L. (1998). A note on the convergence of nonparametric DEA efficiency measures. *Econometric Theory, 14*, 783–793.

Kneip, A., Simar, L., & Wilson, P.W. (2008). Asymptotics and consistent bootstraps for DEA estimators in non-parametric frontier models. *Econometric Theory, 24*, 1663–1697.

Kneip, A., Simar, L., & Wilson, P.W. (2011). A computationally efficient, consistent bootstrap for inference with non-parametric DEA estimators. *Computational Economics*, Institut de Statistique, Université Catholique de Louvain, Louvain-la-Neuve, Belgium.

Korostelev, A., Simar, L., & Tsybakov, A.B. (1995a). Efficient estimation of monotone boundaries. *The Annals of Statistics, 23*, 476–489.

Korostelev, A., Simar, L., & Tsybakov, A.B. (1995b). On estimation of monotone and convex boundaries. *Publications de l'Institut de Statistique de l'Université de Paris XXXIX, 1*, 3–18.

Marsaglia, G. (1972). Choosing a point from the surface of a sphere. *Annals of Mathematical Statistics, 43*, 645–646.

Muller, M.E. (1959). A note on a method for generating points uniformly on n-dimensional spheres. *Communications of the Association for Computing Machinery, 2*, 19–20.

Park, B.U., Simar, L., & Weiner, C. (2000). FDH efficiency scores from a stochastic point of view. *Econometric Theory, 16*, 855–877.

Shephard, R.W. (1970). *Theory of Cost and Production Functions*. Princeton: Princeton University Press.

Simar, L., & Wilson, P.W. (1998). Sensitivity analysis of efficiency scores: How to bootstrap in nonparametric frontier models. *Management Science, 44*, 49–61.

Simar, L., & Wilson, P.W. (2000a). A general methodology for bootstrapping in non-parametric frontier models. *Journal of Applied Statistics, 27*, 779–802.

Simar, L., & Wilson, P.W. (2000b). Statistical inference in nonparametric frontier models: The state of the art. *Journal of Productivity Analysis, 13*, 49–78.

Simar, L., & Wilson, P.W. (2011). Inference by the m out of n Bootstrap in Nonparametric Frontier Models. *Journal of Productivity Analysis, 36*, 33–53.

Wheelock, D.C., & Wilson, P.W. (2008). Non-parametric, unconditional quantile estimation for efficiency analysis with an application to Federal Reserve check processing operations. *Journal of Econometrics, 145*, 209–225.

Wheelock, D.C., & Wilson, P.W. (2009). Robust nonparametric quantile estimation of efficiency and productivity change in U.S. commercial banking, 1985–2004. *Journal of Business and Economic Statistics, 27*, 354–368.

Chapter 7
Explaining Efficiency in Nonparametric Frontier Models: Recent Developments in Statistical Inference

Luiza Bădin and Cinzia Daraio

Abstract The explanation of efficiency differentials is an essential step in any frontier analysis study that aims to measure and compare the performance of decision making units. The conditional efficiency measures that have been introduced in recent years (Daraio and Simar, J. Prod. Anal. 24:93–121, 2005) represent an attractive alternative to two-step approaches, to handle external environmental factors, avoiding additional assumptions such as the separability between the input-output space and the space of external factors. Although affected by the *curse of dimensionality*, nonparametric estimation of conditional measures of efficiency eliminates any potential specification issue associated with parametric approaches. The nonparametric approach requires, however, estimation of a nonstandard conditional distribution function which involves smoothing procedures, and therefore the estimation of a bandwidth parameter. Recently, Bădin et al. (Eur. J. Oper. Res. 201(2):633–640, 2010) proposed a data driven procedure for selecting the optimal bandwidth based on a general result obtained by Hall et al. (J. Am. Stat. Assoc. 99(486):1015–1026, 2004) for estimating conditional probability densities. The method employs least squares cross-validation (LSCV) to determine the optimal bandwidth with respect to a weighted integrated squared error (WISE) criterion.

This paper revisits some of the recent advances in the literature on handling external factors in the nonparametric frontier framework. Following the Bădin et al. (Eur.

L. Bădin (✉)
Department of Applied Mathematics, Bucharest Academy of Economic Studies, Piata Romana nr. 6, 010374 Bucharest, Romania

Department of Statistical Inference, Gh. Mihoc - C. Iacob Institute of Mathematical Statistics and Applied Mathematics, Calea 13 Septembrie nr. 13, Bucharest, Romania
e-mail: luiza.badin@csie.ase.ro

C. Daraio
Department of Management, CIEG - Centro Studi di Ingegneria Economico-Gestionale, University of Bologna, Via Umberto Terracini 28, 40131 Bologna, Italy
e-mail: cinzia.daraio@unibo.it

I. van Keilegom and P.W. Wilson (eds.), *Exploring Research Frontiers in Contemporary Statistics and Econometrics*, DOI 10.1007/978-3-7908-2349-3_7,

J. Oper. Res. 201(2):633–640, 2010) approach, we provide a detailed description of optimal bandwidth selection in nonparametric conditional efficiency estimation, when mixed continuous and discrete external factors are available. We further propose an heterogeneous bootstrap which allows improving the detection of the impact of the external factors on the production process, by computing pointwise confidence intervals on the ratios of conditional to unconditional efficiency measures.

We illustrate these extensions through some simulated data and an empirical application using the sample of U.S. mutual funds previously analyzed in Daraio and Simar (J. Prod. Anal. 24:93–121, 2005; Eur. J. Oper. Res. 175(1):516–542, 2006; Advanced Robust and Nonparametric Methods in Efficiency Analysis: Methodology and Applications, Springer, New York, 2007a).

7.1 Introduction

The nonparametric envelopment estimators of frontiers and technical efficiency, namely Data Envelopment Analysis (DEA, see Farrell 1957; Charnes et al. 1978) and Free Disposal Hull (FDH, see Deprins et al. 1984) have been extensively used in empirical studies aiming to estimate the efficiency scores of producers from a wide variety of fields. The success of these nonparametric envelopment methods is mainly due to the few assumptions required for specifying the Data Generating Process (DGP). However, the nonparametric approach presents also several limitations, namely the difficulty in carrying out statistical inference, the *curse of dimensionality* and the influence of extreme values and outliers. Recent advances on robust nonparametric efficiency estimation provide useful tools for overcoming the main drawbacks of traditional nonparametric efficiency estimators, such as FDH and DEA.

An increasing number of recent studies on efficiency and productivity include external, environmental factors in the analysis of the performances of economic producers. These environmental factors can be considered as exogenous variables, which may influence the production process, but, unlike the inputs and the outputs, are not under the control of the production unit. Including such factors in the analysis can help not only explaining the differences in efficiency, but also improving the management of the analyzed units.

A fully nonparametric setting which includes external variables in the frontier model and permits use of conditional Debreu–Farrell efficiency scores and their nonparametric estimators was developed by Daraio and Simar (2005, 2007a, 2007b), extending previous results from Cazals et al. (2002). Conditional efficiency measures and their nonparametric estimators, including conditional FDH, conditional DEA, conditional order$-m$ and conditional order-α have proved to be a useful tool for investigating the impact of external-environmental factors on the performance of production units in a nonparametric framework. However, nonparametric estimators require at some stage the estimation of a nonstandard conditional distribution which depends on a smoothing parameter (bandwidth). So far, the bandwidth selection method was an open issue in this framework. Daraio and Simar (2005) suggested to use the cross-validation or plug-in rules, but those are based on marginal properties of the external, environmental variables and do not consider the influence of these variables on the production process when estimating the bandwidth.

Bădin et al. (2010) adapted the theoretical results from Hall et al. (2004) and Li and Racine (2008) to the particular setting of conditional frontier estimation and proposed a data-driven method for selecting the optimal bandwidth for nonparametric conditional efficiency estimation. Beside providing the optimal bandwidth, the procedure allows detection of the components of the external variables that have no influence on the production process.

This chapter revisits the method proposed by Bădin et al. (2010) and examines the case where the external variables are vectors having both continuous and discrete components, as opposed to the case considered by Bădin et al. (2010) where all environmental variables are continuous. The paper is organized as follows. Section 7.2 gives an overview of recent results in nonparametric frontier estimation when external, environmental factors are included in the model, stressing advantages as well as shortcomings of the different approaches. In Sect. 7.3 we present the probabilistic formulation of the frontier model, the conditional measures of efficiency, and their corresponding nonparametric estimators. Extending the results of Bădin et al. (2010) we describe in detail how to handle mixed continuous and discrete external factors in Sect. 7.4. In Sect. 7.5 we suggest a consistent bootstrap approach to estimate pointwise confidence intervals for the ratios of conditional to unconditional efficiency measures. We investigate the finite sample behavior of the optimal bandwidth through simulated examples and an empirical application using data on U.S. mutual funds. The simulation exercises, with univariate and multivariate external factors, and with different impact of the external factors, illustrate how to operationalize the statistical inference in this conditional frontier setting. The chapter ends with a section discussing conclusions.

Throughout the chapter, we discuss the output orientation focusing on the nonparametric conditional efficiency estimators. Similar results are straightforward to derive for the input orientation case, which we leave as an exercise for the reader.

7.2 Introducing External Factors in Nonparametric Frontier Models: An Overview on the Literature

For the most part, the literature on efficiency analysis focuses on the estimation of production frontiers, which provide benchmarks against which economic producers are evaluated. Recent studies have attempted to explain efficiency differentials by including exogenous variables that cannot be controlled by the producer, but which may influence the production process. From an economic viewpoint it is important to identify the "particularities" of the production process or the economic conditions that might be responsible for inefficiency, as well as to detect and analyze possible influential factors that can determine changes in productivity patterns.

The meaning and the economic role played by external, environmental variables is strictly linked to the economic field firms are operating in. The choice of environmental variables has to be done on a case-by-case basis, requiring knowledge of the production process and its characteristics and by taking into account the economic field of application.

In the efficiency literature, three main approaches have been proposed to explain efficiency differentials by introducing external, environmental variables (denoted below by Z).

First, the *one-stage approach* is based on including in the model the external factors Z as free-disposable inputs or outputs (see, for example, Banker and Morey 1986). The approach involves external variables Z as inputs or outputs in defining the attainable set, but these are not active in the optimization for estimating the efficiency scores. The method has several drawbacks since it requires restrictive assumptions on the free disposability and convexity of the resulted (augmented) attainable set and moreover the *a priori* specification of the favorable or unfavorable role of the exogenous factors, since they may act either as free disposal inputs or as undesired freely available outputs. Moreover, the linear programs involved in defining the corresponding efficiency scores depend on the returns to scale assumption made on the non-discretionary inputs or outputs. All these assumptions look restrictive because quite often the analyst has not a clear view on the possible influence of Z on the production process.

The second traditional approach is the so-called *two-stage approach*. Here, the nonparametric efficiency estimates obtained in a first stage are regressed in a second stage on covariates interpreted as environmental variables (see Färe et al. 1994, Simar and Wilson 2007, 2008 and all the references therein, as well as the DEA's bibliographies by Cooper et al. 2000 and Gattoufi et al. 2004). Most studies using this approach have used either tobit regression or ordinary least squares for estimating the second-stage parametric models. Simar and Wilson (2007) note that "None of the studies that employ two-stage approaches have described the underlying data-generating process. Since the DGP has not been described, there is some doubt about what is being estimated in the two-stage approaches." In addition, DEA estimates are by construction biased estimators of the true efficiency scores. Finally, a more serious problem arises from the fact that DEA efficiency estimates are serially correlated and that the error term in the second stage is correlated with the regressors. As stated by Simar and Wilson (2007) "consequently, standard approaches to inference [...] are invalid."

Simar and Wilson (2007) proposed a *semi-parametric bootstrap-based approach* to overcome the problems of the traditional two-stage approaches outlined above. In particular, Simar and Wilson (2007) proposed two bootstrap-based algorithms to obtain valid, accurate inference in this framework. Still, the two-stage approach has two inconveniences. First, it relies on a *separability* condition between the input-output space and the space of the external factors, assuming that these factors have no influence on the attainable set, affecting only the probability of being more or less efficient, which may not hold in some situations. Second, the regression in the second stage relies on strong *parametric* assumptions (e.g., linear model and truncated normal error term).

More recently, Park et al. (2008), suggested using a nonparametric model for the second stage regression. Unfortunately, this two-stage approach also relies on the separability condition between the input–output space and the space of external factors that was mentioned above. It is important to note that neither Simar and Wilson (2007) nor Park et al. (2008) advocated using the two-stage approach. The goal of Simar and Wilson was to define a statistical model where a second-stage regression would be meaningful, and to provide a method for valid inference in the second-stage regression. Daraio et al. (2010) provide a test of the separability condition that is required for the second stage regression to be meaningful, and also remark that if this condition is not met, the first-stage estimates have no useful meaning; see also Simar and Wilson (2010) for additional discussion.

A third, more general and appealing approach is the *nonparametric conditional approach* proposed by Daraio and Simar (2005) in which conditional efficiency measures are defined and estimated nonparametrically. The approach is based on the extension of the probabilistic formulation of the production process proposed by Cazals et al. (2002). Here the attainable set is interpreted as the support of some probability measure defined on the input-output space and the traditional Debreu–Farrell efficiency scores are defined in terms of a nonstandard conditional survival function. The approach allows a natural extension of the model in the presence of environmental factors, leading to conditional Debreu–Farrell efficiency measures. The nonparametric estimators of conditional efficiency measures are then easily defined by a plug-in rule, providing conditional FDH estimators as in Daraio and Simar (2005) or conditional DEA estimators, as in Daraio and Simar (2007b). Since the conditional efficiency estimators are based on a nonstandard conditional survival function, smoothing procedures and the estimation of a bandwidth parameter are required. Daraio and Simar (2005, 2007a) suggest using bandwidth selection methods, such as cross-validation and plug-in rules related to kernel density estimation for the external variables Z. For estimating the density of Z, they propose the likelihood cross-validation method based on a k-nearest neighbor technique. This method has the merit of providing bandwidths with appropriate asymptotic rate, but as observed by Daraio and Simar (2005), the resulting bandwidth does not possess optimality properties in finite samples. Moreover, these methods do not take into account the influence of the environmental variables on the production process while determining the window size, since they are based only on marginal properties of Z.

Regarding asymptotic properties of the nonparametric conditional estimators, Jeong et al. (2008) proved their asymptotic consistency and derived the limiting sampling distribution of the conditional efficiency estimators. Methods of statistical inference in conditional frontier models continue to be developed.

Estimation of the bandwidth parameter in the nonparametric, conditional approach was until recently an open issue. Bădin et al. (2010) proposed an adaptive data-driven method for selecting the optimal bandwidth by extending theoretical results obtained by Hall et al. (2004) and Li and Racine (2007, 2008) in the context of conditional density or distribution function estimation to the frontier framework. The approach is based on least squares cross-validation, and provides a bandwidth which optimizes, in terms of integrated square error, the estimation of the conditional survival function involved, where the dependence between the output variable and the external factor Z is implicit.

This paper provides further extensions to the nonparametric conditional approach described above. The methodology applies to full conditional FDH as well as to robust conditional estimators (order-m, order-α). To improve the detection of the impact of Z we propose an heterogenous bootstrap for constructing bootstrap confidence intervals for the ratios of conditional to unconditional efficiency measures.

7.3 Conditional Efficiency: Probabilistic Model and Nonparametric Estimation

The activity of production units is usually characterized by the feasible input-output combinations which define the attainable set Ψ, i.e.,

$$\Psi = \{(x, y) \in \mathbb{R}_+^p \times \mathbb{R}_+^q \mid x \text{ can produce } y\}, \tag{7.1}$$

where $x \in \mathbb{R}_+^p$ is the input vector required to produce the output vector $y \in \mathbb{R}_+^q$.

The probabilistic frontier model introduced by Cazals et al. (2002) assumes that the set of observed production units represents a random sample of independent, identically distributed observations from the population of input-output pairs (X, Y) on $\mathbb{R}_+^p \times \mathbb{R}_+^q$. Consequently, the production process can be fully characterized by the joint distribution of X and Y, whose support is the production possibilities set Ψ. The support of the q-variate survivor function $S_{Y|X}(y|x) = \text{Prob}(Y \geq y|X \leq x)$ can be interpreted as the attainable set of output values Y for a producer using an input level x. For $q = 1$ and for any given x, the upper boundary of the support of this conditional survivor function provides the production frontier, as defined by Cazals et al. (2002), i.e.

$$\varphi(x) = \sup\{y \mid S_{Y|X}(y|x) > 0\}. \tag{7.2}$$

In the multiple output case ($q \geq 1$), it is more convenient to use radial distances

$$\lambda(x, y) = \sup\{\lambda \mid S_{Y|X}(\lambda y|x) > 0\} \tag{7.3}$$

for evaluating the efficiency level of a given point. Here $\lambda(x, y)$ represents the output-efficiency measure for a unit operating at level (x, y) and is equivalent, under free disposability of inputs and outputs, with the Farrell output-efficiency score (Farrell 1957).

Daraio and Simar (2005) extended the probabilistic model by introducing external environmental factors that are exogenous to the production process, but which may influence the process and productivity patterns. Consider $(X, Y, Z) \in \mathbb{R}_+^p \times \mathbb{R}_+^q \times \mathbb{R}^t$, where $Z \in \mathbb{R}^t$ represents a vector of external variables and define the following nonstandard conditional distribution:

$$H(x, y|z) = \text{Prob}(X \leq x, Y \geq y|Z = z). \tag{7.4}$$

Using the decomposition

$$\begin{aligned}\text{Prob}(X \leq x, Y \geq y|Z = z) &= \text{Prob}(Y \geq y \mid X \leq x, Z = z)\,\text{Prob}(X \leq x|Z = z)\\ &= S_{Y|X,Z}(y|x,z)F_{X|Z}(x|z),\end{aligned}$$

where $S_{Y|X,Z}(y|x,z) = H(x,y|z)/H(x,0|z)$ and $F_{X|Z}(x|z) = H(x,0|z)$, the conditional output-efficiency measure can be defined by

$$\lambda(x, y|z) = \sup\{\lambda|S_{Y|X,Z}(\lambda y|x,z) > 0\} = \sup\{\lambda|H(x,\lambda y|z) > 0\}. \tag{7.5}$$

The conditional order-m output efficiency measure can be introduced as follows. For a given level of inputs x, draw m i.i.d. random variables Y_i, $i = 1, \ldots, m$ from $S_{Y|X,Z}(y|x,z)$ and define the set

$$\Psi_m^z(x) = \{(u, v) \in \mathcal{R}_+^{p+q} \mid u \leq x, Y_i \leq v, i = 1, \ldots, n\}. \tag{7.6}$$

This is a random set which depends on z since the Y_i are generated through $S_{Y|X,Z}(y|x,z)$. When the environmental variable Z takes the value z, the conditional survivor of X and Y given $Z = z$ defines the data generating process which depends on the exogenous environment represented by Z. The conditional order-m output efficiency measure is defined as

$$\lambda_m(x, y|z) = E_{Y|X,Z}(\tilde{\lambda}_m^z(x, y) \mid X \leq x, Z = z), \tag{7.7}$$

where $\tilde{\lambda}_m^z(x, y) = \sup\{\lambda \mid (x, \lambda y) \in \Psi_m^z(x)\}$. This can be computed as

$$\lambda_m(x, y|z) = \int_0^\infty [1 - (1 - S_{Y|X,Z}(uy|x,z))^m]du, \tag{7.8}$$

where $\lim_{m\to\infty} \lambda_m(x, y|z) = \lambda(x, y|z)$.

An alternative to the order-m partial frontier is the order-α quantile-type frontier defined by Daouia and Simar (2007) and extended to conditional measures by the same authors. The conditional order-α output efficiency measure given that $Z = z$ is defined as

$$\lambda_\alpha(x, y|z) = \sup\{\lambda \mid S_{Y|X,Z}(\lambda y|x,z) > 1 - \alpha\}. \tag{7.9}$$

The nonparametric conditional output-efficiency estimator is computed by simply replacing the unknown theoretical quantities in (7.5) by their empirical analogs to obtain

$$\widehat{\lambda}_n(x, y|z) = \sup\{\lambda|\widehat{S}_{Y|X,Z}(\lambda y|x,z) > 0\} = \sup\{\lambda|\widehat{H}(x,\lambda y|z) > 0\}. \tag{7.10}$$

A nonparametric estimator of $\lambda_m(x, y|z)$ is given by

$$\widehat{\lambda}_{n,m}(x, y|z) = \int_0^\infty \left[1 - (1 - \widehat{S}_{Y|X,Z}(uy|x, z))^m\right] du$$
$$= \widehat{\lambda}_n(x, y|z) - \int_0^{\widehat{\lambda}_n(x,y|z)} (1 - \widehat{S}_{Y|X,Z}(uy|x, z))^m du. \quad (7.11)$$

Since $\widehat{\lambda}_{n,m}(x, y|z) \to \widehat{\lambda}_n(x, y|z)$ when $m \to \infty$, the order-m conditional efficiency score can again be viewed as a robust estimator of the conditional efficiency score $\lambda(x, y|z)$ when choosing $m = m(n) \to \infty$ with $n \to \infty$. For finite m, the corresponding frontier estimator will be more robust to outliers, since it will not envelop all the data points.

A nonparametric estimator of $\lambda_\alpha(x, y|z)$ is provided in a similar way, by plugging in its formula, the nonparametric estimator of $S_{Y|X,Z}(y|x, z)$. Formally, it is defined as

$$\widehat{\lambda}_{n,\alpha}(x, y|z) = \sup\{\lambda \mid \widehat{S}_{Y|X,Z}(\lambda y|x, z) > 1 - \alpha\}. \quad (7.12)$$

Here also we have $\lim_{\alpha\to 1} \widehat{\lambda}_{n,\alpha}(x, y|z) = \widehat{\lambda}_n(x, y|z)$.

Estimation of the conditional distribution and survivor function involved in (7.10), (7.11) and (7.12) requires smoothing procedures and consequently estimation of a bandwidth parameter, which is crucial for providing a reliable estimator of the conditional efficiency score. Daraio and Simar (2005, 2007a) propose selecting the bandwidth by a cross-validation rule which is based on the estimation of the marginal density of Z, using a nearest-neighbor technique. Although the resulting bandwidth is of appropriate asymptotic order, it is not optimal for finite samples; moreover it does not consider the influence that Z might have on the behavior of Y given that $X \leq x$. For the conditional FDH case, with an univariate baseline bandwidth h, we have

$$\widehat{\lambda}_n(x, y|z) - \lambda(x, y|z) = O_p\big((nh^t)^{-1/(p+q)}\big), \quad (7.13)$$

as $h \to 0$ with $nh^t \to \infty$ when $n \to \infty$ (Jeong et al. 2008). For the unconditional FDH, according to Park et al. (2000) we have

$$\widehat{\lambda}_n(x, y) - \lambda(x, y) = O_p\big(n^{-1/(p+q)}\big). \quad (7.14)$$

Bădin et al. (2010) proposed an adaptive, data-driven method which optimizes the estimation of the conditional survivor function $S_{Y|X,Z}(y|x, z)$, where the dependence between Y and Z for $X \leq x$ is implicit. The optimal order of the bandwidths for estimating the conditional survivor function $S_{Y|X,Z}(y|x, z)$ with the kernel estimator $\widehat{S}_{Y|X,Z}(y|x, z)$ is $h \approx n^{-1/(t+4)}$ (see Li and Racine 2007; for details see Bădin et al. 2010, the extension of Li and Racine's results to the conditional frontier framework). Therefore, as observed by Jeong et al. (2008), the rate of convergence of the FDH conditional efficiency estimators deteriorates to $n^{4/((t+4)(p+q))}$. Similar results are available for the order-m and order-α as well as for the DEA version of the estimators.

7.4 Handling Mixed Continuous and Discrete External Variables

In attempting to explaining efficiency differentials by analyzing the impact of external factors, one is often confronted with situations where mixed continuous and categorical data types are available. These may include, for example, the type of economic environment, ownership and property rights, quality factors, or environmental factors that might be represented by either continuous or discrete variables. A general approach involves developing estimation procedures which allow inclusion of both continuous and discrete exogenous variables. In the following, we describe in detail how to handle multivariate mixed discrete and continuous external variables when selecting the optimal bandwidth for estimating the conditional survival function on which the conditional efficiency estimators are based.

Denote by $Z = (Z^c, Z^d)$ the vector of external variables, where $Z^c \in \mathbb{R}^r$ is the r-dimensional subset of the continuous components of Z and $Z^d \in D^s$ denotes the subset of discrete components of Z.

Consider the conditional pdf of Y given $X \leq x$ and $Z = z$, defined as

$$g(y|X \leq x, z) = \frac{f(y, X \leq x, z)}{g_1(X \leq x, z)} = \frac{f(y, X \leq x, z)}{g_1^c(X \leq x, z^c|z^d)\text{Prob}(Z^d = z^d)}, \quad (7.15)$$

where f and g_1 are respectively the joint and the marginal density in y and z and $g_1^c(X \leq x, z^c|z^d)$ denotes the density of Z^c given $Z^d = z^d$.

Let $\mathcal{X} = \{(X_i, Y_i, Z_i) \mid i = 1, \ldots, n\}$ be the set of observed production units, which can be viewed as a random sample of i.i.d. observations drawn from the joint distribution of $(X, Y, Z) \in \mathbb{R}_+^p \times \mathbb{R}_+^q \times \mathbb{R}^t$, with $t = r + s$. The estimate of the conditional density in (7.15) can be written as

$$\widehat{g}(y|X \leq x, Z = z) = \frac{\widehat{f}(y, X \leq x, z)}{\widehat{g}_1(X \leq x, z)} = \frac{n^{-1}\sum_{i=1}^n \mathbb{I}(X_i \leq x)K_h(Z_i, z^c)K_\lambda(Z_i, z^d)L_{h_y}(Y_i, y)}{n^{-1}\sum_{i=1}^n \mathbb{I}(X_i \leq x)K_h(Z_i, z^c)K_\lambda(Z_i, z^d)}, \quad (7.16)$$

where L_{h_y}, K_h and K_λ are nonnegative generalized kernels, with bandwidths h_y, h and λ respectively; these will be described in more detail below.

Note that the number of observations used in the estimation above is $N_x = \sum_{i=1}^{n} I\!I(X_i \leq x)$, *i.e.* the number of observations in the sample where $X_i \leq x$. Consequently, the selected bandwidths will be determined for a specific value of x, but in order to simplify the notation, we will denote the bandwidths simply by (h_y, h, λ).

For the output variables Y we use the product kernel

$$L_{h_y}(Y_i, y) = \prod_{j=1}^{q} \frac{1}{h_{y_j}} L\left(\frac{Y_{ij} - y_j}{h_{y_j}}\right), \tag{7.17}$$

where $L(\cdot)$ is a traditional univariate kernel function. The kernel used for the external factors Z is described in detail in the following section.

7.4.1 Kernel Smoothing with Mixed Continuous and Discrete Data

Since the conditional efficiency estimators are based on the nonparametric estimation of a conditional survival function and smoothing procedures for the external variable Z, it is natural to involve methods which allow smoothing all the components of Z, continuous or discrete. The classical nonparametric frequency-based estimator for discrete distributions is based on splitting the data into cells according to the distinct values taken by the categorical variables. This may result however, in a large number of cells, sometimes exceeding the sample size, making the estimation unfeasible. Many studies have overcome this issue by proposing kernel estimators for categorical data. Perhaps the most desirable feature of kernel estimators for discrete data is that there is no curse of dimensionality associated with smoothing discrete probability functions, with the estimators achieving $\sqrt{n}$-consistency similar to most parametric estimators.

The idea of kernel smoothing for discrete probability distributions can be traced back to Aitchison and Aitken (1976), who proposed an extension of the kernel approach from continuous to multivariate binary data as an alternative to the classical frequency-based nonparametric estimator. The method allows extensions for data with more than two categories, ordered or unordered, and can be further generalized to handle mixed discrete and continuous data.

Titterington (1980) studied kernel-based density estimates for categorical data, analyzing also several techniques for computing the smoothing parameter for the case of multivariate, multi-category and even incomplete data. Wang and Van Ryzin (1981) proposed a class of smooth weight function estimators for discrete distributions, and provided a general procedure for choosing the weight function smoothing parameter, which is the analogous to the smoothing parameter of Aitchison and Aitken (1976). Various methods for computing the smoothing parameter have been proposed and compared; e.g., jackknife likelihood or cross-validation, minimizing the mean squared error and pseudo-Bayesian techniques.

Although it is known that kernel smoothing is likely to introduce some finite sample bias, the resulting estimators are superior to the frequency-based estimators, significantly reducing the variance, which leads to an overall reduction in mean square error.

Consider the probability function $p(z^d) = \text{Prob}(Z^d = z^d)$, where Z^d is an unordered discrete variable assuming c distinct values $\{0, 1, 2, \ldots, c-1\}$ corresponding to the possible outcomes of Z^d. Aitchison and Aitken (1976) proposed the following kernel estimator for $p(z^d)$:

$$\widehat{p}(z^d) = \frac{1}{n}\sum_{i=1}^{n} l(Z_i^d = z^d), \tag{7.18}$$

where $l(\cdot)$ is a kernel function defined by

$$l(Z_i^d = z^d) = \begin{cases} 1-\lambda & \text{if } Z_i^d = z^d \\ \lambda/(c-1) & \text{otherwise,} \end{cases} \tag{7.19}$$

with smoothing parameter $\lambda \in [0, (c-1)/c]$ and where $\{Z_1^d, \ldots, Z_n^d\}$ is a random sample on Z^d. The range of λ ensures that the kernel function is a pdf with mode at $z^d = Z^d$. For $\lambda = 0$, $l(Z_i^d = z^d)$ becomes the indicator function and $\widehat{p}(z^d)$ simply estimates the probability function by the corresponding relative frequencies, while for λ assuming its upper bound, we have $l(Z_i^d = z^d) = 1/c$, therefore $\widehat{p}(z^d)$ becomes the uniform discrete distribution, whatever the data. This feature is very appealing since for a uniform distribution, $l(Z_i^d = z^d)$ is invariant with respect to the data and seems reasonable to smooth out the corresponding component, when detecting the irrelevant components of the conditioning variable. One possible shortcoming of the previous kernel function could be that it depends on the support of the data which is required to be known. This may be avoided by using the following alternative:

$$\tilde{l}(Z_i^d = z^d) = \begin{cases} 1 & \text{if } Z_i^d = z^d \\ \lambda & \text{otherwise,} \end{cases} \tag{7.20}$$

where $\lambda \in [0, 1]$. We note again that the indicator function and the uniform weight function are obtained as particular cases for $\lambda = 0$ and $\lambda = 1$, respectively. However the above kernel function does not sum to 1. In order to avoid this problem, the resulting probability estimator must be normalized accordingly. For practical situations which also involve ordinal variables, particular weight functions able to reflect the ordered status are needed, as for example near-neighbor weights. Aitchison and Aitken (1976) suggested the following kernel function:

$$k(Z_i^d = z^d) = \binom{c}{j}\lambda^j(1-\lambda)^{c-1}, \text{when } |Z_i^d - z^d| = j, \text{ for } j \in \{0, \ldots, c-1\}. \tag{7.21}$$

This weight function sums to 1, but for any $c \geq 3$ there is no value of λ such that the weights become constant. Wang and Van Ryzin (1981) proposed the following ordered categorical kernel:

$$\tilde{k}(Z_i^d = z^d) = \begin{cases} 1-\lambda & \text{if } |Z_i^d - z^d| = 0 \\ \dfrac{1-\lambda}{2}\lambda^{|Z_i^d - z^d|} & \text{if } |Z_i^d - z^d| \geq 1. \end{cases} \tag{7.22}$$

For alternative kernels for discrete data, or mixed continuous and discrete, see Ouyang et al. (2006), Li and Racine (2008), and Li et al. (2009).

Using generalized product kernels, one can easily extend this approach to the multivariate setting, allowing for mixed discrete and continuous variables, as well as for ordered discrete variables, by replacing the kernel function with an appropriate one. In determining the optimal bandwidths, the use of generalized kernels requires more computational effort, but has the advantage of providing asymptotically optimal smoothing for the relevant components of Z, while eliminating irrelevant components by oversmoothing (the bandwidths for the continuous irrelevant components converge to infinity and those for the discrete attain their upper extreme values). Therefore we determine an optimal bandwidth for each component of Z and detect in the same time irrelevant components of Z.

We develop below the kernel approach for mixed continuous and discrete data in the conditional frontier framework. For the discrete components of the environmental variable Z, we use the kernel proposed by Aitchison and Aitken (1976) and defined in (7.19); our presentation follows Hall et al. (2004).

Consider the vector $Z_i = (Z_i^c, Z_i^d)$ of continuous and discrete components of each observation, where $Z_i^c = (Z_{i1}^c, \ldots, Z_{ir}^c)$ and $Z_i^d = (Z_{i1}^d, \ldots, Z_{is}^d)$, and assume that Z_{ij}^d takes the values $0, 1, \ldots, c_j - 1$. Now define the product kernel

$$K_h(Z_i^c, z^c) = \prod_{j=1}^{r} \frac{1}{h_j} K\left(\frac{Z_{ij}^c - z_j^c}{h_j}\right), \tag{7.23}$$

where $z^c = (z_1^c, \ldots, z_r^c)$, K is a traditional kernel function, and h_j represents the bandwidth for the continuous component Z_{ij}^c with $0 < h_j < \infty$ and $i = 1, \ldots, n$. For the discrete components, consider

$$K_\lambda(Z_i^d, z^d) = \prod_{j=1}^{s} \left(\frac{\lambda_j}{c_j - 1}\right)^{N_{ij}} (1-\lambda_j)^{1-N_{ij}}, \tag{7.24}$$

where $z^d = (z_1^d, \ldots, z_s^d)$, $N_{ij} = I\!I(Z_{ij}^d \neq z_j^d)$ and $\lambda_1, \ldots, \lambda_s$ are the smoothing parameters for the discrete components with $0 \leq \lambda_j \leq (c_j - 1)/c_j$.

At the end, the kernel used for estimation will be the product of the kernels defined in (7.23) and (7.24), i.e.,

$$K(Z_i, z) = K_h(Z_i^c, z^c) K_\lambda(Z_i^d, z^d). \tag{7.25}$$

7.4.2 Optimal Bandwidth Selection

The least squares cross-validation (LSCV) method is employed to find the optimal values for h_y, h and λ. In terms of conditional efficiency estimates, we are mainly interested in finding the optimal values for $(h,\lambda) = (h_1,\ldots,h_r,\lambda_1,\ldots,\lambda_s)$, since we will not use h_y in estimating the conditional survival $S_{Y|X,Z}(y|x,z)$.

The criterion is based on a weighted integrated squared error ($WISE$). We have

$$WISE = \sum_{z^d}\int \{\widehat{g}(y|X\le x, Z=z) - g(y|X\le x, Z=z)\}^2 \times g_1(X\le x,z)w(z^c)dz^c dy, \quad (7.26)$$

where the integral is over (y,z^c) and the sum is taken over all atoms of the distribution of Z^d. Here, $w(z^c)dz^c$ has the role to avoid dividing by 0, or by numbers close to 0, in the ratio $\widehat{f}(y,X\le x,z)/\widehat{g}_1(X\le x,z)$ in (7.16). By straightforward developments, it can be seen that the part of $WISE$ that depends on the bandwidths (h_y,h,λ) can be expressed as $I_{1n}-2I_{2n}$, where

$$I_{1n} = \sum_{z^d}\int \widehat{g}^2(y|X\le x, Z=z)g_1^c(X\le x, z^c|z^d)\text{Prob}(Z^d=z^d)w(z^c)dz^c dy \quad (7.27)$$

and

$$I_{2n} = \sum_{z^d}\int \widehat{g}(y|X\le x, Z=z)f(y,X\le x,z)w(z^c)dz^c dy. \quad (7.28)$$

With the notation $\widehat{G}(x,z)=\int \widehat{f}^2(y,X\le x,z)dy$, the integrals I_{1n} and I_{2n} become:

$$I_{1n} = \sum_{z^d}\int \widehat{G}(x,z)\frac{g_1^c(X\le x, z^c|z^d)\text{Prob}(Z^d=z^d)}{\widehat{g}_1^2(X\le x,z)}w(z^c)dz^c \quad (7.29)$$

and

$$I_{2n} = \sum_{z^d}\int \widehat{g}(y|X\le x, Z=z)f(y,X\le x,z)w(z^c)dz^c. \quad (7.30)$$

Moreover, since $\widehat{G}(x,z)$ can be expressed as

$$\widehat{G}(x,z) = \frac{1}{n^2}\sum_{i_1=1}^{n}\sum_{i_2=1}^{n} K_h(Z_{i_1},z^c)K_h(Z_{i_2},z^c)K_\lambda(Z_{i_1},z^d)K_\lambda(Z_{i_2},z^d) \times \mathbb{I}(X_{i_1}\le x)\mathbb{I}(X_{i_2}\le x)\int L_{h_y}(Y_{i_1},y)L_{h_y}(Y_{i_2},y)dy, \quad (7.31)$$

we obtain the following cross-validation approximations for I_{1n} and I_{2n}:

$$\widehat{I}_{1n} = \frac{1}{n}\sum_{i=1}^{n}\frac{\widehat{G}_{(i)}(x, Z_i)I\!I(X_i \leq x)w(Z_i^c)}{\widehat{g_1}^2_{(i)}(X \leq x, Z_i)} \tag{7.32}$$

and

$$\widehat{I}_{2n} = \frac{1}{n}\sum_{i=1}^{n}\frac{\widehat{f}_{(i)}(Y_i, X \leq x, Z_i)I\!I(X_i \leq x)w(Z_i^c)}{\widehat{g_1}_{(i)}(X \leq x, Z_i)}, \tag{7.33}$$

where by the subscript (i) we indicate that the corresponding function of the data was calculated by the leave-one-out rule, namely by deleting the ith observation (X_i, Y_i, Z_i) from the sample $\{(X_1, Y_1, Z_1), \ldots, (X_n, Y_n, Z_n)\}$.

The optimal bandwidth (h_y, h, λ) is then obtained by minimizing the cross-validation criterion CV given by

$$CV(h_y, h, \lambda) = \widehat{I}_{1n} - 2\widehat{I}_{2n}. \tag{7.34}$$

In the case of two or more local minima, in order to avoid using too small bandwidths, we select the second smallest of these resulting points, as recommended in Hall et al. (2004).

Hall et al. (2004) and Li and Racine (2008) prove that the LSCV criterion provides optimal bandwidths for the relevant components of Z, while for those irrelevant, the corresponding elements of (h, λ) assume their upper extreme values, and consequently, the respective components of Z are smoothed out. In practice, this results in appropriate smoothing for the relevant components of Z and over-smoothing for the potential irrelevant components.

Finally, we have to correct the resulting bandwidths by an appropriate scaling factor, since the method provides the optimal bandwidths for estimating the conditional pdf, while we are considering the estimation of the conditional cdf (survivor) (for details regarding the correction, see Li and Racine 2007). The scaling factor in our conditional frontier framework is $n^{-q/((q+t+4)(t+4))}$ where q is the dimension of Y and t is the dimension of Z.

These bandwidths can be successfully used to estimate all the conditional measures of efficiency discussed in Sect. 7.3: full, order-m and order-α.

7.5 Improving the Detection of the Impact of Z by Bootstrapping

Daraio and Simar (2005) proposed an effective methodology for detecting the impact (positive, negative, neutral or variable) of the external factors on the performance of the production units. The method is based on analyzing the shape of a nonparametric regression curve of $Q^z = \widehat{\lambda}_n(x, y|z)/\widehat{\lambda}_n(x, y)$, the ratio of the conditional to the unconditional efficiency scores, as a function of the conditioning factor Z. An increasing regression in the input orientation case suggests that Z is an unfavorable variable with negative effect on the production process, which may be interpreted as an undesired output, while a decreasing regression suggests that Z is a favorable variable, having a positive effect on the production process, and may be interpreted as a freely available input. Conversely, for the output orientation, an increasing regression indicates that Z is a favorable variable with a positive effect on the production process, while a decreasing regression suggests that Z is an unfavorable variable, having a negative effect on the production process. This approach provides satisfactory results not only for the case of univariate Z but also for the multivariate case that needs more careful analysis in order to recover the marginal impact of the components of Z to the production process. As extensively explained and illustrated in Daraio and Simar (2007) the investigation of the interaction effects may be very useful in this complex framework. The next subsection and the following numerical examples illustrate this methodology.

7.5.1 A Heterogeneous Bootstrap on Q^z

In this section we propose a bootstrap based procedure to test the impact of some external-environmental factors on the production process, i.e., to test whether the impact is *real*, or merely due to sampling variation. Our aim is to investigate the sampling variation of the surface of Q^z against Z_1 and Z_2. In this multivariate setting, we propose a heterogeneous bootstrap approach which avoids the *strict* assumption of the homogeneous bootstrap. Our approach allows to estimate pointwise confidence intervals on the estimates of the smoothed nonparametric regression of Q^z against Z. It is based on the subsampling approach as suggested by Kneip et al. (2008) and adapted to this complex framework a data driven procedure suggested by Simar and Wilson (2009) for selecting the appropriate size of the subsamples.

The subsampling bootstrap (drawing $\tilde{n} < n$ observations out of the n) with replacement or without replacement have the same behavior with $\tilde{n} \to \infty$ and $\frac{\tilde{n}}{n} \to 0$ when $n \to \infty$. In this paper we apply the subsampling without replacement.

As smoothed nonparametric regression we use the local linear technique (Fan and Gijbels 1996; Li and Racine 2007). We estimate the pointwise derivatives of Q^z against Z. In the case of multivariate Z we estimate the marginal derivatives of Q^z against each component of Z using the same procedure to detect interaction effects described in Daraio and Simar (2007a). Pointwise derivatives capture the varying response coefficients across the values of z, while the average derivatives provide constant (fixed) response coefficients. The pointwise derivative captures the local behavior of the shape of the regression function whereas the average derivative gives an indication on the global behavior.

As also pointed out by Pagan and Ullah (1999, pp. 172–173), an important statistical advantage of an average derivative estimator is its $\sqrt{n}$ consistency and asymptotic normality, which is the usual rate of convergence of parametric estimators. Moreover, the asymptotic rate of convergence is not affected neither by the dimension t of Z, nor by the bandwidth h. On the contrary, pointwise derivative estimators which are based on the data in the neighborhood of a point, have less than $\sqrt{n}$ rate of convergence and this rate depends on h, becoming worse with an increasing t. The implication is that for statistical inference, one may need much smaller samples with average derivative estimators compared to pointwise estimators. Despite this disadvantage, pointwise estimates provide useful detailed information able to characterize the behavior of variables that have a changing behavior not normally distributed as it is usually the case e.g. for mutual funds data that we will illustrate in the following.

We first fix a grid of values for Z_1 and Z_2 and study the sampling variation of the estimator of Q^z on this grid. Then we make a local linear fit on this grid. The bootstrap is done by subsampling on $\mathcal{X} = \{(X_i, Y_i, Z_i) \mid i = 1, \ldots, n\}$, keeping the optimal bandwidth h estimated on the full sample using the method described in the previous sections.

We compute basic bootstrap confidence intervals on the statistics of interest (the ratios Q^z and its derivatives) scaled by their appropriate scaling factors that are:

$$\left(\frac{\tilde{n}}{n}\right)^{\frac{4}{(4+t)(p+q)}} \left(\frac{\tilde{n}}{n}\right)^{\frac{2}{5}}$$

for the estimator of the ratios Q^z and

$$\left(\frac{\tilde{n}}{n}\right)^{\frac{4}{(4+t)(p+q)}} \left(\frac{\tilde{n}}{n}\right)^{\frac{1}{5}}$$

for its derivatives, where $\tilde{n}$ is the subsample size and n is the sample size. The exponent of $\frac{\tilde{n}}{n}$ depends on the rates of convergence (see Simar and Wilson 2008 for more details).

7.5.2 *Simulated Examples with Univariate and Bivariate Z*

For the simulated examples we consider the same convex model as in Simar (2007), Daraio and Simar (2007a) and Bădin et al. (2010), with $p = q = 2$ and additive output. The efficient frontier is defined by:

$$y^{(2)} = 1.0845(x^{(1)})^{0.3}(x^{(2)})^{0.4} - y^{(1)} \tag{7.35}$$

where $y^{(1)}$, $y^{(2)}$, and $x^{(1)}$, $x^{(2)}$ are the components of y and x, respectively. The input and output variables are independent, $X_i^{(j)} \sim U(1, 2)$ and $\tilde{Y}_i^{(j)} \sim U(0.2, 5)$ for $j = 1, 2$. The output efficient frontier is characterized by:

$$Y_{i,eff}^{(1)} = \frac{1.0845(X_i^{(1)})^{0.3}(X_i^{(2)})^{0.4}}{S_i + 1} \tag{7.36}$$

$$Y_{i,eff}^{(2)} = 1.0845(X_i^{(1)})^{0.3}(X_i^{(2)})^{0.4} - Y_{i,eff}^{(1)}. \tag{7.37}$$

where $S_i = \tilde{Y}_i^{(2)}/\tilde{Y}_i^{(1)}$ represent the slopes of the random rays in the output space for $j = 1, 2$.

The efficiencies are then simulated according to $U_i \sim Exp(1/3)$ in the case of univariate Z and $U_i \sim Exp(1/2)$ in the multivariate case and the output variables are defined by $Y_i = Y_{i,eff} * \exp(-U_i)$.

7.5.2.1 Univariate Z

In this first example we introduce an environmental factor Z generated from the Uniform distribution, $Z \sim U(1,4)$, having a negative impact on the production process till a value of 2.5 and with a positive impact above this value.

We simulate a sample of $n = 100$ observations according to the following scenario:

$$Y_i^{(1)} = [1 + (Z - 2.5)^2] * Y_{i,eff}^{(1)} * \exp(-U_i) \tag{7.38}$$

$$Y_i^{(2)} = (1 + |Z - 2.5|) * Y_{i,eff}^{(2)} * \exp(-U_i). \tag{7.39}$$

Table 7.1 presents the various measures of efficiency computed on this simulated data set for 10 randomly selected units.

The robust measures of efficiency were computed using $m = 10$ and $\alpha = 0.95$. In practice, the values for the *tuning* parameters m and α are controlled by their

Table 7.1 Results for 10 random units from the simulated data in the case of univariate Z. N is the number of observations dominating the corresponding unit and N_Z, the number points used for estimation given $Z = z$

Units	N	$\hat{\lambda}_n(x,y)$	$\hat{\lambda}_{n,\alpha}(x,y)$	$\hat{\lambda}_{n,m}(x,y)$	h	N_z	$\hat{\lambda}_n(x,y\|z)$	$\hat{\lambda}_{n,\alpha}(x,y\|z)$	$\hat{\lambda}_{n,m}(x,y\|z)$
42	42	3.3069	3.1244	3.1196	0.4779	20	1.7369	1.4455	1.7369
19	0	1.0000	0.7973	0.8908	0.4964	9	1.0000	1.0000	1.0000
71	1	1.6448	1.6448	1.6448	1.2452	2	1.0000	1.0000	1.0000
3	4	1.2354	1.1248	1.2001	0.5691	3	1.1243	1.1243	1.1243
32	3	1.4312	1.2282	1.3217	0.6142	12	1.0000	1.0000	1.0000
35	20	1.7801	1.5273	1.6283	0.5104	24	1.0530	1.0000	1.0530
79	11	1.6905	1.4427	1.5548	0.4825	13	1.0000	1.0000	1.0000
35	20	1.7801	1.5273	1.6283	0.5104	24	1.0530	1.0000	1.0530
53	7	1.4792	1.2710	1.3463	0.5336	16	1.0000	1.0000	1.0000
39	1	1.1691	1.0000	1.1084	0.4835	7	1.1691	1.1691	1.1691
Mean	5.5	1.4820	1.3194	1.3802	1.1327	8.8	1.1441	1.1194	1.1441

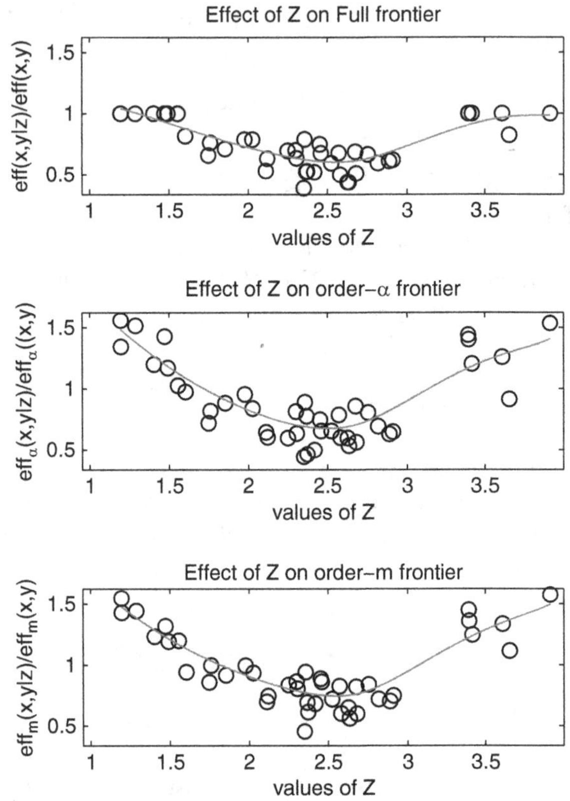

Fig. 7.1 Simulated example with univariate Z. Smoothed nonparametric regression of $\widehat{\lambda}_n(x, y|z)/\widehat{\lambda}_n(x, y)$ on Z (*top panel*), of $\widehat{\lambda}_{n,\alpha}(x, y|z)/\widehat{\lambda}_{n,\alpha}(x, y)$ on Z (*middle panel*) and of $\widehat{\lambda}_{n,m}(x, y|z)/\widehat{\lambda}_{n,m}(x, y)$ on Z (*bottom panel*)

economic interpretation: a benchmark against the best m virtual competitors or against a level of production with a probability $(1-\alpha) \times 100\%$ of being dominated.

For the nonparametric regression, we used a truncated gaussian kernel, but we note that the results remain stable when other kernels with compact support are used. The results are displayed in Fig. 7.1. The three panels (top for the FDH case, middle for the α frontier case and bottom for the m frontier) depict the ratios of conditional to unconditional FDH, α and m efficiency scores respectively, allowing to detect the "$U-$ shape" effect of Z on the production process, as expected.

7.5.2.2 Bivariate Z

We consider now a bivariate external factor $Z = (Z_1, Z_2)$ with two independent components, $Z_j \sim U(1,4)$, $j = 1,2$. The impact of Z on the production process is captured as follows:

$$Y_i^{(1)} = (1 + 2 * |Z_1 - 2.5|^3) * Y_{i,eff}^{(1)} * \exp(-U_i)$$

$$Y_i^{(2)} = (1 + 2 * |Z_1 - 2.5|^3) * Y_{i,eff}^{(2)} * \exp(-U_i).$$

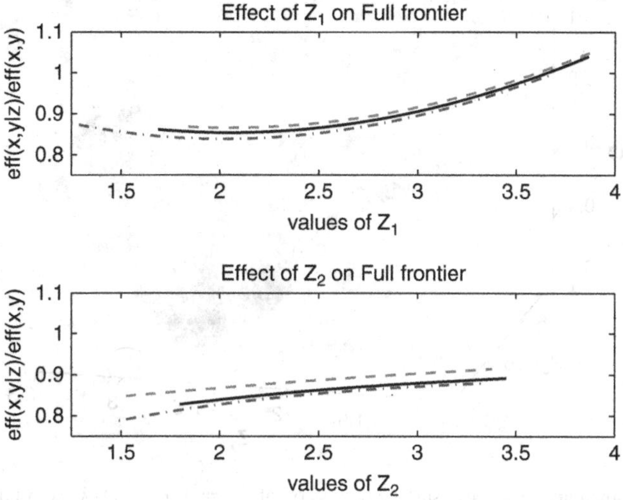

Fig. 7.2 Simulated example with multivariate Z. *Top panel*: smoothed nonparametric regression of $\widehat{\lambda}_n(x,y|z)/\widehat{\lambda}_n(x,y)$ on Z_1 for Z_2's quartiles. *Bottom panel*: smoothed nonparametric regression of $\widehat{\lambda}_n(x,y|z)/\widehat{\lambda}_n(x,y)$ on Z_2 for Z_1's quartiles. The *dashed line* corresponds to the first quartile, the *solid line* to the median and the *dashdot line* to the third quartile

We simulate $n = 100$ observations according to this scenario and we use a multiplicative quartic kernel with a vector of bandwidths $h = (h_1, \ldots, h_r)$ for Z and for Y, a product gaussian kernel with $h_y = h_0 s_y$ where h_0 is an univariate bandwidth and s_y is the vector of empirical standard deviations of Y.

For a more detailed information on the impact of Z on the simulated production process, Fig. 7.2 plots the ratios $\widehat{\lambda}_n(x,y|z)/\widehat{\lambda}_n(x,y)$ against Z_1 (top panel) and Z_2 (bottom panel) at the three quartiles (first, median and third) of the other component of Z. As we expected, we recover a cubic effect of Z_1 and a neutral effect of Z_2. The same effects were obtained also for $\widehat{\lambda}_{n,\alpha}(x,y|z)/\widehat{\lambda}_{n,\alpha}(x,y)$ and for $\widehat{\lambda}_{n,m}(x,y|z)/\widehat{\lambda}_{n,m}(x,y)$ (here we do not have outliers). Here we see that Z_1 has a cubic effect, while Z_2 has no effect on the production process (Fig. 7.3).

The marginal effects can better be viewed in Fig. 7.4 which shows the surface regression evaluated at the observations (X_i, Y_i, Z_i) but represented marginally, as function of each component Z_1 and Z_2 separately. The lines in the pictures represents local average lines of the points to stress the global effect. The cubic effect for Z_1 and the neutral effect for Z_2 are again explicit.

To improve the analysis of the impact of Z we also construct basic bootstrap confidence intervals for Q^z and its derivative and plot the corresponding pointwise intervals against Z_1 and Z_2. For this example we selected $\tilde{n} = 84$ (out of $n = 100$ observations) based on Simar and Wilson (2009). Looking at Fig. 7.5, bottom panel we see that the derivative is negative up to $Z_1 = 2.5$ and positive after this value, suggesting (as also depicted in the upper panel) the shape of Q^z (Fig. 7.6).

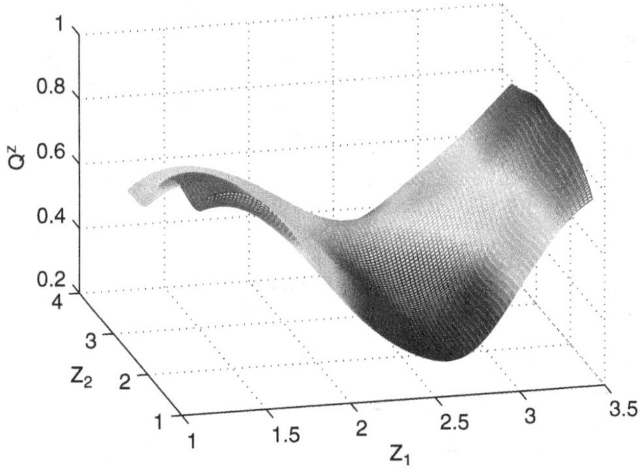

Fig. 7.3 Smoothed nonparametric surface regression of $Q^z = \widehat{\lambda}_n(x, y|z)/\widehat{\lambda}_n(x, y)$ on Z_1 and Z_2. Adapted from Bădin et al. (2010)

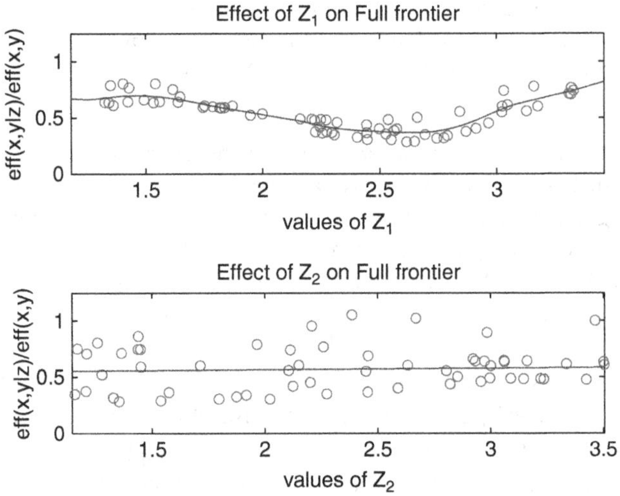

Fig. 7.4 Marginal view of the surface regression of $Q^z = \widehat{\lambda}_n(x, y|z)/\widehat{\lambda}_n(x, y)$ on Z at the observed points (X_i, Y_i, Z_i) viewed as a function of Z_1 (*top panel*) and relative derivatives (*bottom panel*). Adapted from Bădin et al. (2010)

7.5.3 An Illustration on Mutual Funds Data

In this section we apply the heterogeneous bootstrap described in this section to the sample of US Mutual Funds that was previously considered in several papers (see Daraio and Simar 2005, 2006, 2007a; Bădin et al. 2010).

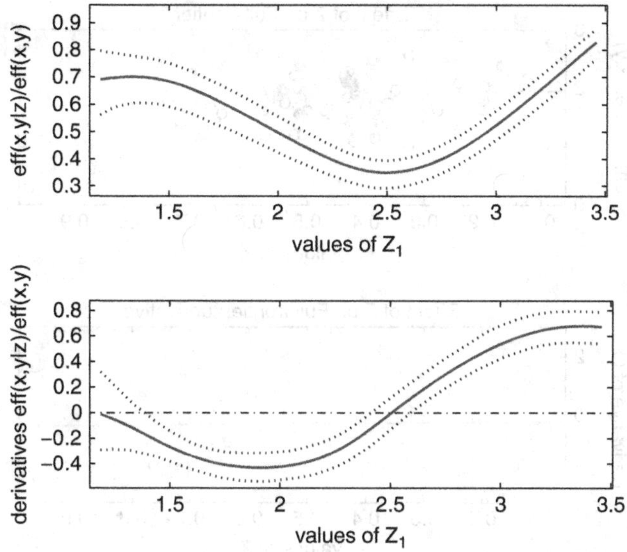

Fig. 7.5 95% BB confidence intervals (*dotted points*) on Q^z (*solid line*) against values of Z_1 (*top panel*) and 95% BB confidence intervals (*dotted points*) on the estimates of the derivatives of Q^z with respect to Z_1 (*solid line*) against values of Z_1 (*bottom panel*)

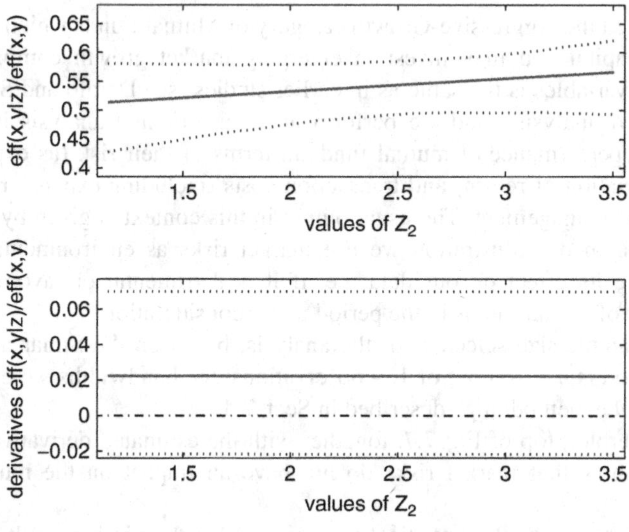

Fig. 7.6 95% BB confidence intervals (*dotted points*) on Q^z (*solid line*) against values of Z_2 (*top panel*) and 95% BB confidence intervals (*dotted points*) on the estimates of the derivatives of Q^z with respect to Z_2 (*solid line*) against values of Z_2 (*bottom panel*)

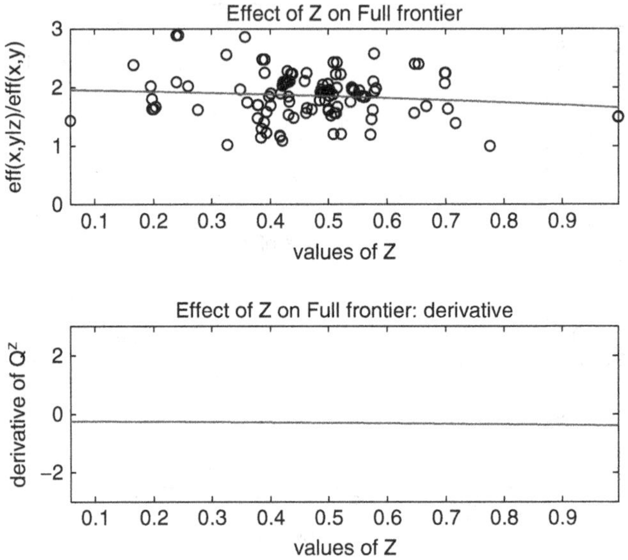

Fig. 7.7 Scatterplot and smoothing nonparametric regression of $Q^z = \widehat{\theta}_n(x, y|z)/\widehat{\theta}_n(x, y)$ on Z (*top panel*) and estimates of the derivatives of Q^z with respect to Z (*bottom panel*)

We selected the Aggressive-Growth category of Mutual Funds which seek rapid growth of capital and may invest in emerging market growth companies. The selection of variables is the same as in earlier studies (see Daraio and Simar 2006 for a detailed analysis), and we perform an input oriented analysis in order to evaluate the performance of mutual funds in terms of their risk (as expressed by standard deviation of return) and transaction costs (including expense ratio, loads and turnover) management. The usual output in this context is given by the return of the funds. In our illustration we use market risks as environmental variable to investigate its effect on our data, i.e. if it is detrimental or favorable to the performance of mutual funds in the period under consideration.

The subsample size selected for the analysis, based on the Simar and Wilson (2009) method is $\tilde{n} = 68$ out of 129 observations; the bandwidths were computed by applying the methodology described in Sect. 7.4.

The scatterplot, top of Fig. 7.7, together with the estimated derivatives, bottom of Fig. 7.7 show that market risks do not have an impact on the mutual funds performance.

Moreover, to assess the statistical meaning of this descriptive result we should look at Fig. 7.8 that shows the basic bootstrap confidence intervals on the estimated Q^z (top panel) as well as the basic bootstrap confidence intervals on the estimated derivatives of Q^z (bottom panel). It appears that there is not a statistical impact of market risks on the performance of mutual funds because the ratios Q^z are flat, the derivatives are constant and the zero is always contained in the pointwise confidence intervals.

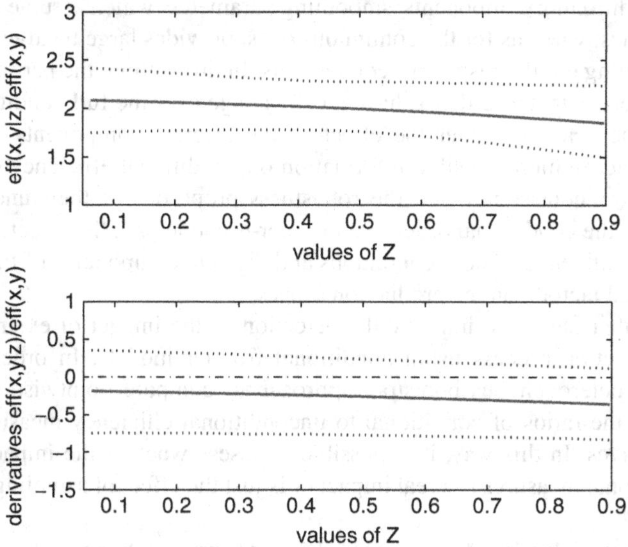

Fig. 7.8 95% BB confidence intervals (*dotted points*) on Q^z (*solid line*) against values of Z (*top panel*) and 95% BB confidence intervals (*dotted points*) on the estimates of the derivatives of Q^z with respect to Z (*solid line*) against values of Z (*bottom panel*)

7.6 Conclusions

The measurement of the technical efficiency of the decision making units is indeed informative as such for making comparisons and informing managers and policy makers on existing differentials and potential improvements across a sample of analyzed units. The step further in the analysis of technical efficiency is to relate the obtained efficiency estimates to some external or environmental variables which may affect the performance evaluation and may help explaining the obtained efficiency differentials. The recently introduced conditional measures of efficiency of full frontiers (FDH, DEA) and of partial frontiers (order-m and order-α) represent attractive fully nonparametric tools to evaluate the efficiency and explain the influence of external factors on the production process.

In this paper we extend the results on nonparametric conditional approach on two directions. On the one hand we focus on bandwidth selection for conditional efficiency estimation when the external factors have both, continuous and discrete components. Based on Bădin et al. (2010) approach for selecting the optimal bandwidth (defined with respect to an integrated square error criterion) through a data-driven method, we illustrate how to handle and implement the estimation in the presence of both categorical and continuous external factors. The approach has the appealing feature of detecting the irrelevant factors, as those conditional components of Z orthogonal to the dependent variable. Based on LSCV, the method automatically detects the irrelevant components, by assigning to those irrelevant discrete conditioning components, smoothing parameters which assume their upper extreme values, whereas for the continuous ones, provides large bandwidth parameters, smoothing out the respective components. In this setting, the partial frontiers, conditional order-m and order-α, have an advantage over the full conditional FDH estimator: they capture better the effect of the discrete components of external factors, providing more sensible information on conditional efficiency estimators. Beside the $\sqrt{n}$-consistency and the robustness properties of their unconditional counterparts, the conditional order-m and order-α encompass the effect, being able to detect the influence of both continuous and discrete components of the external, environmental factors, on the production process.

On the other hand, we improve the detection of the impact of external factors on the production process using conditional frontier models. In order to do so we apply a heterogeneous bootstrap approach to compute pointwise confidence intervals on the ratios of conditional to unconditional efficiency measures and of their derivatives. In this way, it is possible to assess whether the impact detected with conditional measures is a real impact or is just the effect of sampling variation.

Acknowledgements Previous discussions with Léopold Simar as well as his valuable comments and suggestions contributed a lot in improving the quality of the present paper and are gratefully acknowledged. The usual disclaimers apply.

References

Aitchison, J., & Aitken, C.G.G. (1976). Multivariate binary discrimination by the kernel method. *Biometrika*, *63*(3), 413–420.

Banker, R.D., & Morey, R.C. (1986). Efficiency analysis for exogenously fixed inputs and outputs. *Operations Research*, *34*(4), 513–521.

Bădin, L., Daraio, C., & Simar, L. (2010). Optimal bandwidth selection for conditional efficiency measures: a data-driven approach. *European Journal of Operational Research*, *201*(2), 633–640.

Cazals, C., Florens, J.P., & Simar, L. (2002). Nonparametric frontier estimation: a robust approach. *Journal of Econometrics*, *106*, 1–25.

Charnes, A., Cooper, W.W., & Rhodes, E. (1978). Measuring the efficiency of decision making units. *European Journal of Operational Research*, *2*, 429–444.

Cooper, W.W., Seiford, L.M., & Tone, K. (2000). *Data envelopment analysis: A comprehensive text with models, applications, references and DEA-solver software*. Boston: Kluwer Academic Publishers.

Daouia, A., & Simar, L. (2007). Nonparametric efficiency analysis: a multivariate conditional quantile approach. *Journal of Econometrics*, *140*, 375–400.

Daraio, C., & Simar, L. (2005). Introducing environmental variables in nonparametric frontier models: a probabilistic approach. *The Journal of Productivity Analysis*, *24*, 93–121.

Daraio, C., & Simar, L. (2006). A robust nonparametric approach to evaluate and explain the performance of mutual funds. *European Journal of Operational Research*, *175*(1), 516–542.

Daraio, C., & Simar, L. (2007a). *Advanced Robust and Nonparametric Methods in Efficiency Analysis. Methodology and applications*. New York: Springer.

Daraio, C., & Simar, L. (2007b). Conditional nonparametric Frontier models for convex and non convex technologies: A unifying approach. *Journal of Productivity Analysis*, *28*, 13–32.

Daraio, C., Simar, L., & Wilson, P. (2010). Testing whether two-stage estimation is meaningful in nonparametric models of production, Discussion Paper #1030, Institut de Statistique, Université Catholique de Louvain, Louvain-la-Neuve, Belgium.

Deprins, D., Simar, L., & Tulkens, H. (1984). Measuring labor-efficiency in post offices. In M. Marchand, P. Pestieau, & H. Tulkens (eds.) *The Performance of public enterprises – Concepts and Measurement* (pp. 243–267). Amsterdam: North-Holland.

Fan, J., & Gijbels, I. (1996). *Local polinomial modelling and its applications*. London: Chapman and Hall.

Farrell, M.J. (1957). The measurement of the Productive Efficiency. *Journal of the Royal Statistical Society*, Series A, CXX, Part 3, 253–290.

Färe, R., Grosskopf, S., & Lovell, C.A.K. (1994). *Production frontiers*. Cambridge: Cambridge University Press.

Gattoufi, S., Oral, M., Reisman, A. (2004). Data Envelopment Analysis literature: a bibliography update (1951–2001). *Socio-Economic Planning Sciences*, *38*, 159–229.

Hall, P., Racine, J.S., Li, Q. (2004). Cross-validation and the estimation of conditional probability densities. *Journal of the American Statistical Association*, *99*(486), 1015–1026.

Jeong, S.O., Park, B.U., & Simar, L. (2008). Nonparametric conditional efficiency measures: asymptotic properties. *Annals of Operations Research*, doi: 10.1007/s10479-008-0359-5.

Li, Q., & Racine, J. (2007). *Nonparametric econometrics: theory and practice*. New Jersey: Princeton University Press.

Li, Q., & Racine, J. (2008). Nonparametric estimation of conditional CDF and quantile functions with mixed categorical and continuous data. *Journal of Business and Economic Statistics*, *26*(4), 423–434.

Li, Q., Racine, J., Wooldridge, J.M. (2009). Efficient estimation of average treatment effect with mixed categorical and continuous data. *Journal of Business and Economic Statistics*, *26*(4), 423–434.

Pagan, A., & Ullah, A. (1999). *Nonparametric Econometrics*. Cambridge: Cambridge University Press.

Ouyang, D., Li, Q., & Racine, J. (2006). Cross-validation and the estimation of probability distributions with categorical data. *Nonparametric Statistics*, *18*(1), 69–100.

Park, B., Simar, L., & Weiner, C. (2000). The FDH estimator for productivity efficiency scores: asymptotic properties. *Econometric Theory*, *16*, 855–877.

Park, B., Simar, L., & Zelenyuk, V. (2008). Local likelihood estimation of truncated regression and its partial derivatives: Theory and application. *Journal of Econometrics*, *146*(1), 185–198.

Simar, L., & Wilson, P.W. (2007). Estimation and inference in two-stage, semi-parametric models of production processes. *Journal of Econometrics*, *136*(1), 31–64.

Simar, L., & Wilson, P.W. (2008). Statistical inference in nonparametric frontier models: recent developments and perspectives. In Harold O. Fried, C.A. Knox Lovell, & Shelton S. Schmidt (Eds.), *The Measurement of Productive Efficiency*, 2nd edn. Oxford: Oxford University Press.

Simar, L., & Wilson, P.W. (2009). Inference by Subsampling in Nonparametric Frontier Models, Discussion Paper #0933, Institut de Statistique, Université Catholique de Louvain, Louvain-la-Neuve, Belgium.

Simar, L., & Wilson, P.W. (2010). Two-Stage DEA: Caveat Emptor, Discussion Paper #10xx, Institut de Statistique, Université Catholique de Louvain, Louvain-la-Neuve, Belgium.

Titterington, D.M. (1980). A comparative study of kernel-based density estimates for categorical data. *Technometrics*, *22*(2), 259–268.

Wang, M.C., & Van Ryzin, J. (1981). A class of smooth estimators for discrete distributions. *Biometrika*, *68*, 301–309.

Chapter 8
Estimation of a General Parametric Location in Censored Regression

Cédric Heuchenne and Ingrid Van Keilegom

Abstract Consider the random vector (X, Y), where Y represents a response variable and X an explanatory variable. The response Y is subject to random right censoring, whereas X is completely observed. Let $m(x)$ be a conditional location function of Y given $X = x$. In this paper we assume that $m(\cdot)$ belongs to some parametric class $\mathcal{M} = \{m_\theta : \theta \in \Theta\}$ and we propose a new method for estimating the true unknown value θ_0. The method is based on nonparametric imputation for the censored observations. The consistency and asymptotic normality of the proposed estimator are established.

8.1 Introduction

Consider the random vector (X, Y), where Y represents a (possible transformation of a) response variable and X an explanatory variable. This chapter is concerned with the estimation of a location functional of Y given X, when Y is subject to random right censoring and X is completely observed. We suppose that this location functional belongs to some parametric family.

C. Heuchenne (✉)
QuantOM (Centre for Quantitative Methods and Operations Management), HEC-Management School of University of Liége, Rue Louvrex 14, 4000 Liége, Belgium

Institut de statistique, biostatistique et sciences actuarielles, Université catholique de Louvain, Voie du Roman Pays 20, 1348 Louvain-la-Neuve, Belgium
e-mail: cedric.heuchenne@uclouvain.be

I. Van Keilegom
Institut de statistique, biostatistique et sciences actuarielles, Université catholique de Louvain, Voie du Roman Pays 20, 1348 Louvain-la-Neuve, Belgium
e-mail: ingrid.vankeilegom@uclouvain.be

I. van Keilegom and P.W. Wilson (eds.), *Exploring Research Frontiers in Contemporary Statistics and Econometrics*, DOI 10.1007/978-3-7908-2349-3_8,

This problem has been widely studied in the literature when the location functional is the conditional mean (see e.g. Stute (1993), Fygenson and Zhou (1994), Van keilegom and Akritas (2000) among many others) or the conditional median (see e.g. Buchinsky and Hahn (1998), Portnoy (2003), Yin et al. (2008), Wang and Wang (2009) and the references therein). Here we focus attention on L-functionals, given by

$$m(x) = \int_0^1 F^{-1}(s|x)J(s)ds,$$

where $F(y|x) = P(Y \leq y|X = x)$ is the conditional distribution of Y given $X = x$, $F^{-1}(s|x) = \inf\{y : F(y|x) \geq s\}$ is the conditional quantile of order s, and $J(s)$ is a weight function satisfying $J(s) \geq 0$ for all $0 \leq s \leq 1$ and $\int_0^1 J(s)ds = 1$. This type of location functionals includes as special cases the conditional mean, trimmed mean, or any other kind of weighted mean. The conditional median can be regarded as a limiting special case, obtained when $J(s)$ puts all its mass on $s = 1/2$.

Another interesting special case is obtained for $J(s) = I(1 - \delta < s \leq 1)/\delta$ for some $0 < \delta < 1$. Suppose the upper bound of the support of Y given $X = x$ is finite (say equal to τ_x) and one is interested in the estimation of the support curve $x \rightarrow \tau_x$. The above choice of J yields a robust estimator of this curve for small values of δ, and is an interesting alternative to the so-called m-frontiers or α-frontiers, which are based on order statistics of order m or quantiles of order $0 < \alpha < 1$ (see e.g. Cazals et al. (2002) and Aragon et al. (2005)).

We suppose in this chapter that Y is subject to random right censoring, i.e. instead of observing Y we only observe (Z, Δ), where $Z = \min(Y, C)$ is the observed survival time, $\Delta = I(Y \leq C)$ is the censoring indicator, and the random variable C represents the censoring time, which is independent of Y conditionally on X. Let (X_i, Z_i, Δ_i) $(i = 1, \ldots, n)$ be n independent copies of (X, Z, Δ).

In the context of regression with right censored responses it is well known that the nonparametric kernel estimator of the conditional distribution $F(\cdot|x)$ is inconsistent in the right tail. The modeling of the above location functional $m(x)$ is therefore especially attractive in this framework, since particular choices of J-functions enable us to get rid of these inconsistent parts (see Sect. 8.2 for more details).

The goal of this chapter is to propose a new estimation method for $m(x)$, when it is believed that $m(\cdot)$ belongs to the class

$$\mathcal{M} = \{m_\theta : \theta \in \Theta\},$$

consisting of location functionals determined by a finite-dimensional parameter vector $\theta \in \Theta$, where Θ is a compact subset of $I\!R^d$. The class $\mathcal{M}$ can be taken equal to the class of polynomial functions of order $d - 1$, but any other parametric class of "smooth" functions (in θ) can be chosen as well.

The chapter is organized as follows. In the next section we introduce some notation and explain in detail the proposed estimation procedure for θ_0, the true unknown value of θ. Section 8.3 gives the main asymptotic properties of the proposed estimator. In Sect. 8.4 we summarize the results of the chapter and give ideas for future research, whereas the appendix contains the proofs of the main asymptotic results.

8.2 Description of the Method

The estimator θ_n is defined as follows. First, we note that θ_0 can be written as

$$\begin{aligned}\theta_0 &= \text{argmin}_\theta\, E\Big[\Big(\int_0^1 F^{-1}(s|X)J(s)ds - m_\theta(X)\Big)^2\Big]\\ &= \text{argmin}_\theta\, E\Big[(Y - m_\theta(X))^2 J(F(Y|X))\Big]\\ &= \text{argmin}_\theta\, E\Big[(Z - m_\theta(X))^2 J(F(Z|X))\Delta\\ &\quad + \frac{\int_Z^\infty (y - m_\theta(X))^2 J(F(y|X))\, dF(y|X)}{1 - F(Z|X)}(1-\Delta)\Big].\end{aligned}$$

The idea is now to estimate θ_0 by a minimizer θ_n of an empirical version of the above quantity, namely

$$\begin{aligned}\theta_n &= \text{argmin}_{\theta\in\Theta}\, n^{-1}\sum_{i=1}^n \Big[(Z_i - m_\theta(X_i))^2 J(\hat{F}(Z_i|X_i))\Delta_i\\ &\quad + \frac{\int_{Z_i}^\infty (y - m_\theta(X_i))^2 J(\hat{F}(y|X_i))\, d\hat{F}(y|X_i)}{1 - \hat{F}(Z_i|X_i)}(1-\Delta_i)\Big].\end{aligned} \tag{8.1}$$

Here, $\hat{F}(y|x)$ is the nonparametric kernel estimator of the conditional distribution $F(y|x)$ proposed by Beran (1981):

$$\hat{F}(y|x) = 1 - \prod_{Z_i \le y, \Delta_i = 1} \left\{1 - \frac{W_i(x, a_n)}{\sum_{k=1}^n I(Z_k \ge Z_i) W_k(x, a_n)}\right\}, \tag{8.2}$$

(when no ties are present), where

$$W_i(x, a_n) = \frac{K\left(\frac{x - X_i}{a_n}\right)}{\sum_{k=1}^n K\left(\frac{x - X_k}{a_n}\right)},$$

K is a kernel function and $\{a_n\}$ is a bandwidth sequence.

Hence, the estimation procedure for θ_0 can be summarized as follows:

1. First, for fixed θ, estimate the weighted squared error $(Z_i - m_\theta(X_i))^2 J(F(Z_i|X_i))$ of an uncensored observation $(X_i, Y_i, \Delta_i = 1)$ by $(Y_i - m_\theta(X_i))^2 J(\hat{F}(Y_i|X_i))$,

where $\hat{F}$ is defined in (8.2), and of a censored observation $(X_i, C_i, \Delta_i = 0)$ by a nonparametric estimator of $E[(Y - m_\theta(X))^2 J(\hat{F}(Y|X))|X = X_i, Y > C_i]$.
2. Then, estimate θ_0 by minimizing the average of the weighted squared errors obtained under the previous step.

Although the above idea of estimating θ_0 has never been considered in the present context of nonlinear parametric estimation of a general location functional, similar versions of this idea have been applied in other contexts. See e.g. Akritas (1996), who, in the context of polynomial regression, first replaced all observations Z_i (censored and uncensored ones) by a nonparametric estimator $\widehat{m}(X_i)$ of $m(X_i)$, and then applied a classical least squares procedure on the so-obtained 'synthetic' data $(X_i, \widehat{m}(X_i))$. His method has the disadvantage that it is quite sensitive to the choice of the bandwidth, as the bandwidth is playing an important role for both the censored and the uncensored data. Another related methodology is given in Heuchenne and Van Keilegom (2007), who consider the estimation of the conditional mean of Y given X when the relation between Y and X is given by a nonparametric location-scale model. They also replace the censored observations by some kind of synthetic data estimated under the assumed location-scale model. Also see Pardo-Fernandez et al. (2007) for a goodness-of-fit test in parametric censored regression.

For the presentation of the asymptotic results in the next section, we need to introduce the following notation. Let $H(y|x) = P(Z \le y|X = x)$, $H_\delta(y|x) = P(Z \le y, \Delta = \delta|X = x)$ $(\delta = 0, 1)$, $F_\varepsilon(y|x) = P(\varepsilon \le y|X = x)$ and $F_X(x) = P(X \le x)$. The probability density functions of the above distribution functions will be denoted by lower case letters. Also, let

$$\epsilon_\theta(x, z, \delta, F) = \left\{ \delta(z - m_\theta(x))^2 J(F(z|x)) + (1 - \delta) \frac{\int_z^{+\infty} (y - m_\theta(x))^2 J(F(y|x)) dF(y|x)}{1 - F(z|x)} \right\}$$

(where $E[\epsilon_\theta(X, Z, \Delta, F)] = E[(Y - m_\theta(X))^2 J(F(Y|X))]$), which implies that the estimator θ_n can be written as

$$\theta_n = \operatorname{argmin}_{\theta \in \Theta} \sum_{i=1}^{n} \epsilon_\theta(X_i, Z_i, \Delta_i, \hat{F}),$$

where $\hat{F}$ is the Beran estimator defined in (8.2).

8.3 Asymptotic Results

We start by showing the convergence in probability of θ_n and of the least squares criterion function. This will allow us to develop an asymptotic representation for $\theta_{nj} - \theta_{0j}$ $(j = 1, \ldots, d)$, which in turn will give rise to the asymptotic normality of these estimators. The assumptions used in the results below, as well as the proof of the two first results, are given in the appendix.

Theorem 8.3.1. *Assume (A1), (A2) (i), (A4) (i), (iv), (v) and (A7). Moreover, assume that, J is continuously differentiable, $\int_0^1 J(s)ds = 1$, $J(s) \geq 0$ for all $0 \leq s \leq 1$, F_X is two times continuously differentiable,* $\inf_{x \in R_X} f_X(x) > 0$, *$\Theta$ is compact, θ_0 is an interior point of Θ, and $m_\theta(x)$ is continuous in (x, θ) for all x and θ. Let*

$$S_n(\theta) = \frac{1}{n}\sum_{i=1}^{n} \epsilon_\theta(X_i, Z_i, \Delta_i, \hat{F}).$$

Then

$$\theta_n - \theta_0 = o_P(1),$$

and

$$S_n(\theta_n) = E[\epsilon_{\theta_0}(X, Z, \Delta, F)] + o_P(1).$$

The next result decomposes the difference $\theta_n - \theta_0$ into a sum of i.i.d. terms and a negligible term of lower order. This decomposition will be crucial for obtaining the asymptotic normality of θ_n.

Theorem 8.3.2. *Assume (A1)–(A7). Then,*

$$\theta_n - \theta_0 = \Omega^{-1} n^{-1} \sum_{i=1}^{n} \rho(X_i, Z_i, \Delta_i) + \begin{pmatrix} o_P(n^{-1/2}) \\ \vdots \\ o_P(n^{-1/2}) \end{pmatrix},$$

where $\Omega = (\Omega_{jk})$ $(j, k = 1, \ldots, d)$,

$$\Omega_{jk} = E\left[\frac{\partial m_{\theta_0}(X)}{\partial \theta_j}\frac{\partial m_{\theta_0}(X)}{\partial \theta_k}\right],$$

$\rho = (\rho_1, \ldots, \rho_d)^T$, and for any $j = 1, \ldots, d$ and $i = 1, \ldots, n$,

$$\begin{aligned}\rho_j(X_i, Z_i, \Delta_i) = \frac{\partial m_{\theta_0}(X_i)}{\partial \theta_j} \Big\{ & \Delta_i (Z_i - m_{\theta_0}(X_i)) J(F(Z_i|X_i)) \\ & + (1 - \Delta_i) \frac{\int_{Z_i}^{+\infty} (y - m_{\theta_0}(X_i)) J(F(y|X_i)) dF(y|X_i)}{1 - F(Z_i|X_i)} \Big\} \\ & + f_X(X_i) \sum_{\delta=0,1} \int \chi_j((X_i, z, \delta), (Z_i, \Delta_i)) dH_\delta(z|X_i),\end{aligned}$$

where the function χ_j is defined in the appendix.

We are now ready to state the asymptotic normality of θ_n.

Theorem 8.3.3. *Under the assumptions of Theorem 8.3.2, $n^{1/2}(\theta_n - \theta_0) \xrightarrow{d} N(0, \Sigma)$, where*

$$\Sigma = \Omega^{-1} E[\rho(X, Z, \Delta)\rho^T(X, Z, \Delta)]\Omega^{-1}.$$

The proof of this result follows readily from Theorem 8.3.2.

8.4 Summary and Future Research

In this chapter we have proposed a new method to estimate the coefficients of a parametric conditional location function, when the response is subject to random right censoring. The proposed estimator is a least squares type estimator, for which the censored observations are replaced by nonparametrically imputed values. The consistency and asymptotic normality of the estimator are established.

In the future, it would be interesting to compare the proposed method with other estimators that have been proposed in the literature, for instance, when the conditional location is the conditional mean. The least squares estimators obtained in this chapter can be introduced in a test statistic to test the validity of the assumed parametric model, and it would be interesting to work out the asymptotic theory for that test statistic. Finally, extensions of the current work to semiparametric models (like the partial linear or single index model) can also be worked out based on the results in this chapter.

Appendix

We first introduce the following functions, which are needed in the statement of the asymptotic results given in Sect. 8.3:

$$\xi(z, \delta, y|x) = (1 - F(y|x)) \left\{ -\int_{-\infty}^{y \wedge z} \frac{dH_1(s|x)}{(1 - H(s|x))^2} + \frac{I(z \le y, \delta = 1)}{1 - H(z|x)} \right\},$$

$$\begin{aligned}
\chi_j(v^1, z^2, \delta^2) = \frac{\partial m_{\theta_0}(x^1)}{\partial \theta_j} \Bigg\{ \Bigg[& \delta^1 (z^1 - m_{\theta_0}(x^1)) J'(F(z^1|x^1)) \\
& + (1-\delta^1) \frac{\int_{z^1}^{+\infty} (y - m_{\theta_0}(x^1)) J(F(y|x^1)) dF(y|x^1)}{(1 - F(z^1|x^1))^2} \Bigg] \xi(z^2, \delta^2, z^1|x^1) \\
& + (1 - \delta^1) \Bigg[\frac{\int_{z^1}^{+\infty} (y - m_{\theta_0}(x^1)) J(F(y|x^1)) d\xi(z^2, \delta^2, y|x^1)}{1 - F(z^1|x^1)} \\
& + \frac{\int_{z^1}^{+\infty} (y - m_{\theta_0}(x^1)) J'(F(y|x^1)) \xi(z^2, \delta^2, y|x^1) dF(y|x^1)}{1 - F(z^1|x^1)} \Bigg] \Bigg\},
\end{aligned}$$

$j = 1, \ldots, d$, where $v^1 = (x^1, z^1, \delta^1)$.

Let T_x be any value less than the upper bound of the support of $H(\cdot|x)$ such that $\inf_{x \in R_X}(1 - H(T_x|x)) > 0$. For a (sub)distribution function $L(y|x)$ we will use the notations $l(y|x) = L'(y|x) = (\partial/\partial y)L(y|x)$, $\dot{L}(y|x) = (\partial/\partial x)L(y|x)$ and similar notations will be used for higher order derivatives.

The assumptions needed for the results of Sect. 8.3 are listed below.

($A1$) (i) $na_n^3(\log n)^{-3} \to \infty$ and $na_n^4 \to 0$.
(ii) The support R_X of X is a compact interval.
(iii) K is a density with compact support, $\int uK(u)du = 0$ and K is twice continuously differentiable.
(iv) Ω is non-singular.

($A2$) (i) There exist $0 \le s_0 \le s_1 \le 1$ such that $s_1 \le \inf_x F(T_x|x)$, $s_0 \le \inf\{s \in [0,1]; J(s) \neq 0\}$, $s_1 \ge \sup\{s \in [0,1]; J(s) \neq 0\}$ and $\inf_{x \in R_X} \inf_{s_0 \le s \le s_1} f(F^{-1}(s|x)|x) > 0$.
(ii) J is three times continuously differentiable, $\int_0^1 J(s)ds = 1$ and $J(s) \ge 0$ for all $0 \le s \le 1$.

($A3$) F_X is three times continuously differentiable and $\inf_{x \in R_X} f_X(x) > 0$.

($A4$) (i) $L(y|x)$ is continuous,
(ii) $L'(y|x) = l(y|x)$ exists, is continuous in (x, y) and $\sup_{x,y} |yL'(y|x)| < \infty$,
(iii) $L''(y|x)$ exists, is continuous in (x, y) and $\sup_{x,y} |y^2 L''(y|x)| < \infty$,
(iv) $\dot{L}(y|x)$ exists, is continuous in (x, y) and $\sup_{x,y} |y\dot{L}(y|x)| < \infty$,
(v) $\ddot{L}(y|x)$ exists, is continuous in (x, y) and $\sup_{x,y} |y^2\ddot{L}(y|x)| < \infty$,
(vi) $\dot{L}'(y|x)$ exists, is continuous in (x, y) and $\sup_{x,y} |y\dot{L}'(y|x)| < \infty$, for $L(y|x) = H(y|x)$ and $H_1(y|x)$.

($A5$) For the density $f_{X|Z,\Delta}(x|z,\delta)$ of X given (Z, Δ), $\sup_{x,z} |f_{X|Z,\Delta}(x|z,\delta)| < \infty$, $\sup_{x,z} |\dot{f}_{X|Z,\Delta}(x|z,\delta)| < \infty$, $\sup_{x,z} |\ddot{f}_{X|Z,\Delta}(x|z,\delta)| < \infty$ ($\delta = 0, 1$).
($A6$) Θ is compact and θ_0 is an interior point of Θ. All partial derivatives of $m_\theta(x)$ with respect to the components of θ up to order three exist and are continuous in (x, θ) for all x and θ.
($A7$) The function $E[(Y - m_\theta(X))^2 J(F(Y|X))]$ has a unique minimum in $\theta = \theta_0$.

Proof of Theorem 8.3.1. We prove the consistency of θ_n by verifying the conditions of Theorem 5.7 in van der Vaart (1998, p. 45). From the definition of θ_n and condition ($A7$), it follows that it suffices to show that

$$\sup_\theta |S_n(\theta) - S_0(\theta)| \to_P 0, \tag{8.1}$$

where $S_0(\theta) = E[(Y - m_\theta(X))^2 J(F(Y|X))] = E[\varepsilon^2 J(F_\varepsilon(\varepsilon|X))]$. The second statement of Theorem 8.3.1 then follows immediately from (8.1) together with the consistency of θ_n. (8.1) is obtained by using (A2), the uniform consistency of the

Beran (1981) estimator (given in Proposition A.3 of Van keilegom and Akritas (1999)) and Theorem 2 of Jennrich (1969). □

Proof of Theorem 8.3.2. For some θ_{1n} between θ_n and θ_0,

$$\theta_n - \theta_0 = -\left\{\frac{\partial^2 S_n(\theta_{1n})}{\partial\theta\partial\theta^T}\right\}^{-1}\frac{\partial S_n(\theta_0)}{\partial\theta} = -R_{1n}^{-1}R_{2n}.$$

First, we treat R_{2n}.

$$\begin{aligned}
R_{2nk} &= -\frac{2}{n}\sum_{i=1}^{n}\frac{\partial m_{\theta_0}(X_i)}{\partial\theta_k}\Bigg\{\Delta_i(Z_i - m_{\theta_0}(X_i))J'(F(Z_i|X_i))(\hat{F}(Z_i|X_i) - F(Z_i|X_i)) \\
&\quad + (1-\Delta_i)\Bigg[\frac{\int_{Z_i}^{+\infty}(y - m_{\theta_0}(X_i))J'(F(y|X_i))(\hat{F}(y|X_i) - F(y|X_i))dF(y|X_i)}{1 - F(Z_i|X_i)} \\
&\quad + \frac{\int_{Z_i}^{+\infty}(y - m_{\theta_0}(X_i))J(F(y|X_i))dF(y|X_i)}{(1 - F(Z_i|X_i))^2}(\hat{F}(Z_i|X_i) - F(Z_i|X_i)) \\
&\quad + \frac{\int_{Z_i}^{+\infty}(y - m_{\theta_0}(X_i))J(F(y|X_i))d(\hat{F}(y|X_i) - F(y|X_i))}{1 - F(Z_i|X_i)}\Bigg]\Bigg\} \\
&\quad - \frac{2}{n}\sum_{i=1}^{n}\frac{\partial m_{\theta_0}(X_i)}{\partial\theta_k}\Bigg\{\Delta_i(Z_i - m_{\theta_0}(X_i))J(F(Z_i|X_i)) \\
&\quad + (1-\Delta_i)\frac{\int_{Z_i}^{+\infty}(y - m_{\theta_0}(X_i))J(F(y|X_i))dF(y|X_i)}{1 - F(Z_i|X_i)}\Bigg\} + o_P(n^{-1/2}) \\
&= R_{21nk} + R_{22nk} + o_P(n^{-1/2}),
\end{aligned}$$

$k = 1, \ldots, d$. Developing R_{21nk} leads to

$$R_{21nk} = -\frac{2}{n^2 a_n}\sum_{i\neq j}K\Big(\frac{X_i - X_j}{a_n}\Big)\chi_k(V_i, Z_j, \Delta_j) + o_P(n^{-1/2}),$$

where $V_i = (X_i, Z_i, \Delta_i)$. Next, we rewrite R_{21nk} as

$$\begin{aligned}
R_{21nk} &= \frac{-2}{n^2 a_n}\sum_{i\neq j}\{A_k^*(V_i, V_j) + E[A_k(V_i, V_j)|V_i] + E[A_k(V_i, V_j)|V_j] \\
&\quad - E[A_k(V_i, V_j)]\} + o_P(n^{-1/2}) \\
&= T_{1,k}^n + T_{2,k}^n + T_{3,k}^n + T_{4,k}^n + o_P(n^{-1/2}),
\end{aligned}$$

where

$$A_k(V_i, V_j) = K\Big(\frac{X_i - X_j}{a_n}\Big)\chi_k(V_i, Z_j, \Delta_j)$$

and $A_k^*(V_i, V_j) = A_k(V_i, V_j) - E[A_k(V_i, V_j)|V_i] - E[A_k(V_i, V_j)|V_j] + E[A_k(V_i, V_j)]$. Consider

$$\begin{aligned}
E[A_k(V_i, V_j)|V_i] &= \sum_{\delta=0,1} \iint \chi_k(V_i, z, \delta) K\Big(\frac{X_i - x}{a_n}\Big) h_\delta(z|x) f_X(x)\, dz\, dx \\
&= a_n \sum_{\delta=0,1} \iint \chi_k(V_i, z, \delta) K(u)(h_\delta(z|X_i) - a_n u \dot{h}_\delta(z|X_i) + O(a_n^2)) \\
&\quad \times (f_X(X_i) - a_n u f_X'(X_i) + O(a_n^2))\, dz\, du \\
&= a_n f_X(X_i) \sum_{\delta=0,1} \int \chi_k(V_i, z, \delta) h_\delta(z|X_i)\, dz + O(a_n^3) = O(a_n^3)
\end{aligned} \tag{8.2}$$

for $i = 1, \dots, n$, since

$$\sum_{\delta=0,1} \int \xi(z, \delta, y|x) h_\delta(z|x) dz = 0$$

for all $x \in R_X$ and $y \leq T_x$. Hence, we also have $E[A_k(V_i, V_j)] = O(a_n^3)$. In a similar way, using three Taylor expansions of order 2, we get

$$\begin{aligned}
E[A_k(V_i, V_j)|V_j] = a_n f_X(X_j) \sum_{\delta=0,1} \int \chi_k((X_j, z, \delta), (Z_j, \Delta_j))\, dH_\delta(z|X_j) \\
+ O(a_n^3).
\end{aligned} \tag{8.3}$$

Note that for $T_{1,k}^n$, $E[T_{1,k}^n] = 0$, resulting, by Chebyshev's inequality, in

$$\begin{aligned}
P(|T_{1,k}^n| > K(na_n)^{-1}) &\leq K^{-2}(na_n)^2 E[(T_{1,k}^n)^2] \\
&= 4K^{-2} n^{-2} \sum_{j \neq i} \sum_{m \neq l} E[A_k^*(V_i, V_j) A_k^*(V_l, V_m)],
\end{aligned}$$

for any $K > 0$. Since $E[A_k^*(V_i, V_j)] = 0$, the terms for which $i, j \neq l, m$ are zero. The terms for which either i or j equals l or m and the other differs from l and m, are also zero, because, for example when $i = l$ and $j \neq m$,

$$E[A_k^*(V_i, V_j) E[A_k^*(V_i, V_m)|V_i, V_j]] = 0.$$

Thus, only the $2n(n-1)$ terms for which (i, j) equals (l, m) or (m, l) remain. Since $A_k^*(V_i, V_j)$ is bounded by $CK(\frac{X_i - X_j}{a_n}) + O(a_n)$ for some constant $C > 0$, we have (in the case (i, j) equals (l, m)) that

$$E[A_k^{*^2}(V_i, V_j)] \le C^2 a_n \int f_X^2(x)\,dx \int K^2(u)\,du + O(a_n^2) = O(a_n).$$

The case (i, j) equals (m, l) is treated similarly. It now follows that

$$T_{1,k}^n = o_P(n^{-1}a_n^{-1}), \tag{8.4}$$

which is $o_P(n^{-1/2})$. By (8.2), (8.3), (8.4), we finally obtain

$$\begin{aligned} R_{21nk} &= -\frac{2}{n}\sum_{i=1}^{n}\sum_{\delta=0,1}\int \chi_k((X_i, z, \delta), (Z_i, \Delta_i)) f_X(X_i) dH_\delta(z|X_i) \\ &\quad + o_P(n^{-1/2}),\ k = 1, \ldots, d. \end{aligned}$$

Finally, we treat the term R_{1n}.

$$\begin{aligned} R_{1n} &= -\frac{2}{n}\Bigg\{\sum_{i=1}^{n}\Bigg[\Delta_i(Z_i - m_{\theta_{1n}}(X_i))J(\hat F(Z_i|X_i)) \\ &\quad + (1-\Delta_i)\frac{\int_{Z_i}^{+\infty}(y - m_{\theta_{1n}}(X_i))J(\hat F(y|X_i))d\hat F(y|X_i)}{1 - \hat F(Z_i|X_i)}\Bigg]\frac{\partial^2 m_{\theta_{1n}}(X_i)}{\partial\theta\partial\theta^T} \\ &\quad - \sum_{i=1}^{n}\Bigg[\Delta_i J(\hat F(Z_i|X_i)) + (1-\Delta_i)\frac{\int_{Z_i}^{+\infty} J(\hat F(y|X_i))d\hat F(y|X_i)}{1 - \hat F(Z_i|X_i)}\Bigg] \\ &\quad \times \Big(\frac{\partial m_{\theta_{1n}}(X_i)}{\partial\theta}\Big)\Big(\frac{\partial m_{\theta_{1n}}(X_i)}{\partial\theta^T}\Big)\Bigg\} \\ &= R_{11n}(\theta_{1n}, \hat F) + R_{12n}(\theta_{1n}, \hat F). \end{aligned}$$

Using the uniform consistency of the Beran (1981) estimator together with (A6), it is clear that

$$R_{11n}(\theta_{1n}, \hat F) + R_{12n}(\theta_{1n}, \hat F) = R_{11n}(\theta_0, F) + R_{12n}(\theta_0, F) + o_P(1).$$

Since $E[R_{11n}(\theta_0, F)] = 0$ and

$$E[R_{12n}(\theta_0, F)] = 2E\Big[\Big(\frac{\partial m_{\theta_0}(X)}{\partial\theta}\Big)\Big(\frac{\partial m_{\theta_0}(X)}{\partial\theta^T}\Big)\Big],$$

we obtain that

$$R_{1n} = 2E\Big[\Big(\frac{\partial m_{\theta_0}(X)}{\partial \theta}\Big)\Big(\frac{\partial m_{\theta_0}(X)}{\partial \theta^T}\Big)\Big] + o_P(1).$$

This finishes the proof. □

Acknowledgements Financial support from IAP research network P6/03 of the Belgian Government (Belgian Science Policy) is gratefully acknowledged (Cédric Heuchenne). Financial support from IAP research network P6/03 of the Belgian Government (Belgian Science Policy), and from the European Research Council under the European Community's Seventh Framework Programme (FP7/2007-2013)/ERC Grant agreement No. 203650 is gratefully acknowledged (Ingrid van Keilegom).

References

Akritas, M.G. (1996). On the use of nonparametric regression techniques for fitting parametric regression models. *Biometrics, 52*, 1342–1362.

Aragon, Y., Daouia, A., & Thomas-Agnan, C. (2005). Nonparametric frontier estimation: a conditional quantile-based approach. *Econometric Theory, 21*, 358–389.

Beran, R. (1981). *Nonparametric regression with randomly censored survival data.* Technical Report, University California, Berkeley.

Buchinsky, M., & Hahn, J. (1998). An alternative estimator for the censored quantile regression model. *Econometrica, 66*, 653–671.

Cazals, C., Florens, J.P., & Simar, L. (2002). Nonparametric frontier estimation: a robust approach. *Journal of Econometrics, 106*, 1–25.

Fygenson, M., & Zhou, M. (1994). On using stratification in the analysis of linear regression models with right censoring. *Annals of Statistics, 22*, 747–762.

Heuchenne, C., & Van Keilegom, I. (2007). Nonlinear regression with censored data. *Technometrics, 49*, 34–44.

Jennrich, R.I. (1969). Asymptotic properties of nonlinear least squares estimators. *Annals of Mathematical Statistics, 40*, 633–643.

Pardo-Fernández, J.C., Van Keilegom, I., & González-Manteiga, W. (2007). Goodness-of-fit tests for parametric models in censored regression. *Canadian Journal of Statistics, 35*, 249–264.

Portnoy, S. (2003). Censored regression quantiles. *Journal of the American Statistical Association, 98*, 1001–1012.

Stute, W. (1993). Consistent estimation under random censorship when covariables are present. *Journal of Multivariate Analysis, 45*, 89–103.

Van der Vaart, A.W. (1998). *Asymptotic statistics*. Cambridge: Cambridge University Press.

Van Keilegom, I., & Akritas, M.G. (1999). Transfer of tail information in censored regression models. *Annals of Statistics, 27*, 1745–1784.

Van Keilegom, I., & Akritas, M.G. (2000). The least squares method in heteroscedastic censored regression models. In Puri, M.L.(Ed.) *Asymptotics in Statistics and Probability* (pp. 379–391), VSP.

Yin, G., Zeng, D., & Li, H. (2008). Power-transformed linear quantile regression with censored data. *Journal of the American Statistical Association, 103*, 1214–1224.

Wang, J., & Wang, L. (2009). Locally weighted censored quantile regression. *Journal of the American Statistical Association, 104*, 1117–1128.

Chapter 9
On Convex Boundary Estimation

Seok-Oh Jeong and Byeong U. Park

Abstract Consider a convex set S of the form $S = \{(\boldsymbol{x}, y) \in \mathbb{R}_+^p \times \mathbb{R}_+ \,|\, 0 \le y \le g(\boldsymbol{x})\}$, where the function g stands for the upper boundary of the set S. Suppose that one is interested in estimating the set S (or equivalently, the boundary function g) based on a set of observations laid on S. Then one may think of building the convex-hull of the observations to estimate the set S, and the corresponding estimator of the boundary function g is given by the roof of the constructed convex-hull. In this chapter we give an overview of statistical properties of the convex-hull estimator of the boundary function g. Also, we discuss bias-correction and interval estimation with the convex-hull estimator.

9.1 Convex-Hull Estimation

Consider a convex set $S \subset \mathbb{R}_+^p \times \mathbb{R}_+$ defined by $S = \{(\boldsymbol{x}, y) \in \mathbb{R}_+^p \times \mathbb{R}_+ \,|\, 0 \le y \le g(\boldsymbol{x})\}$, where the function g forms the upper boundary of the set S. Suppose that one is interested in estimating the set S (or equivalently, the boundary function g) based on a set of observations $\mathcal{X}_n = \{(\boldsymbol{X}_1, Y_1), (\boldsymbol{X}_2, Y_2), \cdots, (\boldsymbol{X}_n, Y_n)\}$ laid on S. A natural and popular estimator of S is the convex-hull of $\mathcal{X}_n$ defined by

$$\hat{S} = \left\{(\boldsymbol{x}, y) \in \mathbb{R}_+^p \times \mathbb{R}_+ \,\middle|\, \boldsymbol{x} = \boldsymbol{X}^\top \boldsymbol{\gamma}, y = \boldsymbol{\gamma}^\top \boldsymbol{Y} \text{ for some } \boldsymbol{\gamma} \ge \boldsymbol{0} \text{ such that } \boldsymbol{1}^\top \boldsymbol{\gamma} = 1\right\},$$

S.-O. Jeong (✉)
Department of Statistics, Hankuk University of Foreign Studies, Mo-Hyeon Yong-In, Gyeong-Gi, 449-791, South Korea
e-mail: seokohj@hufs.ac.kr

B.U. Park
Department of Statistics, Seoul National University, Seoul 151-747, South Korea
e-mail: bupark@stats.snu.ac.kr

I. van Keilegom and P.W. Wilson (eds.), *Exploring Research Frontiers in Contemporary Statistics and Econometrics*, DOI 10.1007/978-3-7908-2349-3_9,

where $\boldsymbol{Y} = (Y_1, Y_2, \cdots, Y_n)^\top$, $\boldsymbol{X} = [\boldsymbol{X}_1, \boldsymbol{X}_2, \cdots, \boldsymbol{X}_n]^\top$, $\boldsymbol{\gamma} = (\gamma_1, \gamma_2, \cdots, \gamma_n)^\top$ and $\mathbf{1} = (1, 1, \cdots, 1)^\top$. The corresponding estimator of g at $\boldsymbol{x} \in \mathbb{R}_+^p$ is $\hat{g}_{\text{conv}}(\boldsymbol{x}) = \max\{y : (\boldsymbol{x}, y) \in \hat{S}\}$, or equivalently

$$\hat{g}_{\text{conv}}(\boldsymbol{x}) = \max\left\{\boldsymbol{\gamma}^\top \boldsymbol{Y} \;\middle|\; \boldsymbol{x} = \boldsymbol{X}^\top \boldsymbol{\gamma} \text{ for some } \boldsymbol{\gamma} \geq \mathbf{0} \text{ such that } \mathbf{1}^\top \boldsymbol{\gamma} = 1\right\}. \quad (9.1)$$

The problem arises in an area of econometrics where one is interested in evaluating the performance of an enterprise in terms of technical efficiency, see Gattoufi et al. (2004) for a comprehensive list of the existing literature on this problem.

By construction, the convex-hull estimator $\hat{g}_{\text{conv}}(\boldsymbol{x})$ always underestimates the true boundary function $g(\boldsymbol{x})$ for all $\boldsymbol{x} \in \mathbb{R}^p$. One may be interested in quantifying the deviation $\hat{g}_{\text{conv}}(\boldsymbol{x}) - g(\boldsymbol{x})$. For this, the asymptotic properties of the deviation have been investigated under the assumption that $\mathcal{X}_n$ is a set of iid copies of a random vector $(\boldsymbol{X}, Y)$ of which the distribution is supported on S. Kneip et al. (1998) proved that the convex-hull estimator is a consistent estimator with rate of convergence $n^{-2/(p+2)}$. Gijbels et al. (1999), Jeong (2004), Jeong and Park (2006) and Kneip et al. (2008) derived its sampling distribution. This chapter aims at providing an overview of these works. Also we present a practical guide to utilizing them for further analysis such as correcting the bias and constructing confidence intervals for the boundary function.

9.2 Rate of Convergence

For the convergence rate of the convex-hull estimator $\hat{g}_{\text{conv}}(\boldsymbol{x}_0)$ at a given point $\boldsymbol{x}_0 \in \mathbb{R}_+^p$, we assume that

(A1) The boundary function g is twice continuously differentiable in a neighborhood of $\boldsymbol{x}_0$, and its Hessian matrix $\nabla^2 g(\boldsymbol{x}_0)$ is negative definite.
(A2) The density f of $(\boldsymbol{X}_1, Y_1)$ on $\{(\boldsymbol{x}, y) \in S : \|(\boldsymbol{x}, y) - (\boldsymbol{x}_0, g(\boldsymbol{x}_0))\| \leq \epsilon\}$ for some $\epsilon > 0$ is bounded away from zero, where $\|\cdot\|$ is the Euclidean norm.

Theorem 1. *Under the assumptions (A1) and (A2), we have*

$$\hat{g}(\boldsymbol{x}_0) - g(\boldsymbol{x}_0) = O_P(n^{-2/(p+2)}). \quad (9.2)$$

Proof. The proof is basically a special case of the proof of Theorem 1 in Kneip et al. (1998). Define $B_p(\boldsymbol{t}, r) = \{\boldsymbol{x} \in \mathbb{R}^p : \|\boldsymbol{x} - \boldsymbol{t}\| \leq r\}$ and

$$C_r = B(\boldsymbol{x}_0^{(r)}, h/2) \times \mathbb{R}, \quad r = 1, \ldots, 2p,$$

where $\boldsymbol{x}_0^{(2j-1)} = \boldsymbol{x}_0 - h\boldsymbol{e}_j$ and $\boldsymbol{x}_0^{(2j)} = \boldsymbol{x}_0 + h\boldsymbol{e}_j$ for $1 \leq j \leq p$, $\boldsymbol{e}_j$ is the unit p-vector with the j-th element equal to 1, and h is a positive number whose size will be determined later. Let $\boldsymbol{U}_r$ denote the input observation $\boldsymbol{X}_i \in B(\boldsymbol{x}_0^{(r)}, h/2)$

whose output observation Y_i, denoted by V_r, takes the maximal value among those $Y_{i'}$ such that $(\boldsymbol{X}_{i'}, Y_{i'}) \in C_r$, i.e.,

$$(\boldsymbol{U}_r, V_r) = \arg\max \{Y_i : (\boldsymbol{X}_i, Y_i) \in C_r, i = 1, \ldots, n\}, \quad r = 1, \ldots, 2p.$$

Then, we have

$$\hat{g}_{\text{conv}}(\boldsymbol{U}_r) \geq V_r = g(\boldsymbol{U}_r) + O_P(n^{-1}h^{-p}), \quad r = 1, \ldots, 2p. \tag{9.3}$$

In the above claim, the inequality follows from the definition of $\hat{g}_{\text{conv}}$, and the equality from (A2) since then the number of points in $\mathcal{X}_n$ falling into $S \cap C_r$ is proportional to nh^p. Since the convex-hull of $\{\boldsymbol{U}_r : 1 \leq r \leq 2p\}$ contains $\boldsymbol{x}_0$, one may find $w_r \geq 0$, $1 \leq r \leq 2p$, such that $\boldsymbol{x}_0 = \sum_{r=1}^{2p} w_r \boldsymbol{U}_r$ and $\sum_{r=1}^{2p} w_r = 1$. This with (A1) gives

$$\begin{aligned}\sum_{r=1}^{2p} w_r g(\boldsymbol{U}_r) &= \sum_{r=1}^{2p} w_r \left[g(\boldsymbol{x}_0) + g'(\boldsymbol{x}_0)^\top (\boldsymbol{U}_r - \boldsymbol{x}_0)\right] + O_P(h^2) \\ &= g(\boldsymbol{x}_0) + O_P(h^2), \quad r = 1, \ldots, 2p. \end{aligned} \tag{9.4}$$

The results (9.3) and (9.4) together with the concavity of the function $\hat{g}_{\text{conv}}$ entail

$$\begin{aligned} g(\boldsymbol{x}_0) &\geq \hat{g}_{\text{conv}}(\boldsymbol{x}_0) \\ &= \hat{g}_{\text{conv}}\left(\sum_{r=1}^{2p} w_r \boldsymbol{U}_r\right) \geq \sum_{r=1}^{2p} w_r \hat{g}_{\text{conv}}(\boldsymbol{U}_r) \\ &\geq \sum_{r=1}^{2p} w_r V_r = \sum_{r=1}^{2p} w_r g(\boldsymbol{U}_r) + O_P(n^{-1}h^{-p}) \\ &= g(\boldsymbol{x}_0) + O_P(h^2) + O_P(n^{-1}h^{-p}). \end{aligned}$$

Taking $h = n^{-1/(p+2)}$ completes the proof of the theorem. □

9.3 Asymptotic Distribution

The asymptotic distribution of $\hat{g}_{\text{conv}}$ was studied by Jeong and Park (2006). Here, we outline their results, and discuss bias-correction and interval estimation based on the results.

Under the assumption (A1), the matrix $-\nabla^2 g(\boldsymbol{x}_0)/2$ is positive definite. Hence we have a spectral decomposition for this matrix: $-\nabla^2 g(\boldsymbol{x}_0)/2 = P\Lambda P^T$, where Λ is the diagonal matrix whose diagonal elements are the eigenvalues of $-\nabla^2 g(\boldsymbol{x}_0)/2$

and P is the orthonormal matrix formed by the associated orthonormal eigenvectors. Consider a canonical transform: $(\boldsymbol{X}_i, Y_i) \mapsto (\boldsymbol{X}_i^*, Y_i^*)$ with

$$\boldsymbol{X}_i^* = n^{1/(p+2)} \Lambda^{1/2} P^T (\boldsymbol{X}_i - \boldsymbol{x}_0);$$
$$Y_i^* = n^{2/(p+2)} \left\{ Y_i - g(\boldsymbol{x}_0) - \nabla g(\boldsymbol{x}_0)^T (\boldsymbol{X}_i - \boldsymbol{x}_0) \right\}$$

and write $\mathcal{X}_n^* = \{(\boldsymbol{X}_1^*, Y_1^*), (\boldsymbol{X}_2^*, Y_2^*), \cdots, (\boldsymbol{X}_n^*, Y_n^*)\}$. Let $Z_{\text{conv}}^*(\cdot)$ be the roof of the convex-hull of the transformed data $\mathcal{X}_n^*$:

$$Z_{\text{conv}}^*(\boldsymbol{x}^*) = \max \left\{ \boldsymbol{\gamma}^\top \boldsymbol{Y}^* : \boldsymbol{x}^* = \boldsymbol{X}^{*\top} \boldsymbol{\gamma} \text{ for some } \boldsymbol{\gamma} \geq \boldsymbol{0} \text{ such that } \boldsymbol{1}^\top \boldsymbol{\gamma} = 1 \right\}$$

with $\boldsymbol{Y}^* = (Y_1^*, Y_2^*, \cdots, Y_n^*)^\top$ and $\boldsymbol{X}^* = [\boldsymbol{X}_1^*, \boldsymbol{X}_2^*, \cdots, \boldsymbol{X}_n^*]^\top$. Due to Lemma 1 of Jeong and Park (2006), it follows that, with probability tending to one as n goes to infinity,

$$Z_{\text{conv}}^*(\boldsymbol{0}) = n^{2/(p+2)} \left\{ \hat{g}_{\text{conv}}(\boldsymbol{x}_0) - g(\boldsymbol{x}_0) \right\}. \tag{9.5}$$

Under (A1), one can show that the new boundary function g_n^* for $(\boldsymbol{X}_i^*, Y_i^*)$ can be locally approximated by $y^* = -\boldsymbol{x}^{*\top} \boldsymbol{x}^*$. Also, putting

$$\kappa_{\text{conv}} = \left\{ \sqrt{\det \Lambda} / f(\boldsymbol{x}_0, g(\boldsymbol{x}_0)) \right\}^{2/(p+2)}, \tag{9.6}$$

it holds that the density of $(\boldsymbol{X}_i^*, Y_i^*)$ in the transformed coordinate system, denoted by f_n^*, is uniformly approximated by a constant in a neighborhood of the origin: for any $\epsilon \downarrow 0$ as $n \to \infty$,

$$\sup\{ n | f_n^*(\boldsymbol{x}^*, y^*) - \kappa_{\text{conv}}^{-(p+2)/2} | : \|\boldsymbol{x}^*\| \leq \epsilon n^{1/(p+2)}, -\boldsymbol{x}^{*\top} \boldsymbol{x}^*$$
$$- \epsilon n^{2/(p+2)} \leq y^* \leq -\boldsymbol{x}^{*\top} \boldsymbol{x}^* \}$$

goes to zero as $n \to \infty$. Now consider a new random sample

$$\tilde{\mathcal{X}}_n^* = \{(\tilde{\boldsymbol{X}}_1^*, \tilde{Y}_1^*), (\tilde{\boldsymbol{X}}_2^*, \tilde{Y}_2^*), \cdots, (\tilde{\boldsymbol{X}}_n^*, \tilde{Y}_n^*)\}$$

from the uniform distribution on $\mathcal{C}_{n,\kappa_{\text{conv}}}$ where

$$\mathcal{C}_{n,\kappa_{\text{conv}}} = \left\{ (\boldsymbol{x}^*, y^*) : |x_j^*| \leq n^{1/(p+2)} \frac{\sqrt{\kappa_{\text{conv}}}}{2} \text{ for } 1 \leq j \leq p, \right.$$
$$\left. -\boldsymbol{x}^{*\top} \boldsymbol{x}^* - n^{2/(p+2)} \kappa_{\text{conv}} \leq y^* \leq -\boldsymbol{x}^{*\top} \boldsymbol{x}^* \right\}.$$

Note that the uniform density on $\mathcal{C}_{n,\kappa_{\text{conv}}}$ is $n^{-1} \kappa_{\text{conv}}^{-(p+2)/2}$. Let $\tilde{Z}_{\text{conv}}^*$ be the version of Z_{conv}^* constructed from $\tilde{\mathcal{X}}_n^*$. Lemma 2 of Jeong and Park (2006) asserts that the asymptotic distributions of $\tilde{Z}_{\text{conv}}^*$ and Z_{conv}^* are the same.

Theorem 2. (Jeong and Park 2006). *Under (A1) and (A2), we have for $z \le 0$:*

$$P(\tilde{Z}^*_{\mathrm{conv}}(\mathbf{0}) \le z) - P(Z^*_{\mathrm{conv}}(\mathbf{0}) \le z) \longrightarrow 0$$

as $n \to \infty$, and hence, by (9.5), $\tilde{Z}^*_{\mathrm{conv}}(\mathbf{0})$ and $n^{2/(p+2)}\{\hat{g}_{\mathrm{conv}}(\boldsymbol{x}_0) - g(\boldsymbol{x}_0)\}$ have the same limit distribution.

Below in this section we discuss bias-correction for $\hat{g}_{\mathrm{conv}}(\boldsymbol{x}_0)$ and interval estimation based on the estimator. For the moment, suppose that the constant κ_{conv} is available. Consider $\boldsymbol{U}_i \sim \mathrm{Uniform}[-1/2, 1/2]^p$ and $W_i = -\boldsymbol{U}_i^\top \boldsymbol{U}_i - V_i$, where $V_i \sim \mathrm{Uniform}[0,1]$ is independent of $\boldsymbol{U}_i$. Then, we have

$$\{(\tilde{\boldsymbol{X}}_i^*, \tilde{Y}_i^*)\} \overset{d}{=} \left\{\left(n^{1/(p+2)}\sqrt{\kappa_{\mathrm{conv}}} \cdot \boldsymbol{U}_i,\ n^{2/(p+2)}\kappa_{\mathrm{conv}} \cdot W_i\right)\right\}.$$

Let $\tilde{Z}^{**}_{\mathrm{conv}}(\cdot)$ be the version of $\tilde{Z}^*_{\mathrm{conv}}(\cdot)$ constructed from the $(\boldsymbol{U}_i, W_i)$'s. Then,

$$\begin{aligned}\tilde{Z}^*_{\mathrm{conv}}(\mathbf{0}) &= \max\left\{\boldsymbol{\gamma}^\top \tilde{\boldsymbol{Y}}^* : \mathbf{0} = \tilde{\boldsymbol{X}}^{*\top}\boldsymbol{\gamma} \text{ for some } \boldsymbol{\gamma} \ge \mathbf{0} \text{ such that } \mathbf{1}^\top\boldsymbol{\gamma} = 1\right\} \\ &\overset{d}{=} n^{2/(p+2)}\kappa_{\mathrm{conv}} \cdot \tilde{Z}^{**}_{\mathrm{conv}}(\mathbf{0}),\end{aligned}$$

where $\tilde{\boldsymbol{X}}^* = [\tilde{\boldsymbol{X}}_1^*, \tilde{\boldsymbol{X}}_2^*, \cdots, \tilde{\boldsymbol{X}}_n^*]^\top$, $\tilde{\boldsymbol{Y}}^* = (\tilde{Y}_1^*, \tilde{Y}_2^*, \cdots, \tilde{Y}_n^*)^\top$, $\boldsymbol{W} = (W_1, W_2, \cdots, W_n)^\top$ and $\boldsymbol{U} = [\boldsymbol{U}_1, \boldsymbol{U}_2, \cdots, \boldsymbol{U}_n]^\top$. Thus, one may approximate the limiting distribution of $n^{2/(p+2)}\{\hat{g}_{\mathrm{conv}}(\boldsymbol{x}_0) - g(\boldsymbol{x}_0)\}$ by the convex-hull of $(\boldsymbol{U}_i, W_i)$. One can easily obtain an empirical distribution of $\tilde{Z}^{**}_{\mathrm{conv}}(\mathbf{0})$ from a Monte Carlo experiment as follows.

Algorithm to get an empirical distribution of $\tilde{Z}^{**}_{\mathrm{conv}}(\mathbf{0})$.

I. Put $b = 1$.
II. Generate a random sample $\{(\boldsymbol{U}_1^{(b)}, W_1^{(b)}), \ldots, (\boldsymbol{U}_n^{(b)}, W_n^{(b)})\}$ as follows:

$$\begin{aligned}&\boldsymbol{U}_i^{(b)} \sim \mathrm{Uniform}\left[-1/2, 1/2\right]^p, \quad V_i^{(b)} \sim \mathrm{Uniform}\left[0, 1\right], \\ &W_i^{(b)} = -\boldsymbol{U}_i^{(b)\top}\boldsymbol{U}_i^{(b)} - V_i^{(b)}.\end{aligned}$$

III. Compute $\tilde{Z}^{**(b)}_{\mathrm{conv}}(\mathbf{0})$ using $\{(\boldsymbol{U}_1^{(b)}, W_1^{(b)}), \ldots, (\boldsymbol{U}_n^{(b)}, W_n^{(b)})\}$.
IV. Repeat II and III for $b = 2, \cdots, B$.
V. Compute the empirical distribution of $\tilde{Z}^{**(b)}_{\mathrm{conv}}(\mathbf{0})$, $b = 1, \ldots, B$.

The constant κ_{conv} in (9.6) depends on the unknown quantities Λ and $f(\boldsymbol{x}_0, g(\boldsymbol{x}_0))$. We will discuss the estimation of κ_{conv} in Sect. 9.4. The asymptotic bias of the convex-hull estimator $\hat{g}_{\mathrm{conv}}(\boldsymbol{x}_0)$ is approximated by

$$E[\hat{g}_{\mathrm{conv}}(\boldsymbol{x}_0)] - g(\boldsymbol{x}_0) \approx n^{-2/(p+2)}E[\tilde{Z}^*_{\mathrm{conv}}(\mathbf{0})] = \kappa_{\mathrm{conv}}E[\tilde{Z}^{**}_{\mathrm{conv}}(\mathbf{0})],$$

so that a bias-corrected estimator of $g(\boldsymbol{x}_0)$ is given by

$$\hat{g}^{\mathrm{bc}}_{\mathrm{conv}}(\boldsymbol{x}_0) = \hat{g}_{\mathrm{conv}}(\boldsymbol{x}_0) - \hat{\kappa}_{\mathrm{conv}} E[\tilde{Z}^{**}_{\mathrm{conv}}(\mathbf{0})],$$

where $\hat{\kappa}_{\mathrm{conv}}$ is a consistent estimator of κ_{conv}. The expected value $E[\tilde{Z}^{**}_{\mathrm{conv}}(\mathbf{0})]$ may be replaced by the Monte Carlo average of $\tilde{Z}^{**(b)}_{\mathrm{conv}}(\mathbf{0})$ over $1 \le b \le B$. A $100\times(1-\alpha)\%$ confidence interval for $g(\boldsymbol{x}_0)$ can also be obtained. Let Q_α denote the α-quantile of the distribution of $\tilde{Z}^{**}_{\mathrm{conv}}(\mathbf{0})$, i.e., $P[\tilde{Z}^{**}_{\mathrm{conv}}(\mathbf{0}) \le Q_\alpha] = \alpha$. Then, a $100 \times (1-\alpha)\%$ confidence interval for $g(\boldsymbol{x}_0)$ is given by

$$\left[\hat{g}_{\mathrm{conv}}(\boldsymbol{x}_0) - \hat{\kappa}_{\mathrm{conv}} Q_{1-\alpha/2},\ \hat{g}_{\mathrm{conv}}(\boldsymbol{x}_0) - \hat{\kappa}_{\mathrm{conv}} Q_{\alpha/2}\right].$$

In practice, one may replace Q_α by the corresponding empirical quantile of $\{\tilde{Z}^{**(b)}_{\mathrm{conv}}(\mathbf{0})\}_{b=1}^{B}$.

9.4 Estimation of κ_{conv} Via Subsampling

The bias correction and the interval estimation discussed in the previous section require a consistent estimation of the unknown constant κ_{conv}. Jeong and Park (2006) and Park et al. (2010) discussed some methods for estimating the constant. Instead of estimating κ_{conv}, Kneip et al. (2008) considered a subsampling technique for approximating the asymptotic distribution of the convex-hull estimator. As noted there and in other works, a naive bootstrap does not work in the case of boundary estimation. Hall and Park (2002) suggested a bootstrap method based on translations of an initial boundary estimator. Here we discuss the subsampling idea for boundary estimation. The idea of subsampling as a way of estimating a sampling distribution dates back to Wu (1990), and discussed further by Politis and Romano (1994), Bickel et al. (1997), and Bickel (2003) among others. For a detailed account of this technique and a collection of related works, see Politis et al. (1999).

Let $\hat{g}^{\mathrm{SB}}_{\mathrm{conv}}$ denote a version of $\hat{g}_{\mathrm{conv}}$ based on a subsample of size m drawn randomly from the set of observations $\mathcal{X}_n$. Denote by $\mathcal{F}_n$ the subsampling bootstrap distribution of $m^{2/(p+2)}[\hat{g}^{\mathrm{SB}}_{\mathrm{conv}}(\boldsymbol{x}_0) - \hat{g}_{\mathrm{conv}}(\boldsymbol{x}_0)]$ conditioning on $\mathcal{X}_n$. Also, denote by $\mathcal{F}$ the sampling distribution of $n^{2/(p+2)}[\hat{g}_{\mathrm{conv}}(\boldsymbol{x}_0) - g(\boldsymbol{x}_0)]$. Finally, let d_{KS} denote the Kolmogorov-Smirnov metric for the space of distribution functions. Using the technique of Kneip et al. (2008) and our Theorem 2, one can prove the following theorem.

Theorem 3. *Let $m \equiv m(n) = n^c$ for some constant $0 < c < 1$. Under the conditions of Theorem 2, it follows that $d_{KS}(\mathcal{F}_n, \mathcal{F})$ converges to zero in probability as $n \to \infty$.*

In the implementation of the m-out-of-n technique, the choice of m is a key issue. This problem has been treated in a general context by Bickel (2003), and Bickel and Sakov (2008) for example. Here we give an alternative way of determining m, specific to the context of convex-hull estimation, based on a regression idea. Note that, according to Theorems 2 and 3, the conditional distribution $\mathcal{F}_n$ approximates the distribution of $n^{2/(p+2)}\kappa_{\text{conv}}\tilde{Z}^{**}_{\text{conv}}(\mathbf{0})$, denoted by $\mathcal{F}^{**}$. This implies that, asymptotically, the α-quantile $Q_{n,\alpha}$ of $\mathcal{F}_n$ admits the following relation with the α-quantile Q^{**}_α of $n^{2/(p+2)}\tilde{Z}^{**}_{\text{conv}}(\mathbf{0})$:

$$Q_{n,\alpha} = \kappa_{\text{conv}} Q^{**}_\alpha, \quad 0 < \alpha < 1. \tag{9.7}$$

One may obtain empirical values of $Q_{n,\alpha}$ from repeated subsampling, and those of Q^{**}_α by simulating $\tilde{Z}^{**}_{\text{conv}}(\mathbf{0})$ repeatedly. For a large number B which is the number of repetitions, one may get empirical quantiles $Q_{n,b/B}$ and $Q^{**}_{b/B}$ for $b = 1, \ldots, B-1$. The relation (9.7) suggests fitting a simple regression model with $\{(Q^{**}_{b/B}, Q_{n,b/B})\}_{b=1}^{B-1}$ to estimate κ_{conv} by the fitted regression coefficient. For each given m, one may calculate the value of a criterion that measures how well $\{(Q^{**}_{b/B}, Q_{n,b/B})\}_{b=1}^{B-1}$ fit the regression model. As a criterion one may use the residual sum of squares of the fit, or a measure of the prediction error. Then, one can choose m that minimizes the criterion.

Algorithm to determine the subsample size m.

I. Choose a large number B and do II, III, IV below for a grid of $c \in (0, 1)$.
II. Get $\tilde{Z}^{**(b)}_{\text{conv}}(\mathbf{0})$ for $b = 1, \ldots, B$ by the algorithm in Sect. 8.3.
III. Generate B subsamples of size $m = n^c$ and compute $m^{2/(p+2)}[\hat{g}^{\text{SB}(b)}_{\text{conv}}(\boldsymbol{x}_0) - \hat{g}_{\text{conv}}(\boldsymbol{x}_0)]$ for $b = 1, \ldots, B$.
IV. Identify the empirical quantiles $Q^{**}_{b/B}$ and $Q_{n,b/B}$ from the computed values of $n^{2/(p+2)}\tilde{Z}^{**(b)}_{\text{conv}}(\mathbf{0})$ and $m^{2/(p+2)}[\hat{g}^{\text{SB}(b)}_{\text{conv}}(\boldsymbol{x}_0) - \hat{g}_{\text{conv}}(\boldsymbol{x}_0)]$, respectively.
V. Find c that gives the minimal value of a goodness-of-fit measure for fitting a straight line with the empirical quantiles $Q^{**}_{b/B}$ and $Q_{n,b/B}$.

9.5 Conical-Hull Estimation

Conical-hull estimation arises when the maximal value of y for an $\boldsymbol{x}$-value within the support S admits the so-called "constant returns-to-scale". The latter means that an increase of $\boldsymbol{x}$ by a scalar factor gives an increase of y by the same scalar factor. In other words, S is a convex cone with the boundary function g satisfying

$$g(a\boldsymbol{x}) = ag(\boldsymbol{x}) \quad \text{for all scalar } a > 0. \tag{9.8}$$

Park et al. (2010) developed a sound theory for the conical-hull estimator of the boundary g that satisfies the above constant returns-to-scale property (9.8). In this section, we briefly review their derivation of the asymptotic distribution of the conical estimator. Here, we exclude the trivial case where g represents a hyperplane, i.e., $g(\boldsymbol{x}) = \mathbf{c}^\top \boldsymbol{x}$ for some constant vector $\mathbf{c}$. Also, we assume that, for all $\boldsymbol{x}_1, \boldsymbol{x}_2 \in \mathbb{R}^p$ with $\boldsymbol{x}_1 \neq a\boldsymbol{x}_2$ for any $a > 0$,

$$g(\alpha \boldsymbol{x}_1 + (1-\alpha)\boldsymbol{x}_2) > \alpha g(\boldsymbol{x}_1) + (1-\alpha) g(\boldsymbol{x}_2) \quad \text{for all } \alpha \in (0,1).$$

This means that g restricted to the space of $\boldsymbol{x}$ with $\|\boldsymbol{x}\| = c$ is strictly convex for all $c > 0$.

When one knows that the underlying support S is a convex-cone, and thus the boundary g satisfies (9.8), then one would use the conical hull of $\mathcal{X}_n$ defined below, instead of the convex-hull $\hat{S}$ as defined in Sect. 8.1:

$$\hat{S}_{\text{coni}} = \left\{(\boldsymbol{x}, y) \in \mathbb{R}_+^p \times \mathbb{R}_+ : \boldsymbol{x} = \boldsymbol{X}^\top \boldsymbol{\gamma},\ y = \boldsymbol{\gamma}^\top \boldsymbol{Y} \text{ for some } \boldsymbol{\gamma} \geq \mathbf{0}\right\},$$

where $\boldsymbol{Y} = (Y_1, Y_2, \cdots, Y_n)^\top$, $\boldsymbol{X} = [\boldsymbol{X}_1, \boldsymbol{X}_2, \cdots, \boldsymbol{X}_n]^\top$, $\boldsymbol{\gamma} = (\gamma_1, \gamma_2, \cdots, \gamma_n)^\top$. It can be verified that $\hat{S}_{\text{coni}}$ coincides with the convex-hull of the rays $\boldsymbol{R}_i = \{(a\boldsymbol{X}_i, aY_i) \mid a \geq 0\}$, $i = 1, \ldots, n$. The corresponding estimator of g is the conical-hull estimator given by

$$\hat{g}_{\text{coni}}(\boldsymbol{x}) = \max\left\{\boldsymbol{\gamma}^\top \boldsymbol{Y} \,\middle|\, \boldsymbol{x} = \boldsymbol{X}^\top \boldsymbol{\gamma} \text{ for some } \boldsymbol{\gamma} \geq \mathbf{0}\right\}. \tag{9.9}$$

To give the main idea of the derivation of the limiting distribution of $\hat{g}_{\text{coni}}$, fix $\boldsymbol{x}_0 \in \mathbb{R}_+^p$. Project each data point $(\boldsymbol{X}_i, Y_i)$ along the ray $\boldsymbol{R}_i$, $i = 1, \ldots, n$, onto $\mathcal{P}(\boldsymbol{x}_0) \times \mathbb{R}_+$, where $\mathcal{P}(\boldsymbol{x}_0)$ is the hyperplane in $\mathbb{R}_+^p$ which is perpendicular to $\boldsymbol{x}_0$ and passes through $\boldsymbol{x}_0$, that is, $\mathcal{P}(\boldsymbol{x}_0) = \{\boldsymbol{x} \in \mathbb{R}_+^p \mid \boldsymbol{x}_0^\top(\boldsymbol{x} - \boldsymbol{x}_0) = 0\}$. Then, the projected data point $(\boldsymbol{X}_i^0, Y_i^0)$ is given by

$$(\boldsymbol{X}_i^0, Y_i^0) = \frac{\|\boldsymbol{x}_0\|^2}{\boldsymbol{x}_0^\top \boldsymbol{X}_i}(\boldsymbol{X}_i, Y_i), \quad i = 1, \ldots, n.$$

Define $S^0 = [\mathcal{P}(\boldsymbol{x}_0) \times \mathbb{R}_+] \cap S$, which is a section of S cut by $\mathcal{P}(\boldsymbol{x}_0) \times \mathbb{R}_+$. Then, the upper boundary of S^0 equals the locus of $y = g(\boldsymbol{x})$ for $\boldsymbol{x} \in \mathcal{P}(\boldsymbol{x}_0)$, so that the projected points $(\boldsymbol{X}_i^0, Y_i^0)$ are all laid on S^0. Also, the convex-hull, in $\mathcal{P}(\boldsymbol{x}_0) \times \mathbb{R}$, of the projected points $(\boldsymbol{X}_i^0, Y_i^0)$ coincides with the section of $\hat{S}_{\text{coni}}$ cut by $\mathcal{P}(\boldsymbol{x}_0) \times \mathbb{R}$. If we denote the convex-hull by $\hat{S}_{\text{conv}}^0$, then

$$\hat{S}_{\text{conv}}^0 = [\mathcal{P}(\boldsymbol{x}_0) \times \mathbb{R}] \cap \hat{S}_{\text{coni}}.$$

Thus, it follows that

$$g(\boldsymbol{x}_0) = \max\{y : (\boldsymbol{x}_0, y) \in S^0\},$$
$$\hat{g}_{\text{coni}}(\boldsymbol{x}_0) = \max\{y : (\boldsymbol{x}_0, y) \in \hat{S}_{\text{conv}}^0\}.$$

Let Q be a $p \times (p-1)$ matrix of which the columns form an orthonormal basis for $\boldsymbol{x}_0^\top$, the subspace of $\mathbb{R}^p$ perpendicular to $\boldsymbol{x}_0$. Then, $\mathcal{P}(\boldsymbol{x}_0)$ is nothing else than $[\boldsymbol{x}_0 + (\text{the column space of } Q)] \cap \mathbb{R}_+^p$. Note that

$$\boldsymbol{X}_i^0 = \boldsymbol{x}_0 + Q\left(\frac{\|\boldsymbol{x}_0\|^2}{\boldsymbol{x}_0^\top \boldsymbol{X}_i} Q^\top \boldsymbol{X}_i\right), \quad i = 1, \ldots, n.$$

Consider a transform $(\boldsymbol{x}, y) \in \mathbb{R}_+^p \times \mathbb{R}_+ \mapsto (z, y^0) \in \mathbb{R}^{p-1} \times \mathbb{R}_+$ such that

$$(z, y^0) = \frac{\|\boldsymbol{x}_0\|^2}{\boldsymbol{x}_0^\top \boldsymbol{x}}(Q^\top \boldsymbol{x}, y).$$

In the new coordinate system (z, y^0), the projected data points $(\boldsymbol{X}_i^0, Y_i^0)$ are represented by $(\boldsymbol{Z}_i, Y_i^0)$, where

$$\boldsymbol{Z}_i = \frac{\|\boldsymbol{x}_0\|^2}{\boldsymbol{x}_0^\top \boldsymbol{X}_i} Q^\top \boldsymbol{X}_i, \quad Y_i^0 = \frac{\|\boldsymbol{x}_0\|^2}{\boldsymbol{x}_0^\top \boldsymbol{X}_i} Y_i, \qquad i = 1, \ldots, n.$$

Let $y^0 = g^0(z)$ represent the equation $y = g(\boldsymbol{x})$ in the new coordinate system (z, y^0). It follows that $g^0(z) = g(\boldsymbol{x}_0 + Qz)$. Let $\hat{g}_{\text{conv}}^0$ be a version of $\hat{g}_{\text{conv}}$ constructed from $(\boldsymbol{Z}_i, Y_i^0)$. For $z \in \mathbb{R}^{p-1}$,

$$\hat{g}_{\text{conv}}^0(z) = \max\left\{\boldsymbol{\gamma}^\top \boldsymbol{Y}^0 \,\middle|\, z = \boldsymbol{Z}^\top \boldsymbol{\gamma} \text{ for some } \boldsymbol{\gamma} \geq \boldsymbol{0} \text{ such that } \boldsymbol{1}^\top \boldsymbol{\gamma} = 1\right\},$$

with $\boldsymbol{Y}^0 = (Y_1^0, Y_2^0, \cdots, Y_n^0)^\top$ and $\boldsymbol{Z} = [\boldsymbol{Z}_1, \boldsymbol{Z}_2, \cdots, \boldsymbol{Z}_n]^\top$. Then, since $\boldsymbol{x}_0$ corresponds to $\boldsymbol{0}$ in the new coordinate system, it can be shown that

$$g^0(\boldsymbol{0}) = g(\boldsymbol{x}_0), \quad \hat{g}_{\text{conv}}^0(\boldsymbol{0}) = \hat{g}_{\text{coni}}(\boldsymbol{x}_0).$$

The above arguments imply that studying the statistical properties of the conical-hull estimator of a p-variate boundary function that satisfies (9.8) is equivalent to studying those of the convex-hull estimator of a $(p-1)$-variate convex boundary function. Hence, the convergence rate of the conical-hull estimator equals $n^{-2/((p-1)+1+1)} = n^{-2/(p+1)}$, and the asymptotic distribution of $\hat{g}_{\text{coni}}(\boldsymbol{x}_0)$ may be derived from the results for the convex-hull estimators obtained by Jeong and Park (2006). We refer to Park et al. (2010) for technical details.

9.6 Application to Productivity Analysis

Suppose that one is interested in measuring relative efficiencies of decision making units (DMU's) participating in a production activity. Let S be the set of all technically feasible points $(\boldsymbol{x}, y)$, called *the production set*, where $\boldsymbol{x}$ is the vector of p inputs used for the production activity and y is the produced output. Then, the upper boundary of S, represented by the equation $y = g(\boldsymbol{x})$ for a function g defined on $\mathbb{R}_+^p$, forms the set of best production plans. In other words, the value of $g(\boldsymbol{x})$ stands for the maximal amount of output attainable by using the input level $\boldsymbol{x}$. The set $\{(\boldsymbol{x}, y) : y = g(\boldsymbol{x})\}$ is called *the production frontier*, and one may evaluate the efficiency of a DMU by referencing it. Since each point in S corresponds to a production activity of a specific DMU, $\mathcal{X}_n$ can be considered as a set of observations on input and output levels observed from n DMU's. In practice, since neither the production set nor the production frontier are available, we have to estimate them using $\mathcal{X}_n$.

In economics it is often assumed that the production set S is convex and satisfies

$$(\boldsymbol{x}, y) \in S \;\Rightarrow\; (\boldsymbol{x}', y') \in S \text{ if } \boldsymbol{x}' \geq \boldsymbol{x} \text{ and } y' \leq y. \tag{9.10}$$

The property (9.10) is referred to as *free disposability*, and (strict) convexity as *variable returns-to-scale (VRS)*. Convexity and free disposability of S imply that the function g is monotone increasing and concave in $\boldsymbol{x} \in \mathbb{R}^p$. In this case, the convex-hull estimator defined in Sect. 8.1 is slightly modified to take into account free disposability. Instead of the convex-hull $\hat{S}$ of $\mathcal{X}_n$, one uses as an estimator of S the smallest free disposable set that contains $\hat{S}$, and as an estimator of g

$$\hat{g}_{\text{dea}}^{\text{vrs}}(\boldsymbol{x}) = \max\left\{\boldsymbol{\gamma}^\top \boldsymbol{Y} : \boldsymbol{x} \geq \boldsymbol{X}^\top \boldsymbol{\gamma} \text{ for some } \boldsymbol{\gamma} \geq \boldsymbol{0} \text{ such that } \boldsymbol{1}^\top \boldsymbol{\gamma} = 1\right\}. \tag{9.11}$$

The estimator at (9.11) is called the *data envelopment analysis* (*DEA*) estimator.

In addition to convexity and free disposability, it is often assumed that S satisfies

$$aS \equiv S \quad \text{for all scalar } a > 0,$$

which is referred to as *constant returns-to-scale (CRS)* property. The corresponding production frontier g satisfies the identity $g(a\boldsymbol{x}) = ag(\boldsymbol{x})$ for all scalar $a > 0$. The DEA estimator at (9.11) in this case is modified to

$$\hat{g}_{\text{dea}}^{\text{crs}}(\boldsymbol{x}) = \max\left\{\boldsymbol{\gamma}^\top \boldsymbol{Y} : \boldsymbol{x} \geq \boldsymbol{X}^\top \boldsymbol{\gamma} \text{ for some } \boldsymbol{\gamma} \geq \boldsymbol{0}\right\}. \tag{9.12}$$

For some detailed discussion on the VRS and CRS assumptions from an economic point of view, one may refer to Chaps. 2 and 3 in Coelli et al. (2005) and Chap. 5 in Cooper et al. (2000).

The free disposable version in (9.11) equals $\hat{g}_{\text{conv}}$ in (9.1) with probability tending to one in the region of $\boldsymbol{x}$ where the production frontier function g is strictly increasing. This is Proposition 2 of Jeong and Park (2006). Also, it is easy to see that $\hat{g}_{\text{dea}}^{\text{crs}}$ in (9.12) is identical to the estimator $\hat{g}_{\text{coni}}$ defined at (9.9). Thus, under the free disposability assumption, the limiting distributions of $\hat{g}_{\text{dea}}^{\text{vrs}}$ and $\hat{g}_{\text{dea}}^{\text{crs}}$ are the same as those of the estimators $\hat{g}_{\text{conv}}$ and $\hat{g}_{\text{coni}}$, respectively, studied in Sects. 9.3 and 9.5.

As a final remark, we note that the problem of measuring the radial efficiency in the full multivariate setup (with multiple inputs and multiple outputs) can be translated to that of estimating a convex boundary function g with multiple inputs. This is done by making a canonical transformation on the output space so that the problem with multiple outputs is reduced to the case of a single output. The radial efficiency of an input×output pair $(\boldsymbol{x}_0, \boldsymbol{y}_0)$ is defined by

$$\lambda(\boldsymbol{x}_0, \boldsymbol{y}_0) = \sup\{\lambda > 0 : (\boldsymbol{x}_0, \lambda \boldsymbol{y}_0) \in S\}$$

for a production set S of input×output pairs $(\boldsymbol{x}, \boldsymbol{y}) \in \mathbb{R}_+^p \times \mathbb{R}_+^q$. The canonical map T on the output space introduces a new coordinate system on which the q outputs turn into $(q-1)$ inputs and one output. Specifically, one takes $T(\boldsymbol{y}) = (\boldsymbol{u}, w) \in \mathbb{R}^{q-1} \times \mathbb{R}_+$ where $\boldsymbol{u} = \Gamma^\top \boldsymbol{y}$, $w = \boldsymbol{y}_0^\top \boldsymbol{y}/\|\boldsymbol{y}_0\|$, and Γ is $q \times (q-1)$ matrix whose columns form a normalized basis for $\boldsymbol{y}_0^\perp$. If one defines a function on $\mathbb{R}^{p+q-1}$ by

$$g_T(\boldsymbol{x}, \boldsymbol{u}) = \sup\left\{w > 0 : \left(\boldsymbol{x}, \Gamma\boldsymbol{u} + w\frac{\boldsymbol{y}_0}{\|\boldsymbol{y}_0\|}\right) \in S\right\},$$

then the production set S is transformed to $S_T = \{(\boldsymbol{x}, \boldsymbol{u}, w) : 0 \le w \le g_T(\boldsymbol{x}, \boldsymbol{u})\}$ and $\lambda(\boldsymbol{x}_0, \boldsymbol{y}_0) = g_T(\boldsymbol{x}_0, \boldsymbol{0})/\|\boldsymbol{y}_0\|$. This means that the problem of estimating the radial efficiency $\lambda(\boldsymbol{x}_0, \boldsymbol{y}_0)$ from a set of observations in S reduces to the estimation of g_T at $(\boldsymbol{x}_0, \boldsymbol{0}) \in \mathbb{R}^p \times \mathbb{R}^{q-1}$ from the transformed observations in S_T. Thus, the statistical properties of the DEA estimators of $\lambda(\boldsymbol{x}_0, \boldsymbol{y}_0)$ are readily obtained from those of the DEA estimators of the boundary function g. This technique has been used in several works including Jeong (2004) and Park et al. (2010).

References

Bickel, P.J. (2003). Unorthodox bootstraps. *Journal of the Korean Statistical Society, 32*, 213–224.

Bickel, P.J., & Sakov, A. (2008). On the choice of m in the m out of n bootstrap and confidence bounds for extrema. *Statistica Sinica, 18*, 967–985.

Bickel, P.J., Götze, F., & van Zwet, W.R. (1997). Resampling fewer than n observations: Gains, losses, and remedies for losses. *Statistica Sinica, 7*, 1–31.

Coelli, T.J., Prasada Rao, D.S., O'Donnell, C.J., & Battese, G.E. (2005). *An introduction to efficiency and productivity analysis*, 2nd edn. New York: Springer.

Cooper, W.W., Seiford, L.M., & Tone, K. (2000). *Data envelopment analysis: a comprehensive text with models, applications, references and DEA-solver software*. The Netherlands: Kluwer Academic Publishers.

Gattoufi, S., Oral, M., & Reisman, A. (2004). Data envelopment analysis literature: a bibliography update (1951–2001). *Socio-Economic Planning Sciences, 38*, 159–229.

Gijbels, I., Mammen, E., Park, B.U., & Simar, L. (1999). On estimation of monotone and concave frontier functions. *Journal of the American Statistical Association, 94*, 220–228.

Hall, P., & Park, B.U. (2002). New methods for bias correction at endpoints and boundaries. *Annals of Statistics, 30*, 1460–1479.

Jeong, S.-O. (2004). Asymptotic distribution of DEA efficiency scores. *Journal of the Korean Statistical Society, 33*, 449–458.

Jeong, S.-O., & Park, B.U. (2006). Large sample approximation of the limit distribution of convex-hull estimators of boundaries. *Scandinavian Journal of Statistics*, *33*, 139–151.

Kneip, A., Park, B.U., & Simar, L. (1998). A note on the convergence of nonparametric DEA estimators for production efficiency scores. *Econometric Theory*, *14*, 783–793.

Kneip, A., Simar, L., & Wilson, P. (2008). Asymptotics and consistent bootstraps for DEA estimators in nonparametric frontier models. *Econometric Theory*, *24*, 1663–1697.

Park, B.U., Jeong S.-O., & Simar, L. (2010). Asymptotic distribution of conical-hull estimators of directional edges. *Annals of Statistics*, *38*, 1320–1340.

Politis, D.N., & Romano, J.P. (1994). Large sample confidence regions based on subsamples under minimal assumptions. *Annals of Statistics*, *22*, 2031–2050.

Politis, D.N., Romano, J.P., & Wolf, M. (1999). *Subsampling*. New York: Springer.

Wu, C.F.J. (1990). On the asymptotic properties of the jackknife histogram. *Annals of Statistics*, *18*, 1438–1452.

Chapter 10
The Skewness Issue in Stochastic Frontiers Models: Fact or Fiction?

Pavlos Almanidis and Robin C. Sickles

Abstract Skewness plays an important role in the stochastic frontier model. Since the model was introduced by Aigner et al. (J. Econometric 6:21–37, 1977), Meeusen and van den Broeck (Int. Econ. Rev. 18:435–444, 1997), and Battese and Cora (Aust. J. Agr. Econ. 21:169–179, 1977), researchers have often found that the residuals estimated from these models displayed skewness in the wrong direction. In such cases applied researchers were faced with two main and often overlapping alternatives, either respecifying the model or obtaining a new sample, neither of which are particularly appealing due to inferential problems introduced by such data-mining approaches. Recently, Simar and Wilson (Econometric Rev. 29:62–98, 2010) developed a bootstrap procedure to address the skewness problem in finite samples. Their findings point to the latter alternative as potentially the more appropriate-increase the sample size. That is, the skewness problem is a finite sample one and it often arises in finite samples from a data generating process based on the correct skewness. Thus the researcher should first attempt to increase the sample size instead of changing the model specification if she finds the "wrong" skewness in her empirical analyses. In this chapter we consider an alternative explanation to the "wrong" skewness problem and offer a new solution in cases where this is not a mere finite sample fiction but also a fact. We utilize the Qian and Sickles (Stochastic Frontiers with Bounded Inefficiency, Rice University, Mimeo, 2008) model in which an upper bound to inefficiencies or a lower bound to efficiencies is specified based on a number of alternative one-sided bounded inefficiency distributions. We consider one of the set of specifications considered by Qian and Sickles (Stochastic Frontiers with Bounded Inefficiency, Rice University, Mimeo, 2008) wherein inefficiencies are assumed to be doubly-truncated normal. This allows the least square residuals to display skewness in both directions and

---

P. Almanidis (✉) · R.C. Sickles
Department of Economics - MS 22, Rice University, 6100 S. Main Street, Houston, Texas 77005-1892, USA
e-mail: pa1@rice.edu; rsickles@rice.edu

I. van Keilegom and P.W. Wilson (eds.), *Exploring Research Frontiers in Contemporary Statistics and Econometrics*, DOI 10.1007/978-3-7908-2349-3_10,

nests the standard half-normal and truncated-normal inefficiency models. We show and formally prove that finding incorrect skewness does not necessarily indicate that the stochastic frontier model is misspecified in general. Misspecification instead may arise when the researcher considers the wrong distribution for the bounded inefficiency process. Of course if the canonical stochastic frontier model is the proper specification the residuals still may have the incorrect skew in finite samples but this problem goes away as sample size increases. This point was originally made in Waldman (Estimation in Economic Frontier Functions, Unpublished manuscript, University of North Carolina, Chapel Hill, 1977) and Olson et al. (J. Econometric. 13:67–82, 1980). We also conduct a limited set of Monte Carlo experiments that confirm our general findings. We show that "wrong" skewness can be a large sample issue. There is nothing inherently misspecified about the model were this to be found in large samples if one were to consider the bounded inefficiency approach. In this way the "wrong" skewness, while problematic in standard models, can become a property of samples drawn from distributions of bounded inefficiencies.

10.1 Introduction

The stochastic frontier model (SFM) was introduced by Aigner et al. (1977), Meeusen and van den Broeck (1977), and Battese and Cora (1977). The SFM assumes that a parametric functional form exists between the dependent and independent variables, as opposed to the alternative approaches of data envelopment analysis (DEA) proposed by Charnes et al. (1978) and the free disposal hull (FDH) of Deprins et al. (1984). In the SFM model the error is assumed to be composed of two parts, a one-sided term that captures the effects of inefficiencies relative to the stochastic frontier and a two-sided term that captures random shocks, measurement errors and other statistical noise, and allows random variation of frontiers across firms. This formulation proved to be more realistic than, e.g., the deterministic frontier model proposed by Aigner and Chu (1968), since it acknowledges the fact that deviations from the frontier cannot be attributed solely to technical inefficiency which is under firm's control. Since the SFM was introduced a myriad of papers have emerged in the literature discussing either methodological or practical issues, as well as a series of applications of these models to the wide range of data sets. A detailed discussion of up to date innovations and empirical applications in this area can be found in Greene (2007).

Methodological developments in the SFM have been made in model specification and estimation techniques. There are two main methods of estimation that researchers adopt in general. One is based on traditional stochastic frontier models as they were first formulated and uses maximum likelihood estimation techniques (MLE) or linear regression (OLS). The other employs the Bayesian estimation methods introduced by van de Broeck et al. (1994) and Koop (1994), and Koop et al. (1995, 1997). These Bayesian approaches typically utilize Gibbs sampling algorithms with data augmentation and Markov chain Monte Carlo (MCMC) techniques to estimate the model parameters and individual or mean inefficiencies.

Kim and Schmidt (2000) provide a review and empirical comparison of these two methods in panel data models. Regarding model specification, researchers attempted to relax the most restrictive assumptions of the classical stochastic frontier model, such as nonflexible distribution specifications for inefficiencies and their statistical independence from regressors and the random noise. In this chapter we also pursue issues of model specification, specifically the specification of the distribution of the one-sided inefficiency term. In their pioneering work Aigner et al. (1977) proposed the normal distribution for the random error and a half normal distribution for the inefficiency process, while at the same time and independently Meeusen and van den Broeck (1977) proposed exponential distribution for the latter. These random errors were assumed to be independent and identically distributed across observations and statistically independent of each other. More flexible one-sided densities for the efficiency terms such as the gamma (Greene 1980a,b; Stevenson 1980) were also proposed. Other distributions such as the lognormal and Weibull were considered for the prior distribution of inefficiencies by Deprins and Simar (1989b) and Migon and Medici (2001) in Bayesian applications of the SFM. In subsequent years researchers using both classical and Bayesian methods have developed nonparametric approaches to relax the assumption of a specific distribution for inefficiencies. Researchers have also introduced time varying and firm-specific effects and have allowed for more general correlation structures for the random errors in panel stochastic frontier models. Sickles (2005) analyzed these latter developments in panel stochastic frontier models.

A common problem that arises in fitting stochastic frontier models is that the residuals estimated from the SFM may display skewness in the "wrong" direction. While the theory would predict a negative (positive) skewness in production (cost) frontiers in the population, researchers often discover that the sample residuals are positively (negatively) skewed. However, as Simar and Wilson (2010) point out, as do we in this chapter, in finite samples the skewness statistic of the least squares residuals can have the opposite sign from that the theory would predict. Indeed this is more frequent in cases of low dominance of the inefficiency process over the two-sided noise (see also Carree 2002). This is called "finite sample fiction". Nevertheless, researchers still consider the skewness statistic to be an important indicator of the specification of the stochastic frontier model. Therefore, whenever they find the residuals skewed in the "wrong" direction they tend to believe that the model is misspecified or the data are inconsistent with the SFM paradigm. Two course of actions are oftentimes taken: respecify the model or obtain a new sample which hopefully results in the desired sign of skewness. Instead of respecifying the model, applied researchers also often respecify their interpretation of the results by assuming away inefficiencies and utilizing straightforward least squares regression approaches.[1] This weak point of the stochastic frontier models is emphasized in a series of papers, some of which try to justify that this phenomenon might arise in

[1]This is in particular due to the results that Olson et al. (1980) and Waldman (1982) obtain for stochastic frontier models when half-normal distribution for inefficiencies is specified.

finite samples even for models that are correctly specified (see Greene 2007 and Simar and Wilson 2010 for more discussion on this point).

The above discussion brings us directly to the question raised in the title: Is the "wrong" skewness just a finite sample fiction, or it is fact? If it is a fiction, then the bagging method proposed by Simar and Wilson (2010) should be employed to make inference in stochastic frontier models. This method could also be generalized to the case of the bounded inefficiencies we discuss in this chapter, since the normal/half-normal model is a special case of the normal/doubly-truncated-normal model considered herein. We will show via simulations later in this chapter that the former can be recovered from the latter without imposing any a priori restrictions on model parameters. Our main concern in this chapter, however, is the case where this phenomenon is not a finite sample artifact but fact.

In general, this chapter intends to illustrate how the bounded inefficiency formulation proposed by Qian and Sickles (2008) might overcome the issue of the "wrong" skewness in the stochastic frontier model. We first show that the imposition of an upper bound to inefficiency (lower bound to efficiency) enables the distribution of the one-sided inefficiency process to display positive and negative signs of skewness. This is in particular true for the truncated-normal distribution with strictly positive mean. Imposing a bound on the truncated-normal density function apart from the zero yields both positive or negative skewness depending on the position of the bound in the support of inefficiency distribution, thus justifying the occurrence of the so called "wrong" skewness.[2] We show that the normal/doubly-truncated-normal model is capable of handling and estimating the SFM model with "wrong" skewness and we show that it is also quite reasonable to obtain such a pattern of residuals in large samples. Our analysis can be extended to include the gamma and the Weibull distributions as well. It is also worth noting at this point, and it is shown later in the chapter, that the bound is not set a priori but can be estimated along with the rest of the parameters of the stochastic frontier model via the method of maximum likelihood. Moreover, even though the support of the distribution of the inefficiency term depends on unknown bound the composed error term has unbounded range and thus the regularity conditions for maximum likelihood are not violated unless we consider the full frontier model which assumes no stochastic term, a case not studied in current chapter. We also perform a series of Monte Carlo experiments on a stochastic frontier production function with bounded inefficiency and show that when we have a positively skewed distribution of errors we can still get very reasonable MLE estimates of the disturbance and inefficiency variances and as well as other parameters of the model. An interpretation of our results is that although a potential misspecification may occur if the stochastic frontier model is used and skewness is found to be "wrong" this can be avoided if the stochastic frontier model with bounded inefficiency is specified instead. This chapter

---

[2]The "wrong" skewness is as it seen from applied researcher's point of view. This is the reason why we put this word in quotes. We do not perceive this to be wrong. It is either finite sample artifact as discussed in Simar and Wilson (2010) or fact, which implies that the model itself is misspecified in the sense that the wrong distribution for inefficiencies is considered.

is organized as follows. In Sect. 10.2 the general problem of "wrong" skewness in stochastic frontier models and its implications are discussed, as well as solutions proposed in literature to solve it. Section 10.3 provides the main framework of the stochastic frontier model with bounded inefficiency. In Sect. 10.4 we check the validity of the bounded inefficiency model under the "wrong" skewness using simple models and generalize Waldman's proof to formally support the use of stochastic frontier models under these circumstances. Monte Carlo simulation results and concluding remarks are then discussed in Sect. 10.5.

10.2 Skewness Issue in Stochastic Frontier Analysis

10.2.1 "Wrong" Skewness and Its Importance in Frontier Models

We consider here the single cross-sectional (or potentially time-series) classical stochastic frontier model. We will discuss extension to panel stochastic frontiers briefly in a later section. We will also assume that the functional specification of technology or cost is linear in parameters. In this classical setting, the stochastic specification is $\varepsilon_i = \upsilon_i - u_i$ for production frontiers, or $\varepsilon_i = \upsilon_i + u_i$ for the case of cost frontiers. The stochastic term υ_i represents statistical noise and is usually assumed to be $i.i.d$ $N(0, \sigma_\upsilon^2)$ and $u_i \geq 0$ represents technical inefficiency and is usually assumed to be an $i.i.d$ random variable that follows some one-sided distribution. The error terms υ_i and u_i are also usually assumed to be statistically independent of each other and from the regressors. Given these assumptions, the distribution of the composed error is asymmetric and non-normal implying that simple least squares applied to a linear stochastic frontier model will be inefficient and will not provide us with an estimate of the degree of technical inefficiency. However, least squares does provide consistent estimates of all parameters except the intercept since $E(\varepsilon_i) = -E(u_i) \leq 0$. Moreover,

$$E[(\varepsilon_i - E[\varepsilon_i])^3] = E[(\upsilon_i - u_i + E[u_i])^3] = -E[(u_i - E[u_i])^3] \qquad (10.1)$$

which implies that the negative of the third moment of OLS residuals is a consistent estimator of the skewness of the one-sided error.

The common distributions for inefficiencies that appear in the literature are positively skewed, reflecting the fact that a large portion of the firms are expected to operate relatively close to the frontier. For production frontiers whenever we subtract the positively skewed inefficiency component from the symmetric error the composite error should display negative skewness. We will focus on the production function but clearly all that we say about it can be said about the cost function with a sign change on the one-sided error in the composed error term. Thus researchers find stochastic frontier models inappropriate to model inefficiencies if they obtain residuals skewed in the "wrong" direction. The typical conclusion is that, either the model is misspecified or the data are not compatible with the model. However, there can be a third interpretation as well based on the fact that inefficiencies might have been drawn from a distribution which displays negative skewness. This simply says that if the "wrong" skewness is not a finite sample artifact but fact, then any stochastic frontier model based on inefficiencies that are drawn from positively skewed distributions will be misspecified.

The first formal discussion on skewness problem is found in Olson et al. (1980) in their derivation of modified ordinary least squares (MOLS) estimates as a convenient alternative to maximum likelihood estimates. They explicitly assume half-normal distribution for technical inefficiencies in their formulation. MOLS method estimates the slope parameters by OLS. These are unbiased and consistent under standard assumptions about the regressors and the error terms. On the other hand, the OLS estimate of the constant term is shown to be biased and inconsistent. The bias-corrected estimator of the constant term is obtained by adding $\sqrt{2/\pi}\sigma_u$ term which is the expected value of the composed error term. Of course we do not know σ_u. Estimates of σ_v^2 and σ_u^2 are derived by method of moments using the second and third moments of OLS residuals. These are consistent, although not asymptotically efficient, and are given by

$$\hat{\sigma}_u^2 = \left[\sqrt{\pi/2}\left(\frac{\pi}{\pi-4}\right)\hat{\mu}_3\right]^{2/3} \tag{10.2}$$

and

$$\hat{\sigma}_v^2 = \hat{\mu}_2 - \left(\frac{\pi-2}{\pi}\right)\hat{\sigma}_u^2 \tag{10.3}$$

where $\hat{\mu}_2$ and $\hat{\mu}_3$ are the estimated second and third moments of the OLS residuals, respectively.

It is obvious from (10.2) that a serious flaw in this method occurs whenever $\hat{\mu}_3$ is positive, since the estimated variance of inefficiencies becomes negative. This is referred by Olson et al. (1980) as a "Type I" failure of MOLS estimators. A "Type II" failure arises whenever $\hat{\mu}_2 < (\frac{\pi-2}{\pi})\hat{\sigma}_u^2$. Waldman (1982) proved that MLE estimate of σ_u^2 in this case is zero and that the model parameters can be efficiently estimated by OLS. We will outline the main steps and results of Waldman's proof which are necessary benchmarks and links to our further analysis.

Starting from the log-likelihood function for the normal-half-normal model

$$\log L = n\log(\sqrt{2/\pi}) - n\log(\sigma) + \sum_{i=1}^{n}\log\left[1-\Phi\left(\frac{\varepsilon_i\lambda}{\sigma}\right)\right] - \frac{1}{2\sigma^2}\sum_{i=1}^{n}\varepsilon_i^2 \tag{10.4}$$

where $\varepsilon_i = y_i - x_i\beta$, $\lambda = \frac{\sigma_u}{\sigma_v}$, $\sigma^2 = \sigma_v^2 + \sigma_u^2$, and $\Phi(\bullet)$ denotes the cdf of the standard normal distribution, Waldman notes there are two stationary points that potentially can characterize the log-likelihood function. Defining the parameter vector $\theta = (\beta', \sigma^2, \lambda)$, the first stationary point would be the one for which the first derivatives of the log-likelihood function are zero while the second is the OLS solution for θ wherein the parameter λ is set to zero. The superiority of these two stationary points is then compared in cases of the wrong skewness. One way to do this is to examine the second-order derivative matrix of log-likelihood function evaluated at these two points. The Hessian matrix evaluated at OLS solution, $\theta^* = (b', s^2, 0)$, is

$$H(\theta^*) = \begin{bmatrix} -s^{-2}\sum_{i=1}^n x_i x_i' & \sqrt{2/\pi} s^{-1} \sum_{i=1}^n x_i & 0 \\ \sqrt{2/\pi} s^{-1} \sum_{i=1}^n x_i & -2n/\pi & 0 \\ 0 & 0 & -n/2s^4 \end{bmatrix} \tag{10.5}$$

where $b = \left(\sum_{i=1}^n x_i x_i'\right)^{-1} \sum_{i=1}^n x_i y_i$, $s^2 = \frac{1}{n}\sum_{i=1}^n e_i^2$ and e_i is the least squares residual.

This matrix is singular with $k+1$ negative characteristic roots and one zero root. This essentially would require the log-likelihood function to be examined in the direction determined by the characteristic vector associated with this zero root which is given by the vector $z = (s\sqrt{2/\pi}, 1, 0)$. Departing from the point of OLS solution, the term of interest is then the sign of

$$\begin{aligned} \Delta \log L &= \log L(\theta^* + \delta z) - \log L(\theta^*) \\ &= -\delta^2 \frac{n}{\pi} + \sum_{i=1}^n \log[2 - 2\Phi(e_i \delta s^{-1} - \delta^2 \sqrt{2/\pi})] \end{aligned} \tag{10.6}$$

where $\delta > 0$ is an arbitrary small number. If we expand $\Delta \log L$ using a Taylor series it can be shown that

$$\Delta \log L = (\delta^3/6s^3)\sqrt{2/\pi}[(\pi - 4)/\pi] \sum_{i=1}^n \varepsilon_i^3 + O(\delta^4). \tag{10.7}$$

Thus if the term $\sum_{i=1}^n \varepsilon_i^3 > 0$ then the maximum of the log-likelihood function is located at the OLS solution, which is superior to MLE. This result suggests two strategies for practitioners: apply OLS whenever the least squares residuals display positive skewness or increase the sample size, since

$$\text{plim}\left(\frac{1}{n}\sum_{i=1}^n \varepsilon_i^3\right) = \sigma_u^3 \sqrt{2/\pi}[(\pi - 4)/\pi] < 0 \tag{10.8}$$

which implies that asymptotically the sample third moment of least squares residuals converges to its population counterpart by the law of large numbers and thus the problem of the "wrong" skewness goes away.

Undoubtedly, this is true if the inefficiencies are indeed drawn from the half-normal distribution which is positively skewed. What if they are not? What if they are drawn from the distribution which displays negative skewness as well? We will attempt to give answers to these questions.

The problem of the "wrong" skewness is also made apparent and emphasized by the two widely-used computer packages used to estimate stochastic frontiers.

The first package LIMDEP 9.0, which is developed by Greene (2007), calculates and checks the skewness of the OLS residuals just before maximum likelihood estimation begins. In case the sign of the skewness statistic is positive, significantly or not, the message appears that warns the user about the misspecification of the model and suggests using OLS instead of MLE. The second software FRONTIER 4.1, produced by Coelli (1996), also obtains the OLS estimates as a starting values for the grid search of starting value of the γ parameter.[3] If the skewness is positive, the final maximum likelihood value of this parameter is very close to zero, indicating no inefficiencies. More detailed description and comparison of FRONTIER 4.1 and the earlier version 7.0 of LIMDEP can be found in Sena (1999).

Related to these results, several parametric and non-parametric test statistics have been developed to check the skewness of least squares residuals in stochastic frontier models. Schmidt and Lin (1984) proposed the test statistic

$$\sqrt{b_1} = \frac{m_3}{m_2^{3/2}} \tag{10.9}$$

where m_2 and m_3 represent the second and the third moment of the empirical distribution of the least squares residuals. The distribution of $\sqrt{b_1}$ is not standard and the application of this test requires special tables provided by D'Agostino and Pearson (1973). Coelli (1995) proposed an alternative statistic for testing whether the third moment of residuals is greater than or equal to zero

$$\sqrt{b_1^*} = \frac{m_3}{(6m_2^3/N)^{1/2}} \tag{10.10}$$

where N denotes the number of observations in the sample. Under the null hypothesis of zero skewness, the third moment of OLS residuals is asymptotically distributed as a normal random variable with zero mean and variance $6m_2^3/N$. This implies that $\sqrt{b_1^*}$ is asymptotically distributed as a standard normal variable and one can consult the corresponding statistical tables for making an inference. These two tests, although easily computed and implemented, have unknown finite sample properties. Coelli (1995) conducts Monte Carlo experimentations and shows that $\sqrt{b_1^*}$ has correct size and good power in small samples. This is the reason why this test statistic is commonly accepted and used in applications.

10.2.2 Solutions to the "wrong" Skewness

Nonetheless, the standard solutions considered in the case of "wrong" skewness essentially constitute no solutions with regard to the stochastic frontier model. Setting the variance of inefficiency process to be equal to zero based on the skewness of OLS residuals is not a very comforting solution. This solution to the

[3] $\gamma = \frac{\sigma_u}{\sigma_u + \sigma_v}$, which is another reparametrization used in stochastic frontier models

problem would imply that all firms in the industry are fully efficient. Moreover, the estimated standard errors will not be correct if straightforward OLS is applied to the data.[4] Moreover, data-mining techniques will introduce inferential problems and possibly biases in parameters and their standard errors (Leamer 1978). Carree (2002), Greene (2007), and Simar and Wilson (2010) note that in finite samples, even the correctly specified stochastic frontier model is capable of producing least squares residuals with the "wrong" skewness sign with relatively high frequency. Thus another suggested solution is to get more data. Of course the availability of the data in economics is often rather limited and this alternative may not possible in many empirical settings. Another solution is to argue that the inefficiencies are drawn from an efficiency distribution with negative skewness. A major problem with this assumption is that it implies that there is only a very small fraction of the firms that attain a level of productivity close to the frontier. For example, Carree (2002) considers a distribution for inefficiencies that allows for both, negative and positive skewness.[5] He proposes a binomial distribution $b(n, p)$ which for a range of values of the parameter p is negatively skewed.[6] This is a discrete distribution wherein continuous inefficiencies fall into discrete "inefficiency categories". He employs the method-of-moments estimators as in Olson et al. (1980) and Greene (1990) and provides an explanation for how theoretically and empirically the "wrong" skewness issue may arise in stochastic frontier model.[7] Empirically, the use of the binomial distribution can be justified by a model in which the cycle of innovations and imitations occurs. This would suggest that the occurrence of positively skewed residuals would correspond to the cases where very few firms in the industry innovate while the large proportion of firms experience large inefficiencies. In contrast, as it will be shown later, the stochastic frontier model with doubly-truncated normal inefficiencies does not imply such a pattern in firms' inefficiencies, but instead it precludes the probability of occurrence of extreme inefficiencies.

10.3 Stochastic Frontier Model with Bounded Inefficiency

10.3.1 Model

In this section we briefly introduce the stochastic frontier model with bounded inefficiency proposed by Qian and Sickles (2008). The formulation of the model is similar to the traditional stochastic frontier model. The key difference is that an

---

[4]We thank anonymous referee for this point.

[5]Carree (2002) also argues that distributions with bounded range can be negatively skewed but further development of these is not pursued by the author.

[6]Other authors also considered distributions with negative skew (see Johnson et al. 1992, 1994).

[7]The shortcoming of this approach is that method-of-moments estimators may not be defined for some empirical values of the higher sample moments of the least squares residuals

upper bound to inefficiencies or a lower bound to efficiencies is specified. This amounts to imposing a second truncation point other than zero to the distribution of the inefficiency process. Thus, if the model is $y_i = x_i\beta + \varepsilon_i$, the composed error is $\varepsilon_i = \upsilon_i - u_i$ where $\upsilon_i \sim^{iid} N(0, \sigma_\upsilon^2)$ is statistical noise and υ_i and u_i are assumed to be statistically independent from each other and from regressors, we assume that the u_i, which represent the unobserved technical inefficiencies, are non-negative random variables with doubly-truncated-normal distribution density $f(\cdot)$ defined on the positive domain. The doubly-truncated normal distribution nests the truncated-normal, the truncated-half-normal and the half-normal distribution.

The initial purpose of the bounded inefficiency model was to introduce a stochastic frontier model in which the bound can be used for gauging the tolerance for or ruthlessness against the inefficient firms and thus to serve as an index of competitiveness of an industry. It can be thought as naturally instituted by the market, competitive in most of the cases, and precludes the existence of extremely inefficient firms. Another purpose was to introduce (in a panel data setting) another time-varying technical efficiency model in the literature. However, in this chapter we note another useful feature of this model, which is reflected in the flexibility of the one-sided distribution of inefficiencies. This flexibility enables the truncated-normal distribution with strictly positive mean to display positive, negative, and zero skewness. This potentially attractive option leads us to take a closer look at the doubly-truncated-normal inefficiencies whose density is given by

$$f_u(x) = \frac{\frac{1}{\sigma_u}\phi(\frac{x-\mu}{\sigma_u})}{\Phi(\frac{B-\mu}{\sigma_u}) - \Phi(\frac{-\mu}{\sigma_u})} I_{[0,B]}(x), \qquad \sigma_u > 0, B > 0 \tag{10.11}$$

where $\Phi(\cdot)$ and $\phi(\cdot)$ are the cdf and pdf of the standard normal distribution respectively, and $I(\cdot)$ denotes the indicator function. Under parametric restrictions, it can be verified that this distribution generalizes the truncated-normal ($B = \infty$), truncated-half-normal ($\mu = 0$), and the half-normal ($B = \infty, \mu = 0$) distributions.

10.3.2 Estimation

Using the parametrization of Aigner et al. (1977), the log-likelihood function for the normal/doubly-truncated normal model composed error for the SFM is given by

$$\begin{aligned} \log(L) = &-n \log\left[\Phi\left(\frac{B-\mu}{\sigma_u(\sigma,\lambda)}\right) - \Phi\left(\frac{-\mu}{\sigma_u(\sigma,\lambda)}\right)\right] \\ &-n\log\sigma - \frac{n}{2}\log(2\pi) - \sum_{i=1}^{n}\frac{(\varepsilon_i+\mu)^2}{2\sigma^2} \\ &+\sum_{i=1}^{n}\log\left\{\Phi\left(\frac{(B+\varepsilon_i)\lambda + (B-\mu)\lambda^{-1}}{\sigma}\right) - \Phi\left(\frac{\varepsilon_i\lambda - \mu\lambda^{-1}}{\sigma}\right)\right\} \end{aligned} \tag{10.12}$$

where $\sigma_u(\sigma,\lambda) = \frac{\sigma}{\sqrt{1+\frac{1}{\lambda^2}}}$, $\lambda = \frac{\sigma_u}{\sigma_\upsilon}$, and $\varepsilon_i = y_i - x_i\beta$.

This log-likelihood function can be maximized to obtain the MLE estimates of the model parameters along with the parameter that determines the bound of the one-sided distribution. The conditional distribution of the inefficiency term $E[u_i|\varepsilon_i = \hat{\varepsilon}_i]$, where $\hat{\varepsilon}_i = y_i - X_i\hat{\beta}$, can be used in the same spirit as in Jondrow et al. (1982) to derive individual and mean technical inefficiencies. It should be noted that, while the support of the distribution of u depends on the bound, the support of the composite error is unbounded. Hence, this regularity condition for MLE is not violated. However, global identifiability of this model fails (Rothenberg 1971), which is also true for the normal/truncated-normal model, and thus some parameters are only identified locally.

10.4 Skewness Statistic Under the Bounded Inefficiencies

10.4.1 Derivation of Skewness and MOLS Estimates with Doubly-Truncated-Normal Inefficiencies

The location parameter of the doubly-truncated-normal distribution, as a function of the bound (B), mean (μ), and variance of the normal distribution (σ_u^2) is given by

$$\psi_1(B, \mu, \sigma_u^2) = E(u) = \mu + \sigma_u \eta \tag{10.13}$$

with

$$\eta \equiv \frac{\phi(\xi_1) - \phi(\xi_2)}{\Phi(\xi_2) - \Phi(\xi_1)} \tag{10.14}$$

where $\xi_1 = \frac{-\mu}{\sigma_u}$, $\xi_2 = \frac{B-\mu}{\sigma_u}$, while $\Phi(\cdot)$ and $\phi(\cdot)$ are the cdf and pdf of the standard normal distribution, respectively. η represents the inverse Mill's ratio and it is equal to $\sqrt{2/\pi}$ in the normal-half-normal model and ξ_1 and ξ_2 are the lower and upper truncation points of the standard normal density, respectively.

The central population moments up through order four as a functions of B, μ, and σ_u^2 are given by

$$\psi_2(B, \mu, \sigma_u^2) = \sigma_u^2 \left(1 - \eta^2 + \frac{\xi_1\phi(\xi_1) - \xi_2\phi(\xi_2)}{\Phi(\xi_2) - \Phi(\xi_1)}\right) \tag{10.15}$$

$$\psi_3(B, \mu, \sigma_u^2) = \sigma_u^3 \left(2\eta^3 - \left[3\left(\frac{\xi_1\phi(\xi_1) - \xi_2\phi(\xi_2)}{\Phi(\xi_2) - \Phi(\xi_1)}\right) + 1\right]\eta + \frac{\xi_1^2\phi(\xi_1) - \xi_2^2\phi(\xi_2)}{\Phi(\xi_2) - \Phi(\xi_1)}\right) \tag{10.16}$$

$$\psi_4(B,\mu,\sigma_u^2) = \sigma_u^4\Bigg(-3\eta^4 + 2\left[3\frac{\xi_1\phi(\xi_1)-\xi_2\phi(\xi_2)}{\Phi(\xi_2)-\Phi(\xi_1)} - 1\right]\eta^2 \qquad (10.17)$$
$$-4\eta\left(\frac{\xi_1^2\phi(\xi_1)-\xi_2^2\phi(\xi_2)}{\Phi(\xi_2)-\Phi(\xi_1)}\right) + 3\left(\frac{\xi_1\phi(\xi_1)-\xi_2\phi(\xi_2)}{\Phi(\xi_2)-\Phi(\xi_1)}\right)$$
$$+\frac{\xi_1^3\phi(\xi_1)-\xi_2^3\phi(\xi_2)}{\Phi(\xi_2)-\Phi(\xi_1)} + 3\Bigg)$$

Three special cases immediately arise from the doubly-truncated-normal distribution. If we let $\xi_1 = -\infty$ and $\xi_2 = \infty$ then η becomes zero and if we additionally use the L'Hospital's Rule, it can be shown that $\lim_{\xi_1\to-\infty}\xi_1\phi(\xi_1) = \lim_{\xi_2\to\infty}\xi_2\phi(\xi_2) = 0$ and $\lim_{\xi_1\to-\infty}\xi_1^2\phi(\xi_1) = \lim_{\xi_2\to\infty}\xi_2^2\phi(\xi_2) = 0$, in which case we obtain exactly the cumulants of the normal distribution. On the other hand, if only the lower truncation point exists and μ is non zero we obtain results for the truncated-normal distribution. Setting μ to zero we obtain the moments of the half-normal distribution.

The skewness of the doubly-truncated-normal distribution is derived from expressions in (10.15) and (10.16) as

$$\gamma_1(B,\mu,\sigma_u^2) = \frac{\psi_3}{\psi_2^{3/2}} = \frac{\psi_3}{\psi_2\sqrt{\psi_2}}$$
$$= \frac{\left(2\eta^3 - \left[3\left(\frac{\xi_1\phi(\xi_1)-\xi_2\phi(\xi_2)}{\Phi(\xi_2)-\Phi(\xi_1)}\right)+1\right]\eta + \frac{\xi_1^2\phi(\xi_1)-\xi_2^2\phi(\xi_2)}{\Phi(\xi_2)-\Phi(\xi_1)}\right)}{\left[1-\eta^2+\frac{\xi_1\phi(\xi_1)-\xi_2\phi(\xi_2)}{\Phi(\xi_2)-\Phi(\xi_1)}\right]^{3/2}}. \qquad (10.18)$$

The skewness parameter describes the shape of the distribution independent of location and scale. Although many non-symmetric distributions have either positive or negative skewness statistic for the doubly-truncated-normal distribution the sign of skewness is ambiguous. It is either positive, whenever $B > 2\mu$, or nonpositive when $B \le 2\mu$ for μ strictly positive. This follows from the fact that the bound B is strictly positive by assumption. We provide a graphical representation of doubly-truncated inefficiencies with $\mu > 0$ in the appendix.[8]

The consequences of both positive and negative skewness of the doubly-truncated-normal distribution in the SFM are not clear. The residuals in a SFM can be skewed in both directions while the variance of inefficiency term is nonzero. Moreover, in finite samples the sampling variability of the skewness statistic

[8]It should be noted that parameter μ is not restricted to be strictly positive in estimation procedure. It takes negative values as well.

could give rise to a positive or negative skewness statistic even if the population skewness parameter was negative or positive. The issues for estimation of the SFM when the inefficiencies are doubly-truncated-normal are similar to those when the inefficiencies are truncated-normal. MOLS estimates can be used to obtain method-of-moments estimates of λ and σ. The second and higher order moments of ε when it is a mixture of a normal and doubly-truncated normal distribution are nonlinear functions of parameters. Due to the two-folded nature of the normal distribution we cannot express the distributional parameters of the model uniquely as a function of the moments of the least squares residuals and the data. Hence, the global identifiability of the model fails (Rothenberg 1971), which is also the case for the normal/truncated-normal model. Identifiability is problematic even for large number of observations for large values of parameter μ. This is because for large μ the truncated-normal distribution converges to the normal distribution resulting in a mixture of normally distributed inefficiency and random components in the composed error. Correspondingly, it can be shown that the second order derivative of the log-likelihood function with respect to μ converges to zero as μ increases without bound, resulting in a nearly flat log-likelihood function in the dimension of μ of course in inferential problems as well. The existence of the bound avoids this problem. We provide a simple proof of the points we make above in the appendix. Greene (1997), and Ritter and Simar (1997) provide more discussion on identification issues in SFM. Local identification of both models can be shown by examining the Fisher's information matrix, which is nonsingular at any point of the parameter space Θ for which λ parameter is strictly greater than zero and is not too large. However, for $\lambda \to 0$ and $\lambda \to \infty$, Wang and Schmidt (2008) show that the distribution of $\hat{u}$ degenerates to a point mass at $E[u]$ or to the distribution of u, respectively. We do not pursue these two limiting cases but rather on intermediate cases for strictly positive and bounded λ.

The second and third central moments of the SFM residuals based on OLS are $\hat{m}_2 = \sigma_v^2 + \sigma_u^2(1-\eta^2+\frac{\xi_1\phi(\xi_1)-\xi_2\phi(\xi_2)}{\Phi(\xi_2)-\Phi(\xi_1)})$ and $\hat{m}_3 = -\sigma_u^3(2\eta^3-[3(\frac{\xi_1\phi(\xi_1)-\xi_2\phi(\xi_2)}{\Phi(\xi_2)-\Phi(\xi_1)})+1]\eta+\frac{\xi_1^2\phi(\xi_1)-\xi_2^2\phi(\xi_2)}{\Phi(\xi_2)-\Phi(\xi_1)})$ respectively. Solving these two equations we can derive the MOLS estimators of σ_v^2 and σ_u^2. The MOLS estimators of the variances of the inefficiency term and the random disturbance are given by

$$\hat{\sigma}_u^2 = \left[\frac{-\hat{m}_3}{2\eta^3 - \left[3\left(\frac{\xi_1\phi(\xi_1)-\xi_2\phi(\xi_2)}{\Phi(\xi_2)-\Phi(\xi_1)}\right)+1\right]\eta + \frac{\xi_1^2\phi(\xi_1)-\xi_2^2\phi(\xi_2)}{\Phi(\xi_2)-\Phi(\xi_1)}}\right]^{2/3} \tag{10.19}$$

and

$$\hat{\sigma}_v^2 = \hat{m}_2 - \hat{\sigma}_u^2\left(1-\eta^2+\frac{\xi_1\phi(\xi_1)-\xi_2\phi(\xi_2)}{\Phi(\xi_2)-\Phi(\xi_1)}\right) \tag{10.20}$$

To illustrate how skewness is indeterminate for the normal/doubly-truncated model we can fix the values of parameters B and μ and calculate σ_v^2 and σ_u^2 from

(10.19) and (10.20). In the normal/half-normal model these values are also fixed ($B = \infty$, $\mu = 0$). Since the negative of the third moment of the OLS residuals is an unbiased and consistent estimator of the skewness of inefficiencies, one can see that the estimate of the σ_u^2 can have positive sign even in the case of positively skewed residuals as opposed to the standard model. Most importantly, the "type I" failure goes away asymptotically since a positive $\hat{m}_3$ would imply that ψ_3 is negative, which occurs whenever $B < 2\mu$. Thus $\hat{\sigma}_u^2$ cannot take on negative values. In cases where we have $B = 2\mu$ the ratio in (10.19) is unidentified. By applying the L'Hospital rule and evaluating the limits it is straightforward to show that the variance of the inefficiency term is a strictly positive number. Only in the case when $B = 0$ is the variance of the inefficiency term zero.

We can test the extent to which the distribution of unobservable inefficiencies can display negative or positive skewness using the observable residuals based on the expression in (10.1). For this purpose we can utilize the adjusted for skewness test statistic proposed by Bera and Premaratne (2001), since the excess kurtosis is not zero. By using the standard test for skewness we will have either over-rejection or under-rejection of the null hypothesis of non-negative skewness and this will depend primarily on the sign of the excess kurtosis. In addition, since there are two points at which the doubly-truncated-normal distribution has zero skewness, the standard tests are not appropriate. Since the standard tests do not distinguish these two cases, application of standard skewness tests may lead researchers to accept the null hypothesis of zero variance when it is false at levels larger than nominal test size would suggest. The modified likelihood ratio statistic (see Lee 1993) is still asymptotically distributed as $\frac{1}{2}\chi^2(0) + \frac{1}{2}\chi^2(1)$ and should still be appropriate in this case.

10.4.2 Generalization of Waldman's Proof

We next examine the consistency and identifiability of the parameters of the normal/doubly-truncated normal bounded inefficiency stochastic frontier model utilizing the same approach as in Waldman (1982). To compare and contrast the problem of the "wrong" skewness with the benchmark case of the normal/half-normal model of Aigner et al. (1977), we set the values of the deep parameters B and μ and consider the scores of parameter vector $\theta = (\beta', \sigma^2, \lambda)$ as a function of these fixed parameters. Note that the normal/half-normal model fixes these values at ∞ and 0, respectively. We begin by examining the second-order derivative matrix evaluated at the OLS solution point, $\theta^* = (b', s^2, 0)$, given by

$$H(\theta^*) = \begin{bmatrix} -s^{-2}\sum_{i=1}^{n} x_i x_i' & -\frac{1}{s^4}\sum_{i=1}^{n}(e_i - \mu)x_i & 0 \\ -\frac{1}{s^4}\sum_{i=1}^{n}(e_i - \mu)x_i & \frac{n}{2s^4} - \frac{1}{s^6}\sum_{i=1}^{n}(e_i - \mu)^2 & 0 \\ 0 & 0 & 0 \end{bmatrix} \qquad (10.21)$$

where $b = (\sum_{i=1}^{n} x_i x_i')^{-1} \sum_{i=1}^{n} x_i y_i$, $s^2 = \frac{1}{n}\sum_{i=1}^{n} e_i^2$ and e_i is the least squares residual.

Obviously, $H(\theta^*)$ is singular with $k+1$ negative characteristic roots and one zero root. The eigenvector associated with this zero root is given by $z = (\mathbf{0}', 0, 1)$. We then need to search the sign of $\Delta \log L = \log L(\theta^* + \delta z) - \log L(\theta^*)$ in the positive direction ($\delta > 0$), since λ is constrained to be non-negative. By expanding the $\Delta \log L$ the first term in the series is 0 since OLS is a stationary point. The second term also vanishes since $|H(\theta^*)| = 0$. Thus, the only relevant point that remains to be considered is the third derivative of the log-likelihood function with respect to parameter λ evaluated at the OLS solution,

$$\frac{1}{6}\delta^3 \frac{\partial^3 \log(\theta^*)}{\partial \lambda^3} \tag{10.22}$$

Substituting for the third derivative and ignoring higher order terms, we can write

$$\Delta \log L \cong \frac{\delta^3}{6s^3}\left\{ -2\varpi^3 + \left[3\left(\frac{\omega_2\phi(\omega_2) - \omega_1\phi(\omega_1)}{\Phi(\omega_1) - \Phi(\omega_2)}\right) + 1\right]\varpi \right. \\ \left. - \frac{\omega_2^2\phi(\omega_2) - \omega_1^2\phi(\omega_1)}{\Phi(\omega_1) - \Phi(\omega_2)}\right\} \sum_{i=1}^{n} e_i^3 \tag{10.23}$$

where $\omega_1 = -\frac{\mu}{s}$, $\omega_2 = \frac{B-\mu}{s}$, and $\varpi = \frac{\phi(\omega_1) - \omega\phi(\omega_2)}{\Phi(\omega_2) - \Phi(\omega_1)}$.

The simple inspection of (10.23) reveals that the third order term of the least squares residuals need not always have the opposite sign of $\Delta \log L$. This will mainly depend on the relationship between the imposed bound B and the mean of the normal distribution μ. For $B < 2\mu$, ϖ is negative and the term in the curly brackets becomes positive. Thus positive skewness would imply the existence of inefficient firms in the sample. The implication of this is that whenever a researcher finds positively skewed residuals it may be the case that the inefficiencies have been drawn from a distribution that has negative skew. For $B = 2\mu$, $\Delta \log L = 0$ and in this case MLE should be employed since it will be more efficient than OLS and will provide us with technical inefficiency estimates. Asymptotically the third order term of OLS residuals and the expression in curly brackets have the same sign since

$$\text{plim}\left(\frac{1}{n}\sum_{i=1}^{n} e_i^3\right) = -\sigma_u^3\left(2\eta^3 - \left[3\left(\frac{\xi_1\phi(\xi_1) - \xi_2\phi(\xi_2)}{\Phi(\xi_2) - \Phi(\xi_1)}\right) + 1\right]\eta \right. \\ \left. + \frac{\xi_1^2\phi(\xi_1) - \xi_2^2\phi(\xi_2)}{\Phi(\xi_2) - \Phi(\xi_1)}\right) \tag{10.24}$$

which implies that we can observe the "wrong" skewness even in large samples. Thus, in our model the problem of the "wrong" skewness is not a just finite sample issue; positive or negative skewness of least squares residuals will always imply a positive variance of the inefficiency process in large samples. In finite samples, anything can happen. We can obtain negatively skewed residuals even if we sample from a negatively skewed distribution of inefficiencies.

10.5 Further Discussion and Conclusions

We also conduct Monte Carlo experiment in the same spirit as in Simar and Wilson (2010) wherein they note that in the finite samples even the correctly specified stochastic frontier model is capable of generating least squares residuals with the "wrong" skewness statistic with relatively high frequency. They calculate the proportion of samples with positively skewed residuals which converges to zero as the sample size grows large. We conduct the same experiment under the modified error specification allowing for bounded inefficiency and display the results in Table 10.1. Without loss of generality, we set the parameter μ to 1 and use the inverse CDF method to sample from a convolution of inefficiency and random noise distributions by varying the bound parameter. We examine the three cases of the skewness sign and compute the proportion of 1,000 samples with positive

Table 10.1 Proportion of samples with positively skewed residuals in normal/doubly-truncated-normal model

	n	$B=1$	$B=2$	$B=5$	$B=10$
	50	0.519	0.505	0.480	0.509
	100	0.481	0.501	0.516	0.520
	200	0.495	0.473	0.514	0.493
	500	0.487	0.503	0.539	0.507
	10^3	0.520	0.516	0.510	0.494
	10^4	0.504	0.483	0.512	0.498
$\lambda = 0.1$	10^5	0.532	0.492	0.437	0.405
	50	0.517	0.485	0.503	0.510
	100	0.545	0.491	0.459	0.479
	200	0.551	0.490	0.486	0.466
	500	0.520	0.488	0.431	0.459
	10^3	0.564	0.514	0.453	0.435
	10^4	0.684	0.491	0.397	0.318
$\lambda = 0.5$	10^5	0.759	0.496	0.107	0.092
	50	0.565	0.536	0.367	0.383
	100	0.524	0.513	0.317	0.335
	200	0.529	0.512	0.224	0.245
	500	0.567	0.514	0.155	0.122
	10^3	0.576	0.524	0.063	0.051
	10^4	0.709	0.501	0	0
$\lambda = 1$	10^5	0.943	0.503	0	0

skewness. The first column shows that the proportion of the samples with the positive ("wrong") skewness increases as the sample size grows larger. It converges to one as the variance of the one-sided inefficiency term becomes larger relative to the variance of two-sided error. In the second column we have the case where $B = 2\mu$. Under this case there is about a 50-50 chance that we generate a sample with positive skewness. In most of the cases, the positive skewness appears to be statistically insignificant. The third and forth columns correspond to the case where the distribution of inefficiencies is positively skewed and as in Simar and Wilson (2010) the proportion decreases as the sample size and parameter λ increase. Our findings again clearly indicate that the skewness issue is also a large sample issue, since for $B < 2\mu$ the proportion of the samples with positive skewness converges to one. This simply would mean that if the true DGP is based on inefficiencies that are drawn from a doubly-truncated-normal distribution, the researcher fails to recognize this and finds a skewness statistic with the "wrong" sign, then she will reject her model. Moreover, if there is the potential for increasing sample size and researcher keeps increasing it and finds continuously positive signs of skewness then she may erroneously conclude that all firms in her sample are super efficient. The flexibility of the bounded inefficiency model avoids this problem. Qian and Sickles (2008) also conduct experiments for other error specification such as truncated-half-normal ($\mu = 0$) and truncated exponential model. It is worth mentioning that in these cases, for certain levels of the bound, the skewness statistic becomes statistically insignificant and thus the null hypothesis of no inefficiency cannot be rejected, even if λ is not zero, when one applies the standard tests. We also note that if the DGP from which a sample of data is drawn has bounded inefficiency but this is ignored, then it will sometimes mask the true skewness. It is often the case in such settings that point estimates of skewness may have the "wrong" sign. However, this may simply due to the weak identifiability of skewness in a stochastic frontier with bounded inefficiency and the "wrong" sign is not "significantly wrong" in a statistical sense.

An important question for applied researchers is, what happens in the case if the true model is the normal/half-normal but we estimate the normal/doubly-truncated-normal model instead and visa-versa? To answer the first question, we conduct an additional Monte Carlo experiment to assess the validity of the model whenever the underlying true data generating process is the one proposed in ALS, that is wherein inefficiencies are drawn from the half-normal distribution. For this purpose, we specify the simple Cobb-Douglas production frontier with two inputs as

$$y_i = \alpha_0 + \alpha_1 \ln x_{1i} + \alpha_2 \ln x_{2i} + \upsilon_i - u_i$$

where $\upsilon_i \sim^{iid} N(0, \sigma_\upsilon^2)$ and $u_i \sim^{iid} N^+(0, \sigma_u^2)$. υ_i and u_i, as previously, are assumed to be independent of each over and from regressors.

Throughout, we set $\alpha_0 = 0.9$, $\alpha_1 = 0.6$, and $\alpha_2 = 0.5$. $\ln x_{ji}|_{j=1,2}$ are drawn from $N(\mu_{xj}, \sigma_{xj}^2)$ with $\mu_{x1} = 1.5$, $\mu_{x2} = 1.8$, and $\sigma_{x1}^2 = \sigma_{x2}^2 = 0.3$. These draws are fixed across Monte Carlo replications. We keep $\sigma_u = 0.3$ and vary the σ_υ in a way that λ^2 takes on values of 1, 10, and 100, while we also vary the sample size

Table 10.2 Monte Carlo results for Half-Normal model. The number of repetitions $M = 1{,}000$

			$n = 100$		$n = 200$		$n = 1{,}000$	
		True	AVE	MSE	AVE	MSE	AVE	MSE
	$\hat{\sigma}$	0.42	0.4307	0.0099	0.4359	0.0079	0.4250	0.0041
	$\hat{\gamma}$	0.5	0.5634	0.1332	0.5605	0.0025	0.5565	0.0382
	$\hat{\mu}$	0.0	−0.0071	0.3069	−0.0351	0.2801	0.0082	0.2003
	$\hat{\alpha}_0$	0.9	0.9799	0.1197	0.9728	0.0728	0.9757	0.0450
	$\hat{\alpha}_1$	0.6	0.5814	0.0160	0.5997	0.0034	0.6021	0.0012
$\lambda^2 = 1$	$\hat{\alpha}_2$	0.5	0.5105	0.0139	0.5043	0.0021	0.4958	0.0013
	$\hat{\sigma}$	0.31	0.3264	0.0076	0.3256	0.0021	0.3148	0.0011
	$\hat{\gamma}$	0.91	0.9404	0.0068	0.9161	0.0006	0.9107	0.0004
	$\hat{\mu}$	0.0	−0.0398	0.1277	−0.0566	0.0439	−0.0115	0.0175
	$\hat{\alpha}_0$	0.9	0.9144	0.0247	0.8982	0.0043	0.9001	0.0026
	$\hat{\alpha}_1$	0.6	0.6079	0.0043	0.6033	0.0006	0.6012	0.0003
$\lambda^2 = 10$	$\hat{\alpha}_2$	0.5	0.5015	0.0036	0.4973	0.0007	0.5005	0.0004
	$\hat{\sigma}$	0.30	0.3064	0.0049	0.3050	0.0013	0.3030	0.0006
	$\hat{\gamma}$	0.99	0.9958	0.0001	0.9912	0.0001	0.9905	0.0001
	$\hat{\mu}$	0.0	−0.0252	0.0559	−0.0092	0.0104	−0.005	0.0050
	$\hat{\alpha}_0$	0.9	0.9102	0.0098	0.8986	0.0014	0.8951	0.0006
	$\hat{\alpha}_1$	0.6	0.5995	0.0016	0.5993	0.0108	0.6003	0.0001
$\lambda^2 = 100$	$\hat{\alpha}_2$	0.5	0.4978	0.0012	0.5016	0.0125	0.5002	0.0001

by 100, 200, and 1,000, respectively. To facilitate the numerical optimization we consider the γ-parametrization instead of the λ-parametrization in the maximum likelihood estimation. γ is contained in the compact set $[0, 1]$ and is related to λ through $\gamma = \lambda^2/(1 + \lambda^2)$. We set the number Monte Carlo replications to 1,000 and examine the performance of normal/doubly-truncated-normal model without imposing any restrictions on model parameters. Table 10.2 reports the averaged values (AVE) of the estimates over the replications and their mean squared errors (MSE). The first case in the first column is where $n = 100$ and $\lambda^2 = 1$. In this case about $1/3$ of the samples will have least squares residuals positively skewed according to Simar and Wilson (2010). The distributional parameters obtained from the DTN model have relatively large mean squared errors. We presume that this is due to the fictitious "wrong" skewness that yields large variances of the estimates because the determinant of Fisher's information matrix is close to zero. DTN cannot provide a remedy in this case and the bagging technique proposed by Simar and Wilson (2010) can be employed to make an inference. As either the sample size or signal-to-noise ratio increase, the fictitious "wrong" skewness goes away, MSE decreases and the ALS model is recovered from the DTN model. It would appear from our intuition and from our simulations that in finite samples the large estimated standard errors of the distributional parameters can serve as an indicator of the presence of "fictitious" wrong skewness in the model.

To answer the second part of the question we consider the conditional mean inefficiencies conditional on the stochastic error in the same spirit as in Jondrow et al. (1982). For the DTN model these are given by

$$E(u_i|\hat{\varepsilon}_i) = \mu_* + \sigma_* \frac{\phi(-\frac{\mu_*}{\sigma_*}) - \phi(\frac{B-\mu_*}{\sigma_*})}{\Phi(\frac{B-\mu_*}{\sigma_*}) - \Phi(-\frac{\mu_*}{\sigma_*})},$$

where $\mu_* = \frac{\mu\sigma_v^2 - \varepsilon\sigma_u^2}{\sigma^2}$ and $\sigma_* = \frac{\sigma_u\sigma_v}{\sigma}$. Ignoring the bound will yield incorrect estimates of the inefficiencies scores.

As we have pointed out in this chapter, most of the distributions for inefficiencies considered in the stochastic frontier models literature are positively skewed. The half-normal distribution is commonly used in the literature and in applications. The doubly-truncated-normal inefficiency distribution generalizes the SFM in a way that allows for negative skewness as well. This implies that finding incorrect skewness does not necessarily indicate that the model is misspecified. A misspecification would arise, however, were the researcher to consider an incorrect distribution for the inefficiency process, which has a skewness that is not properly identified by the least squares residuals. The "wrong" skewness can be a finite sample artifact or fact. In this chapter we have considered the latter case and have shown that the normal/doubly-truncated-normal composed error SFM can still be valid with the "wrong" sign of the skewness statistic using a generalization of Waldman's (1982) proof. Moreover, "wrong" skewness in finite samples does not necessarily preclude its appearance in large samples under our specification. This chapter thus provides a rationale for applied researchers to adopt an additional strategy in cases when this perceived empirical anomaly is found.

Appendix

Truncated-Normal and Doubly-Truncated-Normal Distributions

Four distributions for inefficiencies that are discussed in this chapter. These are the truncated-normal distribution and the doubly-truncated-normal distribution with zero, positive, and negative skewness. The skewness statistic of truncated-normal distribution is given by

$$\gamma_1(B, \mu, \sigma_u^2) = \frac{(2\tilde{\eta}^3 - [3(\frac{\xi_1\phi(\xi_1)}{1-\Phi(\xi_1)}) + 1]\tilde{\eta} + \frac{\xi_1^2\phi(\xi_1)}{1-\Phi(\xi_1)})}{\left[1 - \tilde{\eta}^2 + \frac{\xi_1\phi(\xi_1)}{1-\Phi(\xi_1)}\right]^{3/2}} \tag{10.25}$$

where $\tilde{\eta} \equiv \frac{\phi(\xi_1)}{1-\Phi(\xi_1)}$, which has positive sign implying that γ_1 is positive also. Note that, as μ grows large $\tilde{\eta}$ as well as γ_1 tend to zero and the truncated-normal distribution resembles the bell-shaped normal distribution.

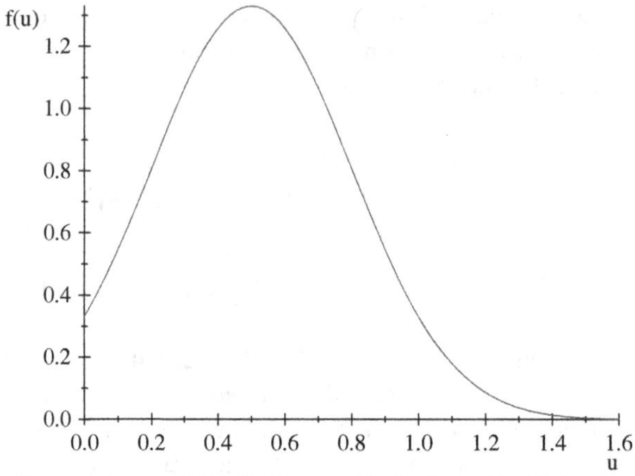

Truncated-normal distribution: positively skewed inefficiencies

In the standard model the variance of the inefficiency term must be assigned a zero value whenever the skewness of the OLS residuals is zero. This leads to the conclusion that there are no inefficient firms in the sample. On the other hand, in the normal/doubly-truncated-normal model we may have the variance of the inefficiency process strictly positive even if the skewness is zero. This case is well illustrated in the next figure where $B = 2\mu$ and $\sigma_u = 0.3$.

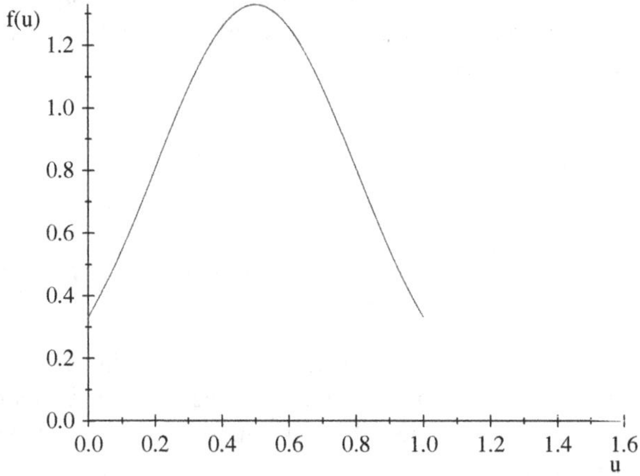

Doubly-truncated-normal distribution: symmetrically distributed inefficiencies ($\gamma_1 = 0$)

The graphs of positively skewed ($B > 2\mu$) and negatively skewed ($B < 2\mu$) inefficiencies are provided in the next set of figures. Again it is clear that negative skewness does not imply a lack of inefficiency.

The last distribution can describe a scenario where only a small fraction of firms attain levels of productivity close to the frontier. This is described as "few stars, most

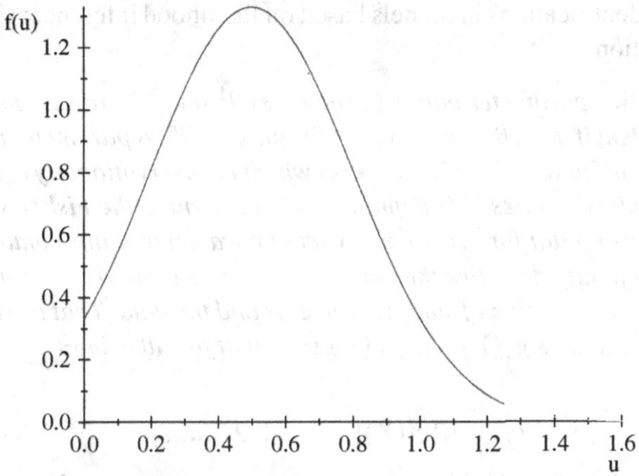

Doubly-truncated-normal distribution: positively skewed inefficiencies ($\gamma_1 = 0.69$)

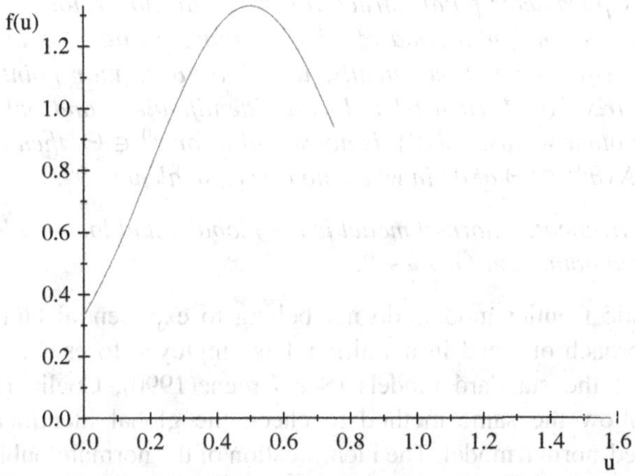

Doubly-truncated-normal distribution: negatively skewed inefficiencies ($\gamma_1 = -1.2$)

dogs" in the business press. The standard models only describe the case where there are "most stars, few dogs". Note that, this is not the only case we can reproduce negatively skewed distribution for inefficiencies. The condition that is required is that $B < 2\mu$ with strictly positive μ.

Identification in Normal/Doubly-Truncated-Normal Model

The classical SFM is likelihood-based and such as requires careful examination of the identification of parameters and measures of inefficiencies, as pointed out by Greene (1997) and Ritter and Simar (1997). Rothenberg (1971), defines two kinds of parametric identifications in models based on likelihood inference: global and the local identification.

Definition 1. *Two parameter points (structures) θ^1 and θ^2 are said to be observationally equivalent if $f(y, \theta^1) = f(y, \theta^2)$ for all y in R^n. A parameter point $\theta^0 \in \Theta$ is said to be identifiable if there is no $\theta \in \Theta$ which is observationally equivalent. For exponential family densities, this definition is equivalent to the Fisher's information matrix to be nonsingular for every convex set of parameter points containing Θ. For nonexponential family densities the condition requires that every parameter can be expressed only as a function of sample moments and the data. That is, suppose there exist p known functions $g_1(Y), \ldots., g_p(Y)$ such that for all θ in Θ*

$$\theta_i = E[g_i(Y)] \quad i = 1, 2, \ldots\ldots, p$$

Then every θ in Θ is identifiable.

Definition 2. *A parameter point (structure) θ^0 is said to be locally identified if there exists an open neighborhood of θ^0 containing no other θ in Θ which is observationally equivalent. Equivalently, let θ^0 be a regular point of Fisher's information matrix $I(\theta)$.[9] Then θ^0 is locally identifiable if and only if $I(\theta^0)$ is nonsingular. In other words, if $I(\theta)$ is nonsingular for $\theta^0 \in \Theta$, then there exists a neighborhood $N(\theta^0) \subset \Theta$ of θ^0 in which no θ is equivalent to θ^0.*

Claim: *Normal/truncated-normal model is not globally and locally identified as μ increases without bound and/or $\lambda \to 0$.*

Proof. Stochastic frontier models do not belong to exponential family and thus the second approach outlined in definition 1 is employed to establish the global identifiability of the standard models (See Greene(1990), Coeli (1995) among others). We follow the same method to check the global identification of the normal/truncated-normal model. The identification of the normal/doubly-truncated-normal deserves special treatment and it is discussed in Almanidis et al. (2010).

Under the assumptions made for the error terms the population central moments of the composed error term, ε up to forth order are given by[10]

$$\mu_1 = E[\varepsilon] = -\mu - \sigma_u \tilde{\eta}$$

$$\mu_2 = E[(\varepsilon - \mu_1)^2] = E[v^2] + E[(u - E[u])^2] = \sigma_v^2 + \sigma_u^2 \left[1 - \tilde{\eta}^2 + \xi_1 \tilde{\eta}\right]$$

$$\mu_3 = E[(\varepsilon - \mu_1)^3] = -E[(u - E[u])^3] = -\sigma_u^3 [2\tilde{\eta}^3 - [3\xi_1 \tilde{\eta} + 1]\tilde{\eta} + \xi_1^2 \tilde{\eta}]$$

[9] see Rothenberg 1971, pp 579 for the relevant definition of the regular point

[10] Note that p in the definition 1 is equal to 4.

$$\begin{aligned}\mu_4 &= E[(\varepsilon-\mu_1)^4] = E[v^4] + 6E[v^2]E[(u-E[u])^2] + E[(u-E[u])^4]\\ &= 3\sigma_v^4 + 6\sigma_v^2\sigma_u^2\left[1-\tilde{\eta}^2+\xi_1\tilde{\eta}\right] + \sigma_u^4[-3\tilde{\eta}^4 + 6\xi_1\tilde{\eta}^3 - 2\tilde{\eta}^2 - 4\xi_1^2\tilde{\eta}^2\\ &\quad + 3\xi_1\tilde{\eta} + \xi_1^3\tilde{\eta} + 3]\end{aligned}$$

where $\tilde{\eta}$ and ξ_1 are as defined above.

In addition it can be shown that

$$\mu_4 - 3\mu_2^2 = E[(u-E[u])^4] - 3\left(E[(u-E[u])^2]\right)^2$$

and after dividing this expression by $\mu_3^{4/3}$ we get

$$g(\xi_1) = \mu_3^{-4/3}(\mu_4 - 3\mu_2^2)$$

which is a function of ξ_1 only. Replacing the population moments by their sample counterparts we can solve for ξ_1 numerically from

$$h(\xi_1) = \frac{4\tilde{\eta}^2 - 3\tilde{\eta}\xi_1 - 6\tilde{\eta}^4 + \tilde{\eta}\xi_1^3 + 12\tilde{\eta}^3\xi_1 - 7\tilde{\eta}^2\xi_1^2}{\left(-[2\tilde{\eta}^3 - [3\xi_1\tilde{\eta}+1]\tilde{\eta} + \xi_1^2\tilde{\eta}]\right)^{-4/3}} - \hat{\mu}_3^{-4/3}(\hat{\mu}_4 - 3\hat{\mu}_2^2) = 0$$

For a range of values of ξ_1 it is shown by implicit function theorem that function $h(\cdot)$ has no unique solution. Hence, the global identifiability fails. However, the local identifiability of the normal/truncated normal model, which is necessary condition, can be established according to definition 2 unless μ is very large and/or $\lambda \to 0$. In this two extreme cases both global and local identification fail since the Fisher's information matrix $I(\theta)$ evaluated at this points is close to singular

$$I(\theta^0) \to \begin{bmatrix} -\frac{1}{\sigma^2}\sum_{i=1}^n x_i x_i' & -\frac{1}{\sigma^4}\sum_{i=1}^n(\varepsilon_i+\mu)x_i & 0 & -\frac{1}{\sigma^2}\sum_{i=1}^n x_i\\ -\frac{1}{\sigma^4}\sum_{i=1}^n(\varepsilon_i+\mu)x_i & \frac{n}{2\sigma^4} - \frac{1}{\sigma^6}\sum_{i=1}^n(\varepsilon_i+\mu)^2 & 0 & -\frac{1}{\sigma^4}\sum_{i=1}^n(\varepsilon_i+\mu)\\ 0 & 0 & 0 & 0\\ -\frac{1}{\sigma^2}\sum_{i=1}^n x_i & -\frac{1}{\sigma^4}\sum_{i=1}^n(\varepsilon_i+\mu) & 0 & -\frac{n}{\sigma^2} \end{bmatrix}$$

The above argument can be made simpler as in Ritter and Simar (1997). The authors note that the distribution of the composite error in the normal/gamma stochastic frontier model tends to the normal distribution as the shape parameter of the gamma distribution increases without bound and the scale parameter remains relatively low. In this case the model parameters and inefficiencies cannot be identified. This is also the case for normal/truncated-normal model, where for relatively small values of parameter λ the distribution resembles the normal distribution as parameter μ becomes relatively large. Therefore, this model fails to be locally identified in this particular case. On the other hand, the bounded inefficiency model is still capable of identifying the model parameters even for large values of μ, since the existence of the bound will distinguish the distribution of inefficiencies from the normal distribution which is assumed for the noise term.

An illustration of these two cases is provided below in the two figures, where the truncated-normal density appears like the normal density for values of parameter μ as low as 1, while keeping the variance of inefficiencies to be one third. On the other hand, the doubly-truncated-normal distribution distinguishes itself from the normal distribution in the sense that its right tail will be shorter than its corresponding left tail. Thus the model will be identified even for large values of the parameter μ. It is also worth mentioning that the empirical distribution of inefficiencies may often appear to be shaped like the normal distribution and, therefore, without imposing the bound it is difficult to identify it from normally distributed noise term.

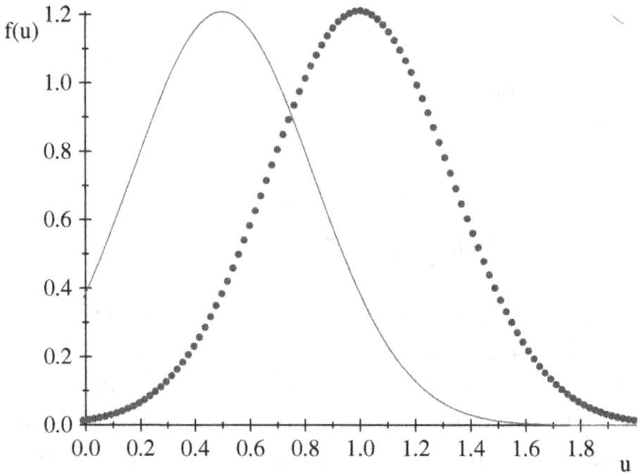

Truncated-normal distribution with $\mu = 0.5$ (*solid line*) and $\mu = 1$ (*dotted line*)

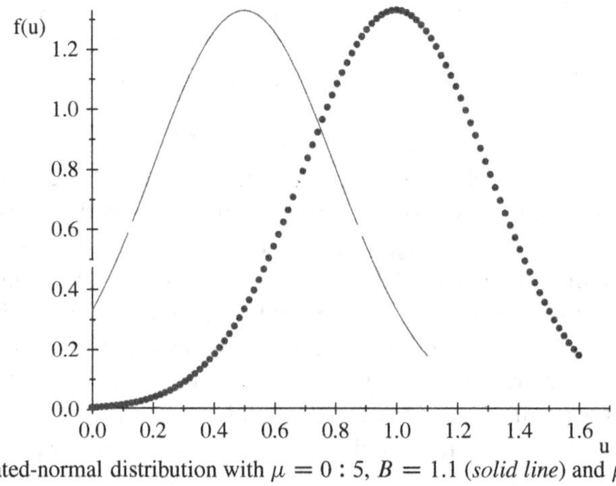

Doubly-truncated-normal distribution with $\mu = 0:5$, $B = 1.1$ (*solid line*) and $\mu = 1$, $B = 1.6$ (*dotted line*)

Both the normal/doubly-truncated normal SFM and normal/truncated-normal SFM fail to be globally identified, but are potentially locally identified. However, if the inefficiencies have a large mean, the latter model fails to be even locally identified. The former can still yield precise and stable estimates of parameters and inefficiencies. It remains unclear how statistical inference can be validated in the locally identified case. Shapiro (1986) and Dasgupta et al. (2007) discuss some cases there valid statistical inference can be obtained. Moreover, Bayesian methods could be utilized to identify the parameters through nonlinear constraints imposed by the skewness condition. The Ritter and Simar (1997) argument can also be avoided if one considers the normal/truncated-gamma stochastic frontier model. We leave these possibilities for the future work.

Acknowledgements The authors would like to thank the participants of the Workshop on Exploring Research Frontiers in Contemporary Statistics and Econometrics in Honor of Leopold Simar, Institute of Statistics of the Université Catholique de Louvain, Louvain-la-Neuve, Belgium, May 14–15, 2009, the 2009 XI European Workshop on Efficiency and Productivity Analysis, Pisa, Italy, and the 2009 Rice University Econometrics Workshop, as well as two anonymous referees and Co-Editor Paul Wilson for comments and criticisms that substantially strengthened this chapter. The usual caveat applies.

References

Aigner, D.J., & Chu, S. (1968). On estimating the industry production function. *American Economic Review, 58*, 826–839.

Aigner, D.J., Lovell, C.A.K., & Schmidt, P. (1977). Formulation and estimation of stochastic frontier models. *Journal of Econometrics, 6*, 21–37.

Almanidis, P., Qian, J., & Sickles, R.C. (2010). Bounded Stochastic Frontiers with an Application to the US Banking Industry: 1984–2009. Working paper.

Battese, G.E., & Cora, G.S. (1977). Estimation of a production frontier model: with application to the pastoral zone of eastern Australia. *Australian Journal of Agricultural Economics, 21*, 169–179.

Bera, A.K., & Premaratne, G. (2001). Adjusting the tests for skewness and kurtosis for distributional misspecifications. Working Paper.

Carree, M.A. (2002). Technological inefficiency and the skewness of the error component in stochastic frontier analysis. *Economics Letter, 77*, 101–107.

Charnes, A., Cooper, W.W., & Rhodes, E.L. (1978). Measuring the efficiency of decision making units. *European Journal of Operational Research, 2*, 429–444.

Coelli, T. (1995). Estimators and hypothesis tests for a stochastic frontier function: A Monte Carlo analysis. *Journal of Productivity Analysis, 6*, 247–268.

Coelli, T. (1996). A guide to FRONTIER version 4.1: A computer program for stochastic frontier production and cost function estimation. CEPA working paper #96/07, Center for Efficiency and Productivity Analysis, University of New England, Arimidale, NSW 2351, Australia.

Dasgupta, A., Steven, G., Selfb, & Das Gupta, S. (2007). Non-identifiable parametric probability models and reparametrization. *Journal of Statistical Planning and Inference, 137*, 3380–3393.

D'Agostino, R.B., & Pearson, E.S. (1973). Tests for departure from normality: Empirical results for the distribution of b_2 and $\sqrt{b_1}$. *Biometrica, 60*, 613–622.

Deprins, D., Simar, L., & Tulkens, H. (1984). Measuring labor inefficiency in post offices. In Marchand, M., Pestieau, P., & Tulkens, H. (Eds.) *The performance of public enterprises: concepts and measurements*, pp. 243–267. Amsterdam: North-Holland.

Deprins, D., & Simar, L. (1989b). Estimating technical inefficiencies with corrections for environmental conditionswith an application to railway companies. *Annals of Public and Cooperative Economics, 60*, 81–102.

Greene, W.H. (1980a). Maximum likelihood estimation of econometric frontier functions. *Journal of Econometrics, 13*, 27–56.

Greene, W.H. (1980b). On the estimation of a flexible frontier production model. *Journal of Econometrics, 13*, 101–115.

Greene, W.H. (1990). A gamma distributed stochastic frontier model. *Journal of Econometrics, 46*, 141–164.

Greene, W.H. (1997). Frontier production functions. In Microeconomics, H., Pesaran, & Schmidt, P. (Eds.) *Handbook of Applied Econometrics*, vol. 2. Oxford: Oxford University Press.

Greene, W.H. (2007a). *LIMDEP Version 9.0 User's Manual*. New York: Econometric Software, Inc.

Greene, W.H. (2007b). The econometric approach to efficiency analysis. In Fried, H.O., Lovell, C.A.K., & Schmidt, S.S. (Eds.) *The measurement of productive efficiency: techniques and applications*. New York: Oxford University Press.

Jondrow, J., Lovell, C.A.K., Materov, I.S., & Schmidt, P. (1982). On the estimation of technical inefficiency in the stochastic frontier production function model. *Journal of Econometrics, 19*, 233–238.

Johnson, N.L., Kotz, S., & Kemp, A.W. (1992). *Univariate discrete distributions*, 2nd edn. New York: Wiley.

Johnson, N.L., Kotz, S., & Balakrishnan, N. (1994). *Continuous univariate distributions*, vol. 1, 2nd edn. New York: Wiley.

Kim, Y., & Schmidt, P. (2000). A review and empirical comparison of Bayesian and classical approaches to inference on efficiency levels in stochastic frontier models with panel data. *Journal of Productivity Analysis, 14*, 91–118.

Koop, G. (1994). Recent progress in applied Bayesian Econometrics. *Journal of Economic Surveys, 8*, 1–34.

Koop, G., Osiewalski, J., & Steel, M.F.J. (1995). Posterior analysis of stochastic frontier models using Gibbs sampling. *Computational Statistics, 10*, 353–373.

Koop, G., Osiewalski, J., & Steel, M.F.J. (1997). Bayesian efficiency analysis through individual effects: Hospital cost frontiers. *Journal of Econometrics, 76*, 77–105.

Leamer, E.E. (1978). *Specification Searches: Ad hoc inference with nonexperimental data*. New York: Wiley.

Lee, L. (1993). Asymptotic distribution of the MLE for a stochastic frontier function with a singular information matrix. *Econometric Theory, 9*, 413–430.

Meeusen, W., & van den Broeck, J. (1977). Efficiency estimation from Cobb-Douglas production functions with composed error. *International Economic Review, 18*, 435–444.

Migon, H.S., & Medici, E. (2001). A Bayesian Approach to Stochastic Production Frontier, Technical Report n. 105, DME-IM/UFRJ.

Olson, J.A., Schmidt, P., & Waldman, D.M. (1980). A Monte Carlo study of estimators of the stochastic frontier production function. *Journal of Econometrics, 13*, 67–82.

Qian, J., & Sickles, R.C. (2008). Stochastic frontiers with bounded inefficiency. Mimeo: Rice University.

Ritter, C., & Simar, L. (1997). Pitfalls of normal-gamma stochastic frontier models. *Journal of Productivity Analysis, 8*, 167–182.

Rothenberg, T.J. (1971). Identification in parametric models. *Econometrica, 39*(3), 577–591.

Sena, V. (1999). Stochastic frontier estimation: A review of the software options. *Journal of Applied Econometrics, 14*, 579–586.

Sickles, R.C. (2005). Panel estimators and the identification of firm-specific efficiency levels in parametric, semiparametric and nonparametric settings. *Journal of Econometrics, 126*, 305–334.

Simar, L., & Wilson, P.W. (2010). Estimation and inference in cross-sectional, stochastic frontier models. *Econometric Reviews, 29*, 62–98.

Schmidt, P., & Lin, T. (1984). Simple tests of alternative specifications in stochastic frontier models. *Journal of Econometrics, 24*, 349–361.

Shapiro, A. (1986). Asymptotic theory of overparameterized structural models. *Journal of American Statistical Association, 81*, 142–149.

Stevenson, R.E. (1980). Likelihood functions for generalized stochastic frontier estimation. *Journal of Econometrics, 13*, 57–66.

van den Broeck, J., Koop, G., & Osiewalski, J. (1994). Stochastic frontier models: a Bayesian perspective. *Journal of Econometrics, 61*, 273–303.

Waldman, D. (1977). Estimation in economic frontier functions. Unpublished manuscript, University of North Carolina, Chapel Hill.

Waldman, D. (1982). A stationary point for the stochastic frontier likelihood. *Journal of Econometrics, 18*, 275–279.

Wang, W.S., & Schmidt, P. (2008). On the distribution of estimated technical efficiency in stochastic frontier models. *Journal of Econometrics, 148*, 36–45.

Chapter 11
Optimal Smoothing for a Computationally and Statistically Efficient Single Index Estimator

Yingcun Xia, Wolfgang Karl Härdle, and Oliver Linton

Abstract In semiparametric models it is a common approach to under-smooth the nonparametric functions in order that estimators of the finite dimensional parameters can achieve root-n consistency. The requirement of under-smoothing may result, as we show, from inefficient estimation methods or technical difficulties. Xia et al. (J. Roy. Statist. Soc. B. 64:363–410, 2002) proposed an adaptive method for the multiple-index model, called MAVE. In this chapter we further refine the estimation method. Under some conditions, our estimator of the single-index is asymptotically normal and most efficient in the semi-parametric sense. Moreover, we derive higher-order expansions for our estimator and use them to define an optimal bandwidth for the purposes of index estimation. As a result we obtain a practically more relevant method and we show its superior performance in a variety of applications.

11.1 Introduction

Single index models (SIMs) are widely used in the applied quantitative sciences. Although the context of applications for SIMs almost never prescribes the functional or distributional form of the involved statistical error, the SIM is commonly fitted

---

Y. Xia (✉)
Department of Statistics and Applied Probability and Risk Management Institute, National University of Singapore, Singapore
e-mail: staxyc@nus.edu.sg

W.K. Härdle
C.A.S.E. Centre for Applied Statistics and Economics, School of Business and Economics, Humboldt-Universität zu Berlin, Unter den Linden 6, 10099 Berlin, Germany
e-mail: haerdle@wiwi.hu-berlin.de

O. Linton
Faculty of Economics, Austin Robinson Building, Sidgwick Avenue, Cambridge, CB3 9DD, UK
e-mail: obl20@cam.ac.uk

I. van Keilegom and P.W. Wilson (eds.), *Exploring Research Frontiers in Contemporary Statistics and Econometrics*, DOI 10.1007/978-3-7908-2349-3_11,

with (low dimensional) likelihood principles. Both from a theoretical and practical point of view such fitting approach has been criticized and has led to semiparametric modelling. This approach involves high dimensional parameters (nonparametric functions) and a finite dimensional index parameter. Consider the following single-index model,

$$Y = g(\theta_0^\top X) + \varepsilon, \tag{11.1}$$

where $E(\varepsilon|X) = 0$ almost surely, g is an unknown link function, and θ_0 is a single-index parameter with length one and first element positive for identification. In this model there is a single linear combination of covariates X that can capture most information about the relation between response variable Y and covariates X, thereby avoiding the "curse of dimensionality". Estimation of the single-index model is very attractive both in theory and in practice. In the last decade a series of papers has considered estimation of the parametric index and the nonparametric part with focus on root-n estimability and efficiency issues, see Carroll et al. (1997) and Delecroix et al. (2003, 2006) for an overview. There are numerous methods proposed or can be used for the estimation of the model. Amongst them, the most popular ones are the average derivative estimation (ADE) method investigated by Härdle and Stoker (1989), the sliced inverse regression (SIR) method proposed by Li (1991), the semiparametric least squares (SLS) method of Ichimura (1993) and the simultaneous minimization method of Härdle et al. (1993).

The existing estimation methods are all subject to some or other of the following four critiques: (1) *Heavy computational burden:* see, for example, Härdle et al. (1993), Delecroix et al. (2003), Xia and Li (1999) and Xia et al. (2002). These methods include complicated optimization techniques (iteration between bandwidth choice and parameter estimation) for which no simple and effective algorithm is available up to now. (2) *Strong restrictions on link functions or design of covariates X:* Li (1991) required the covariate to have a symmetric distribution; Härdle and Stoker (1989) and Hristache et al. (2001) needed a non-symmetric structure for the link function, i.e., $|Eg'(\theta_0^\top X)|$ is bounded away from 0. If these conditions are violated, the corresponding methods are inconsistent. (3) *Inefficiency:* The ADE method of Härdle and Stoker (1989) or the improved ADE method of Hristache et al. (2001) is not asymptotically efficient in the semi-parametric sense, Bickel et al. (1993). Nishiyama and Robinson (2000, 2005) considered the Edgeworth correction to the ADE methods. Härdle and Tsybakov (1993) discussed the sensitivity of the ADE. Since this method involves high dimensional smoothing and derivative estimation, its higher order properties are poor. (4) *Under-smoothing:* Let h_g^{opt} be the optimal bandwidth in the sense of MISE for the estimation of the link function g and let h_θ be the bandwidth used for the estimation of θ_0. Most of the methods mentioned above require the bandwidth h_θ to be much smaller than the bandwidth h_g^{opt}, i.e. $h_\theta / h_g^{opt} \to 0$ as $n \to \infty$, in order that estimators of θ_0 can achieve root-n consistency, see, Härdle and Stoker (1989), Hristache et al. (2001), Robinson (1988), Hall (1989) and Carroll et al. (1997) among others. Due to technical complexities, there are few investigations about how to select the bandwidth h_θ for the estimation of the single-index. Thus it could be the case that even if $h_\theta = h_g^{opt}$ allows for root-n consistent estimation of θ, that $h_\theta^{opt} / h_g^{opt} \to 0$ or $h_g^{opt} / h_\theta^{opt} \to 0$, where h_θ^{opt} is the optimal bandwidth for estimation of θ. This would mean that using a single bandwidth h_g^{opt} would result in suboptimal performance for the estimator of θ. Higher order properties of other semiparametric procedures have been studied in Linton (1995) *inter alia*.

Because the estimation of θ_0 is based on the estimation of the link function g, we might expect that a good bandwidth for the link function should be a good bandwidth for the single-index, i.e., under-smoothing should be unnecessary. Unfortunately, most of the existing estimation methods involve for technical reasons "under-smoothing" the link function in order to obtain a root-n consistent estimator of θ_0. See, for example, Härdle and Stoker (1989), Hristache et al. (2001, 2002), Carroll et al. (1997) and Xia and Li (1999). Härdle et al. (1993) investigated this problem for the first time and proved that the optimal bandwidth for the estimation of the link function in the sense of MISE can be used for the estimation of the single-index to achieve root-n consistency. As mentioned above, for its computational complexity the method of Härdle et al. (1993) is hard to implement in practice.

This chapter refines the method in Xia et al. (2002) and Xia (2006). It avoids undersmoothing and the computational complexity of former procedures and achieves the semiparametric efficiency bound. It is based on the MAVE method of Xia et al. (2002), which we outline in the next section. Using local linear approximation and global minimization, we give a very simple iterative algorithm. The proposed method has the following advantages: (a) the algorithm only involves one-dimensional smoothing and is proved to converge at a geometric rate; (b) with normal errors in the model, the estimator of θ_0 is asymptotically normal and efficient in the semiparametric sense; (c) the optimal bandwidth for the estimation of the link function in the sense of MISE can be used to estimate θ_0 with root-n consistency; (d) by a second order expansion, we further show that the optimal bandwidth for the estimation of the single-index θ_0, h_θ^{opt}, is of the same magnitude as h_g^{opt}.

Therefore, the commonly used "under-smoothing" approach is inefficient in the sense of second order approximation. Powell and Stoker (1996) investigated bandwidth selection for the ADE methods. We also propose an automatic bandwidth selection method for our estimator of θ. Xia (2006) has recently shown the first order asymptotic properties of this method. Our theoretical results are proven under weak moment conditions.

In section 3 we present our main results. We show the speed of convergence, give the asymptotic estimation and derive a smoothing parameter selection procedure. In the following section we investigate the proposed estimator in simulation and application. Technical details are deferred to the appendix.

11.2 The MAVE Method

Suppose that $\{(X_i, Y_i) : i = 1, 2, \ldots, n\}$ is a random sample from model (11.1). The basic idea of our estimation method is to linearly approximate the smooth link function g and to estimate θ_0 by minimizing the overall approximation errors. Xia et al. (2002) proposed a procedure via the so called minimum average conditional variance estimation (MAVE). The single index model (11.1) is a special case of what they considered, and we can estimate it as follows. Assuming function g and parameter θ_0 are known, then the Taylor expansion of $g(\theta_0^\top X_i)$ at $g(\theta_0^\top x)$ is

$$g(\theta_0^\top X_i) \approx a + d\theta_0^\top (X_i - x),$$

where $a = g(\theta_0^\top x)$ and $d = g'(\theta_0^\top x)$. With fixed θ, the local estimator of the conditional variance is then

$$\sigma_\theta^2(x) = \min_{a,d} \left\{ n\hat{f}_\theta(x)\}^{-1} \sum_{i=1}^n [Y_i - \{a + d\theta^\top (X_i - x)\}]^2 K_h\{\theta^\top (X_i - x)\right\},$$

where $\hat{f}_\theta(x) = n^{-1}\sum_{i=1}^n K_h\{\theta^\top (X_i - x)\}$, K is a univariate density function, h is the bandwidth and $K_h(u) = K(u/h)/h$; see Fan and Gijbels (1996). The value $\sigma_\theta^2(x)$ can also be understood as the local departure of Y_i with X_i close to x from a local linear model with given θ. Obviously, the best approximation of θ should minimize the overall departure at all $x = X_j, j = 1, \cdots, n$. Thus, our estimator of θ_0 is to minimize

$$Q_n(\theta) = \sum_{j=1}^n \sigma_\theta^2(X_j) \tag{11.2}$$

with respect to $\theta : |\theta| = 1$. This is the so-called minimum average conditional variance estimation (MAVE) in Xia et al. (2002). In practice it is necessary to include some trimming in covariate regions where density is low, so we weight $\sigma_\theta^2(x)$ by a sequence $\hat{\rho}_j^\theta$, where $\hat{\rho}_j^\theta = \rho_n\{\hat{f}_\theta(X_j)\}$, that is discussed further below.

The corresponding algorithm can be stated as follows. Suppose θ_1 is an initial estimate of θ_0. Set the number iteration $\tau = 1$ and bandwidth h_1. We also set a final bandwidth h. Let $X_{ij} = X_i - X_j$.

Step 1: With bandwidth h_τ and $\theta = \theta_\tau$, calculate $\hat{f}_\theta(X_j) = n^{-1}\sum_{i=1}^n K_{h_\tau}(\theta^\top X_{ij})$ and the solutions of a_j and d_j to the inner problem in (11.2):

$$\begin{pmatrix} a_j^\theta \\ d_j^\theta h_\tau \end{pmatrix} = \left\{ \sum_{i=1}^n K_{h_\tau}(\theta^\top X_{ij}) \begin{pmatrix} 1 \\ \theta^\top X_{ij}/h_\tau \end{pmatrix} \begin{pmatrix} 1 \\ \theta^\top X_{ij}/h_\tau \end{pmatrix}^\top \right\}^{-1}$$

$$\times \sum_{i=1}^n K_{h_\tau}(\theta^\top X_{ij}) \begin{pmatrix} 1 \\ \theta^\top X_{ij}/h_\tau \end{pmatrix} Y_i.$$

Step 2: Fix the weight $K_{h_\tau}(\theta^\top X_{ij})$, $f_\theta(X_j)$, a_j^θ and d_j^θ. Calculate the solution of θ to (11.2):

$$\theta = \left\{ \sum_{i,j=1}^{n} K_{h_\tau}(\theta^\top X_{ij}) \hat{\rho}_j^\theta \{d_\theta(X_j)\}^2 X_{ij} X_{ij}^\top \hat{f}_\theta(\theta^\top X_j) \right\}^{-1}$$

$$\sum_{i,j=1}^{n} K_{h_\tau}(\theta^\top X_{ij}) \hat{\rho}_j^\theta d_\theta(X_j) X_{ij} (Y_i - a_j^\theta) / \hat{f}_\theta(\theta^\top X_j),$$

where $\hat{\rho}_j^\theta = \rho_n\{\hat{f}_\theta(X_j)\}$.

Step 3: Set $\tau = \tau + 1$, $\theta_\tau := \theta/|\theta|$ and $h_\tau := \max\{h, h_\tau/\sqrt{2}\}$, go to Step 1.

Repeat steps 1 and 2 until convergence.

The iteration can be stopped by a common rule. For example, if the calculated θ's are stable at a certain direction, we can stop the iteration. The final vector $\theta := \theta/|\theta|$ is the MAVE estimator of θ_0, denoted by $\hat{\theta}$. Note that these steps are an explicit algorithm of the Xia et al. (2002) method for the single-index model with some version of what they called 'refined kernel weighting' and boundary trimming. Similar to the other direct estimation methods, the calculation above is easy to implement. See Horowitz and Härdle (1996) for more discussions. After θ is estimated, the link function can then be estimated by the local linear smoother as $\hat{g}^{\hat{\theta}}(v)$, where

$$\hat{g}^\theta(v) = [n\{s_2^\theta(v)s_0^\theta(v) - (s_1^\theta(v))^2\}]^{-1}$$
$$\times \sum_{i=1}^{n} \{s_2^\theta(v) - s_1^\theta(v)(\theta^\top X_i - v)/h_\tau\} K_{h_\tau}(\theta^\top X_i - v) Y_i, \qquad (11.3)$$

and $s_k^\theta(v) = n^{-1}\sum_{i=1}^n K_{h_\tau}(\theta^\top X_i - v)\{(\theta^\top X_i - v)/h_\tau\}^k$ for $k = 0, 1, 2$. Actually, $\hat{g}^{\hat{\theta}}(v)$ is the final value of a_j^θ in Step 1 with $\theta^\top X_j$ replaced by v.

In the algorithm, $\rho_n(.)$ is a trimming function employed to handle the boundary points. There are many choices for the estimator to achieve the root-n consistency; see e.g. Härdle and Stoker (1989) and Härdle et al. (1993). However, to achieve the efficiency bound, $\rho_n(v)$ must tend to 1 for all v. In this chapter, we take $\rho_n(v)$ as a bounded function with third order derivatives on $\mathbb{R}$ such that $\rho_n(v) = 1$ if $v > 2c_0 n^{-\varsigma}$; $\rho_n(v) = 0$ if $v \le c_0 n^{-\varsigma}$ for some constants $\varsigma > 0$ and $c_0 > 0$. As an example, we can take

$$\rho_n(v) = \begin{cases} 1, & \text{if } v \ge 2c_0 n^{-\varsigma}, \\ \frac{\exp\{(2c_0 n^{-\varsigma} - v)^{-1}\}}{\exp\{(2c_0 n^{-\varsigma} - v)^{-1}\} + \exp\{(v - c_0 n^{-\varsigma})^{-1}\}}, & \text{if } 2c_0 n^{-\varsigma} > v > c_0 n^{-\varsigma}, \\ 0, & \text{if } v \le c_0 n^{-\varsigma}. \end{cases} \qquad (11.4)$$

The choice of ς will be given below. The trimming function is selected to be smooth and to include all sample space as $n \to \infty$.

11.3 Main Results

We impose the following conditions to obtain the asymptotics of the estimators.

(C1) [Initial estimator] The initial estimator is in $\Theta_n = \{\theta : |\theta - \theta_0| \le Cn^{-\alpha}\}$ for some $C > 0$ and $0 < \alpha < 1/2$.

(C2) [Design] The density function $f_\theta(v)$ of $\theta^\top X$ and its derivatives up to 6th order are bounded on $\mathbb{R}$ for all $\theta \in \Theta_n$, $E|X|^6 < \infty$ and $E|Y|^3 < \infty$. Furthermore, $\sup_{v\in R,\theta\in\Theta_n} |f_\theta(v) - f_{\theta_0}(v)| \le c|\theta - \theta_0|$ for some constant $c > 0$.

(C3) [Link function] The conditional mean $g_\theta(v) = E(Y|\theta^\top X = v)$, $E(X|\theta^\top X = v)$, $E(XX^\top|\theta^\top X = v)$ and their derivatives up to 6th order are bounded for all θ such that $|\theta - \theta_0| < \delta$ for some $\delta > 0$.

(C4) [Kernel function] $K(v)$ is a symmetric density function with finite moments of all orders.

(C5) [Bandwidth and trimming parameter] The trimming parameter ς is bounded by 1/20 and the bandwidth h is proportional to $n^{-\rho}$ for some ρ with $1/5 - \epsilon \le \rho \le 1/5 + \epsilon$ for some $\epsilon > 0$.

Assumption (C1) is feasible because such an initial estimate is obtainable using existing methods, such as Härdle and Stoker (1989), Powell et al. (1989) and Horowitz and Härdle (1996). Actually, Härdle et al. (1993) even assumed that the initial value is in a root-n neighborhood of θ_0, $\{\theta : |\theta - \theta_0| \le C_0 n^{-1/2}\}$. Assumption (C2) means that X may have discrete components providing that $\theta^\top X$ is continuous for θ in a small neighborhood of θ_0; see also Ichimura (1993). The moment requirement on X is not strong. Härdle et al. (1993) obtained their estimator in a bounded area of $\mathbb{R}^p$, which is equivalent to assume that X is bounded; see also Härdle and Stoker (1989). We impose slightly higher order moment requirements than the existence of the second moment for Y to ensure the optimal bandwidth in (C5) can be used in applying Lemma 11.8.1 in section 6. The smoothness requirements on the link function in (C3) can be relaxed to the existence of a bounded second order derivative at the cost of more complicated proofs and smaller bandwidth. Assumption (C4) includes the Gaussian kernel and the quadratic kernel. Assumption (C5) includes the commonly used optimal bandwidth in both the estimation of the link function and the estimation of the index θ_0. Actually, imposing these constraints on the bandwidth is for ease of exposition in the proofs.

Let $\mu_\theta(x) = E(X|\theta^\top X = \theta^\top x)$, $\nu_\theta(x) = \mu_\theta(x) - x$, $w_\theta(x) = E(XX^\top|\theta^\top X = \theta^\top x)$, $W_0(x) = \nu_{\theta_0}(x)\nu_{\theta_0}^\top(x)$. Let A^+ denote the Moore-Penrose inverse of a symmetric matrix A. Recall that K is a symmetric density function. Thus, $\int K(v)dv = 1$ and $\int vK(v)dv = 0$. For ease of exposition, we further assume that $\mu_2 = \int v^2 K(v)dv = 1$. Otherwise, we can redefine $K(v) := \mu_2^{1/2} K(\mu_2^{1/2} v)$.

We have the following asymptotic results for the estimators.

Theorem 11.3.1 (Speed of algorithm). *Let θ_τ be the value calculated in Step 3 after τ iterations. Suppose assumptions (C1)-(C5) hold. If $h_\tau \to 0$ and $|\theta_\tau - \theta_0|/h_\tau^2 \to 0$, we have*

$$\theta_{\tau+1} - \theta_0 = \frac{1}{2}\{(I - \theta_0\theta_0^\top) + o(1)\}(\theta_\tau - \theta_0) + \frac{1}{2\sqrt{n}}N_n + O(n^{2\varsigma}h_\tau^4)$$

almost surely, where $N_n = [E\{g'(\theta_0^\top X)^2 W_0(X)\}]^+ n^{-1/2}\sum_{i=1}^n g'(\theta_0^\top X_i)\nu_{\theta_0}(X_i)$ $\varepsilon_i = O_p(n^{-1/2})$.

Theorem 11.3.1 indicates that the algorithm converges at a geometric rate, i.e. after each iteration, the estimation error reduces by half approximately. By Theorem 11.3.1 and the bandwidth requirement in the algorithm, we have

$$|\theta_{\tau+1} - \theta_0| = \left\{\frac{1}{2} + o(1)\right\} |\theta_\tau - \theta_0| + O(n^{-1/2} + n^{2\varsigma}h_\tau^4).$$

Starting with $|\theta_1 - \theta_0| = Cn^{-\alpha}$, in order to achieve root-n consistency, say $|\theta_k - \theta_0| \le cn^{-1/2}$ i.e. $2^{-k}Cn^{-\alpha} \le cn^{-1/2}$, the number of iterations k can be calculated roughly by

$$k = \left\{\left(\frac{1}{2} - \alpha\right)\log n + \log(C/c)\right\} \Big/ \log 2. \tag{11.5}$$

Based on Theorem 11.3.1, we immediately have the following limiting distribution.

Theorem 11.3.2 (Efficiency of estimator). *Under the conditions (C1)–(C5), we have*

$$\sqrt{n}(\hat\theta - \theta_0) \xrightarrow{\mathcal{L}} N(0, \Sigma_0),$$

where $\Sigma_0 = [E\{g'(\theta_0^\top X)^2 W_0(X)\}]^+ E\{g'(\theta_0^\top X)^2 W_0(X)\varepsilon^2\}[E\{g'(\theta_0^\top X)^2 W_0(X)\}]^+$.

By choosing a similar trimming function, the estimators in Härdle et al. (1993) and Ichimura (1993) have the same asymptotic covariance matrix as Theorem 11.3.2. If we further assume that the conditional distribution of Y given X belongs to a canonical exponential family

$$f_{Y|X}(y|x) = \exp\{y\eta(x) - \mathcal{B}(\eta(x)) + \mathcal{C}(y)\}$$

for some known functions $\mathcal{B}$, $\mathcal{C}$ and η, then Σ_0 is the lower information bound in the semiparametric sense (Bickel et al. 1993). See also the proofs in Carroll et al. (1997) and Härdle et al. (1993). In other words, our estimator is the most efficient one in the semiparametric sense.

For the estimation of the single-index model, it was generally believed that undersmoothing the link function must be employed in order to allow the estimator of the parameters to achieve root-n consistency. However, Härdle et al. (1993) established that undersmoothing the link function is not necessary. They derived an asymptotic expansion of the sum of squared residuals. We also derive an asymptotic expansion but of the estimator $\widehat{\theta}$ itself. This allows us to measure the higher order cost of estimating the link function. We use the expansion to propose an automatic bandwidth selection procedure for the index. Let $f_{\theta_0}(.)$ be the density function of $\theta_0^\top X$.

Theorem 11.3.3 (Higher order expansion). *Under conditions (C1)-(C5) and if ε_i is independent of X_i, we have almost surely*

$$\hat{\theta} - \theta_0 = \mathcal{E}_n + \frac{c_{1,n}}{nh} + c_{2,n}h^4 + \mathcal{H}_n + O\{n^{2\varsigma}\gamma_n^3\},$$

where $\gamma_n = h^2 + (nh/\log n)^{-1/2}$,

$$\mathcal{E}_n = (W_n)^+ \sum_{i=1}^n \rho_n\{f_{\theta_0}(X_j)\}g'(\theta_0^\top X_i)\nu_{\theta_0}(\theta_0^\top X_i)\varepsilon_i,$$

with $W_n = n^{-1}\sum_{j=1}^n \rho_n\{f_{\theta_0}(X_j)\}(g'(\theta_0^\top X_i))^2\nu_{\theta_0}(X_j)\nu_{\theta_0}^\top(X_j)$, $\mathcal{H}_n = O\{n^{-1/2}\gamma_n + n^{-1}h^{-1/2}\}$ *with* $E\{\mathcal{H}_n\mathcal{E}_n\} = o\{(nh)^{-2} + h^8\}$ *and*

$$c_{1,n} = \int K^2(v)v^2 dv\sigma^2(nW_n)^{-1}\sum_{j=1}^n \rho_n\{f_\theta(X_j)\}\{\nu'_{\theta_0}(X_j) + f'_0(X_j)\nu_{\theta_0}(X_j)/f_{\theta_0}(X_j)\},$$

$$c_{2,n} = \frac{1}{4}\left(\int K(v)v^4 dv - 1\right)(nW_n)^{-1}\sum_{j=1}^n \rho_n\{f_\theta(X_j)\}g'(\theta_0^\top X_j)g''(\theta_0^\top X_j)\nu''_{\theta_0}(X_j).$$

Because $K(v)$ is a density function and we constrain that $\int v^2 K(v) = 1$, it follows that $\mu_4 = \int K(v)v^4 dv > 1$. In the expansion of $\hat{\theta} - \theta_0$, the first term $\mathcal{E}_n$ does not depend on h. The second and third terms are the leading term among the remainders. The higher order properties of this estimator are better than those of the AD method, see Nishiyama and Robinson (2000), and indeed do not reflect a curse of dimensionality.

To minimize the stochastic expansion, it is easy to see that the bandwidth should be proportional to $n^{-1/5}$. Moreover, by Theorem 11.3.2 we consider the Mahalanobis distance

$$(\hat{\theta} - \theta_0)^\top \Sigma_0^+ (\hat{\theta} - \theta_0) = T_n + o\{h^8 + (nh)^{-2}\},$$

where

$$T_n = \left(\mathcal{E}_n + \frac{c_{1,n}}{nh} + c_{2,n}h^4 + \mathcal{H}_n\right)^\top \Sigma_0^+ \left(\mathcal{E}_n + \frac{c_{1,n}}{nh} + c_{2,n}h^4 + \mathcal{H}_n\right)$$

is the leading term. We have by Theorem 11.3.3 that

$$ET_n = E(\mathcal{E}_n^\top \Sigma_0^+ \mathcal{E}_n) + \left(\frac{c_1}{nh} + c_2 h^4\right)^\top \Sigma_0^+ \left(\frac{c_1}{nh} + c_2 h^4\right) + o\{h^8 + (nh)^{-2}\},$$

where $c_1 = \int K^2(v)v^2 dv \sigma^2 W_0^+ E\{v_0'(X) + f^{-1}(X)f'(X)v_0(X)\}$, $W_0 = E\{(g'(\theta_0^\top X))^2 v_{\theta_0}(X)v_{\theta_0}^\top(X)\}$ and

$$c_2 = \frac{1}{4}\left(\int K(v)v^4 dv - 1\right) W_0^+ E[g'(\theta_0^\top X)g''(\theta_0^\top X)v_{\theta_0}''(X)].$$

Note that $E(\mathcal{E}_n^\top \Sigma_0^+ \mathcal{E}_n)$ does not depend on h. By minimizing ET_n with respective to h, the optimal bandwidth should be

$$h_\theta = \left\{\frac{(9r_2^2 + 16r_1)^{1/2} - 3r_2}{8}\right\}^{1/5} n^{-1/5},$$

where $r_1 = c_1^\top \Sigma_0^+ c_1/(c_2^\top \Sigma_0^+ c_2)$ and $r_2 = c_1^\top \Sigma_0^+ c_2 / c_2^\top \Sigma_0^+ c_2$. As a comparison, we consider the optimal bandwidth for the estimation of the link function g. By Lemma 11.5.1 and Theorem 11.3.2, if $f_{\theta_0}(v) > 0$ we have

$$\hat{g}(v) = g(v) + \frac{1}{2}g''(v)^2 h^2 + \frac{1}{nf_{\theta_0}(v)}\sum_{i=1}^{n} K_h(\theta_0^\top X_i - v)\varepsilon_i + O_P(n^{-1/2} + h^2\gamma_n). \tag{11.6}$$

In other words, the link function can be estimated with the efficiency as if the index parameter vector is known. A brief proof for (11.6) is given in section 5. It follows that

$$|\hat{g}(v) - g(v)|^2 = S_n(v) + O_P\{(n^{-1/2} + h^2\gamma_n)\gamma_n\},$$

where the leading term is $S_n(v) = [\frac{1}{2}g''(v)^2 + \{nf_{\theta_0}(v)\}^{-1}\sum_{i=1}^n K_h(\theta_0^\top X_i - v)\varepsilon_i]^2$. Suppose we are interested in a constant bandwidth in the region $[a,b]$ with weight $w(v)$. Minimizing $\int_{[a,b]} ES_n(v)w(v)dv$ with respect to h, we have that the optimal bandwidth for the estimation of the link function is

$$h_g = \left[\frac{\int K^2(v)dv \int_{[a,b]} f_{\theta_0}^{-1}(v)\sigma_{\theta_0}^2(v)w(v)dv}{\int_{[a,b]} g''(v)^2 w(v)dv}\right]^{1/5} n^{-1/5}.$$

It is noticeable that the optimal bandwidth for the estimation of the parameter vector θ_0 is of the same order as that for the estimation of the link function. In other words, under-smoothing may lose efficiency for the estimation of θ_0 in the higher order sense. These optimal bandwidths h_θ^{opt} and h_g^{opt} can be consistently estimated by plug-in methods; see Ruppert et al. (1995).

Although the optimal bandwidth for the estimation of θ is different from that for the link function, its estimation such as the plug-in method may be very unstable because of the estimation of second order derivatives. Moreover, its estimation needs another pilot parameter which is again hard to choose. In practice it is convenient to apply h_g^{opt} for h_θ^{opt} directly, and since h_g^{opt} and h_θ^{opt} have the same order, the loss of efficiency in doing so should be small. For the former, there are a number of estimation methods such as CV and GCV methods. If the CV method is used, in each iteration with the latest estimator θ, the bandwidth is selected by minimizing

$$\hat{h}_g = \underset{h}{\text{argmin}}\, n^{-1} \sum_{j=1}^{n} \{Y_j - \hat{g}_j^{\theta}(\theta^\top X_j)\}^2,$$

where $\hat{g}_j^{\theta}(v)$ is the delete-one-observation estimator of the link function, i.e. the estimator of $\hat{g}^{\theta}(v)$ in (11.3) using data $\{(X_i, Y_i), i \neq j\}$. Another advantage of this approach is that we can also obtain the estimator for the link function.

11.4 Numerical Results

In the following calculation, the Gaussian kernel function and the trimming function (11.4) with $\varsigma = 1/20$ and $c_0 = 0.01$ are used. A MATLAB code rMAVE.m for the calculations below is available at

http://www.stat.nus.edu.sg/%7Estaxyc

In the first example, we check the behavior of the bandwidths h_g and h_θ. We consider two sets of simulations to investigate the finite performance of our estimation method, and to compare the bandwidths for the estimation of the link function g and the single-index θ_0. Our models are

$$\text{model A: } Y = (\theta_0^\top X)^2 + 0.2\varepsilon, \quad \text{model B: } Y = \cos(\theta_0^\top X) + 0.2\varepsilon,$$

where $\theta_0 = (3, 2, 2, 1, 0, 0, -1, -2, -2, -3)^\top/6$, $X \sim N_{10}(0, I)$, and $\varepsilon \sim N(0, 1)$ is independent of X. The OPG method was used to choose the initial value of θ. With different sample sizes n and bandwidths h, we estimate the model and calculate estimation errors

$$err_\theta = \{1 - |\theta_0^\top \hat{\theta}|\}^{1/2}, \quad err_g = \frac{1}{n} \sum_{j=1}^{n} \rho_n\{\hat{f}_{\hat{\theta}}(\hat{\theta}^\top X_j)\} |\hat{g}^{\hat{\theta}}(\hat{\theta}^\top X_j) - g(\theta_0^\top X_j)|,$$

where $\hat{g}^{\hat{\theta}}(\hat{\theta}^\top X_j)$ is defined in (11.3). With 200 replications, we calculate the mean errors $mean(err_\theta)$ and $mean(err_g)$. The results are shown in Fig. 11.1.

With the outer product of gradients estimator as the initial value (Xia et al. 2002), we have the following observations. (1) Notice that $n^{1/2} mean(err_\theta)$ tends to decrease as n increases, which means the estimation error err_θ enjoys a root-n consistency (and slightly faster for finite sample size). (2) Notice that the U-shape curves of err_θ have a wider bottom than those of err_g. Thus, the estimation of θ_0 is more robust to the bandwidth than the estimation of g. (3) Let $h_\theta^{opt} = \arg\min_h mean(err_\theta)$ and $h_g^{opt} = \arg\min_h mean(err_g)$. Then h_θ^{opt} and h_g^{opt}

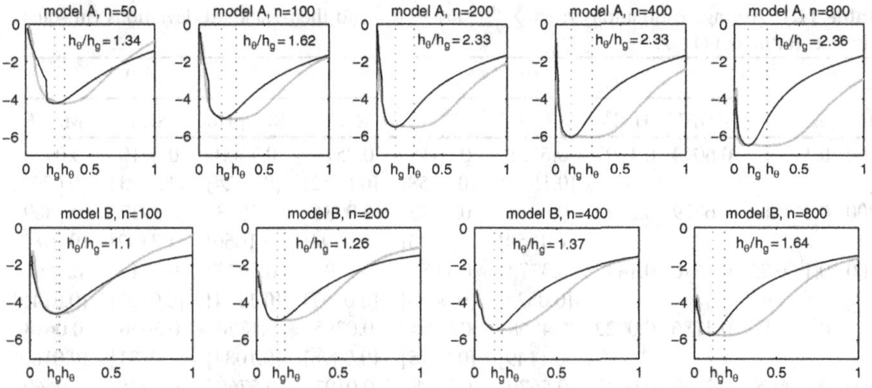

Fig. 11.1 The *wide solid lines* are the values of $\log\{n^{1/2} mean(err_\theta)\}$ and the *narrow lines* are the values of $\log\{n^{1/2} mean(err_g)\}$ (re-scaled for easier visualisation). The *dotted vertical lines* correspond to the bandwidths h_θ and h_g respectively

represent the best bandwidths respectively for the estimation of the link function g and the single-index θ_0. Notice that $h_\theta^{opt} / h_g^{opt}$ tends to increase as n increases, but seems to converge to a constant. Thus the under-smoothing bandwidth is not optimal.

Next, we compare our method with some of the existing estimation methods including ADE in Härdle and Stocker (1993), MAVE, the method in Hristache et al. (2001), called HJS hereafter, the SIR and pHd methods in Li (1991, 1992) and SLS in Ichimura (1993). For SLS, we use the algorithm in Friedman (1984) in the calculation. The algorithm has best performance among those proposed for the minimization of SLS, such as Weisberg and Welsh (1994) and Fan and Yao (2003). We consider the following model used in Hristache et al. (2001),

$$Y = (\theta_0^\top X)^2 \exp(a\theta_0^\top X) + \sigma\varepsilon, \tag{11.7}$$

where $X = (\mathbf{x}_1, \cdots, \mathbf{x}_{10})^\top$, $\theta_0 = (1, 2, 0, \ldots, 0)^\top / \sqrt{5}$, $\mathbf{x}_1, \cdots, \mathbf{x}_{10}, \varepsilon$ are independent and $\varepsilon \sim N(0, 1)$. For the covariates X: $(\mathbf{x}_k + 1)/2 \sim Beta(\tau, 1)$ for $k = 1, \cdots, p$. The parameter a is introduced to control the shape of the function. If $a = 0$, the structure is symmetric; the bigger it is, the more monotone the function is.

Following Hristache et al. (2001), we use the absolute deviation $\sum_{j=1}^p |\hat{\theta}_j - \theta_j|$ to measure the estimation errors. The calculation results for different σ and τ based on 250 replications are shown in Table 11.1. We have the following observations from Table 11.1. Our methods has much better performance than ADE and the method of Hristache et al. (2001). For each simulation, the better one of SIR and pHd is reported in Table 11.1, suggesting that these methods are not so competitive. Actually the main application of SIR and pHd is not in the estimation of single-index models. See Li (1991, 1992). For SLS, its performance depends much on the data and the model. If the model is easy to estimate (such as monotone and having big signal/noise ratio), it performs quite well. But overall SLS is still not so good

Table 11.1 Average estimation errors $\sum_{j=1}^{p}|\hat{\theta}_j - \theta_j|$ and their standard deviations (in square bracket) for model (11.7)

					$a = 1$			$a = 0$		
n	σ	τ	ADE*	HJS*	SIR/pHd	SLS	MAVE	SIR/pHd	SLS	MAVE
200	0.1	1	0.6094	0.1397	0.6521	0.0645	0.0514	0.7500	0.6910	0.0936
					[0.1569]	[0.0258]	[0.0152]	[0.1524]	[1.2491]	[0.0255]
200	0.2	1	0.6729	0.2773	0.6976	0.1070	0.0934	0.7833	0.8937	0.1809
					[0.1759]	[0.0375]	[0.0294]	[0.1666]	[1.3192]	[0.0483]
400	0.1	0.75	0.7670	0.1447	0.3778	0.1151	0.0701	0.6037	0.0742	0.0562
					[0.0835]	[0.0410]	[0.0197]	[0.1134]	[0.0193]	[0.0146]
400	0.1	1	0.4186	0.0822	0.4868	0.0384	0.0295	0.5820	0.5056	0.0613
					[0.1149]	[0.0125]	[0.0096]	[0.1084]	[1.0831]	[0.0167]
400	0.1	1.5	0.2482	0.0412	0.5670	0.0208	0.0197	0.5760	0.0923	0.0669
					[0.1524]	[0.0063]	[0.0056]	[0.1215]	[0.0257]	[0.0175]
400	0.2	1	0.4665	0.1659	0.5249	0.0654	0.0607	0.6084	0.7467	0.1229
					[0.1353]	[0.0207]	[0.0178]	[0.1064]	[1.2655]	[0.0357]
400	0.4	1	0.5016	0.3287	0.6328	0.1262	0.1120	0.6994	0.9977	0.2648
					[0.1386]	[0.0406]	[0.0339]	[0.1370]	[1.2991]	[0.1880]

*The values are adopted from Hristache et al. (2001)

as MAVE. The proposed method has the best performance in all the simulations we have done.

11.5 Proof of Theorems

Let $f_\theta(v)$ be the density function of $\theta^\top X$ and $\Lambda_n = \{x : |x| < n^c, f_\theta(x) > n^{-2\varsigma}, \theta \in \Theta_n\}$ where $c > 1/3$ and $\varsigma > 0$ is defined in (C5). Suppose A_n is a random matrix depending on x and θ. By $A_n = \mathcal{O}(a_n)$ (or $A_n = o(a_n)$) we mean that all elements in A_n are $O_{a.s.}(a_n)$ (or $o_{a.s.}(a_n)$) uniformly for $\theta \in \Theta_n$ and $x \in \Lambda_n$. Let $\delta_n = (nh/\log n)^{-1/2}$, $\gamma_n = h^2 + \delta_n$ and $\delta_\theta = |\theta - \theta_0|$. For any vector $V(v)$ of functions of v, we define $(V(v))' = dV(v)/dv$.

Suppose $(X_i, Z_i), i = 1, 2, \ldots, n$, are i.i.d. samples from (X, Z). Let $X_{ix} = X_i - x$,

$$s_k^\theta(x) = n^{-1}\sum_{i=1}^{n} K_h(\theta^\top X_{ix})\{\theta^\top X_{ix}/h\}^k,$$

$$t_k^\theta(x) = n^{-1}\sum_{i=1}^{n} K_h(\theta^\top X_{ix})\{\theta^\top X_{ix}/h\}^k X_i,$$

$$w_k^\theta(x) = n^{-1}\sum_{i=1}^{n} K_h(\theta^\top X_{ix})\{\theta^\top X_{ix}/h\}^k X_i X_i^\top,$$

$$e_k^\theta(x) = n^{-1}\sum_{i=1}^{n} K_h(\theta^\top X_{ix})\{\theta^\top X_{ix}/h\}^k \varepsilon_i,$$

$\epsilon_k^\theta = s_k^\theta(x) - Es_k^\theta(x)$, $\xi_k^\theta = t_k^\theta(x) - Et_k^\theta(x)$, $D_{n,k}^\theta(x) = s_2^\theta(x)s_k^\theta(x) - s_1^\theta(x)s_{k+1}^\theta(x)$, $E_{n,k}^\theta = s_0^\theta(x)s_{k+1}^\theta(x) - s_1^\theta(x)s_k^\theta(x)$ for $k = 1, 2, \ldots$. For any random variable Z and its random observations $Z_i, i = 1, \ldots, n$, let

$$T_{n,k}^\theta(Z|x) = s_2^\theta(x)n^{-1}\sum_{i=1}^n K_h^\theta(X_{ix})(\theta^\top X_{ix}/h)^k Z_i - s_1^\theta(x)n^{-1} \times \sum_{i=1}^n K_h^\theta(X_{ix})(\theta^\top X_{ix}/h)^{k+1} Z_i,$$

$$S_{n,k}^\theta(Z|x) = s_0^\theta(x)n^{-1}\sum_{i=1}^n K_h^\theta(X_{ix})(\theta^\top X_{ix}/h)^{k+1} Z_i - s_1^\theta(x)n^{-1} \times \sum_{i=1}^n K_h^\theta(X_{ix})(\theta^\top X_{ix}/h)^k Z_i.$$

By the Taylor expansion of $g(\theta_0^\top X_i)$ at $\theta_0^\top x$, we have

$$\begin{aligned} g(\theta_0^\top X_i) &= g(\theta_0^\top x) + \sum_{k=1}^5 \frac{1}{k!} g^{(k)}(\theta_0^\top x)\{\theta^\top X_{ix} + (\theta_0 - \theta)^\top X_{ix}\}^k \\ &\quad + O(\{\theta^\top X_{ix} + (\theta_0 - \theta)^\top X_{ix}\}^6) \\ &= g(\theta_0^\top x) + A^\theta(x, X_i) + B^\theta(x, X_i)(\theta_0 - \theta) \\ &\quad + O\{(\theta^\top X_{ix})^6 + \delta_\theta^3(|X_i|^6 + |x|^6)\}, \end{aligned} \tag{11.8}$$

where $A^\theta(x, X_i) = \sum_{\ell=1}^5 (k!)^{-1} g^{(k)}(\theta_0^\top x)(\theta^\top X_{ix})^k$ and

$$B^\theta(x, X_i) = \sum_{k=1}^5 \frac{1}{(k-1)!} g^{(k)}(\theta_0^\top x)(\theta^\top X_{ix})^{k-1} X_{ix}^\top + \frac{1}{2} g''(\theta_0^\top x)(\theta - \theta_0)^\top X_{ix} X_{ix}^\top.$$

For ease of exposition, we simplify the notation and abbreviate g for $g(\theta_0^\top x)$ and g', g'', g''' for $g'(\theta_0^\top x)$, $g''(\theta_0^\top x)$, $g'''(\theta_0^\top x)$ respectively. Without causing confusion, we write $f_\theta(\theta^\top x)$ as f_θ, $f_\theta(\theta^\top X_j)$ as $f_\theta(X_j)$ and $K_h(\theta^\top X_{ij})$ as $K_h^\theta(X_{ij})$. Similar notations are used for the other functions.

Lemma 11.5.1 (Link function). *Let*

$$\Sigma_n^\theta(x) = n^{-1}\sum_{i=1}^n K_h^\theta(X_{ix}) \begin{pmatrix} 1 \\ \theta^\top X_{ix}/h \end{pmatrix} \begin{pmatrix} 1 \\ \theta^\top X_{ix}/h \end{pmatrix}^\top$$

and

$$\begin{pmatrix} a_\theta(x) \\ d_\theta(x)h \end{pmatrix} = \{n\Sigma_n^\theta(x)\}^{-1} \sum_{i=1}^n K_h^\theta(X_{ix}) \begin{pmatrix} 1 \\ \theta^\top X_{ix}/h \end{pmatrix} Y_i.$$

Under assumptions (C2)–(C5), we have

$$a_\theta(x) = g(\theta_0^\top x) + A_n^\theta(x)h^2 + B_n^\theta(x)(\theta_0 - \theta) + V_n^\theta(x) + \mathcal{O}(h^2\gamma_n^2 + \delta_\theta^3)(1 + |x|^6),$$

$$d_\theta(x)h = g'(\theta_0^\top x)h + \tilde{A}_n^\theta(x)h^2 + \tilde{B}_n^\theta(x)(\theta_0 - \theta)h + \tilde{V}_n^\theta(x) + \mathcal{O}(h^2\gamma_n^2 + \delta_\theta^3)(1 + |x|^6),$$

where

$$A_n^\theta(x) = \frac{1}{2}g'' + \frac{1}{4}\left\{(\mu_4 - 1)g'' f_\theta^{-2}(f_\theta f_\theta'' - 2(f_\theta')^2) + \frac{1}{24}\mu_4 g^{(4)}\right\} h^2 + H_{1,n}^\theta(x),$$

$$\tilde{A}_n^\theta(x) = \frac{1}{2}g''(\mu_4 - 1) f_\theta^{-1} f_\theta' h + \frac{1}{6}g^{(3)}\mu_4 h + \frac{1}{2}g'' f_\theta^{-1}(\epsilon_3^\theta - \epsilon_1^\theta) + \mathcal{O}(h\gamma_n),$$

$$B_n^\theta(x) = g'\nu_\theta + \mathcal{O}(\gamma_n + \delta_\theta), \quad \tilde{B}_n^\theta(x) = g'(\theta_0^\top x) f_\theta^{-1}\{f_\theta \nu_\theta(x)\}' + \mathcal{O}(\gamma_n),$$

where $H_{1,n}^\theta(x) = \frac{1}{2}g''(\theta_0^\top x)\{f_\theta^{-1}(\epsilon_2^\theta - \epsilon_0^\theta) + (2 - \mu_4) f_\theta^{-2} f_\theta' h\epsilon_1^\theta - f_\theta^{-2} f_\theta' h\epsilon_3^\theta\} + \frac{1}{6} f_\theta^{-1} g''' h\epsilon_3^\theta$ *and* $V_n^\theta(x) = f_\theta^{-1} e_0^\theta - f_\theta^{-2} f_\theta' h e_1^\theta + \mu_4 f_\theta^{-2} f_\theta'' h^2 e_0^\theta/2 + f_\theta^{-2}(e_0^\theta \epsilon_2^\theta - e_1^\theta \epsilon_1^\theta) - \mu_4 f_\theta^{-2} f_\theta''' h^3 e_1^\theta + \{f_\theta^{-2}(f_\theta')^2 - (\mu_4 + 1) f_\theta^{-1} f_\theta''\}\{f_\theta^{-1} h^2 e_0^\theta - f_\theta^{-2} f_\theta' h^3 e_1^\theta\} - f_\theta^{-1}(\epsilon_0^\theta + \epsilon_1^\theta)\{f_\theta^{-1} e_0^\theta - f_\theta^{-2} f_\theta' e_1^\theta\} + 2 f_\theta^{-2} f_\theta' h\epsilon_1^\theta f_\theta^{-1} e_0^\theta$ *and* $\tilde{V}_n^\theta(x) = f_\theta^{-1} e_1^\theta + f_\theta^{-2} f_\theta'' h^2 e_1^\theta/2 + f_\theta^{-2}(\epsilon_0^\theta e_1^\theta - \epsilon_1^\theta e_0^\theta) - f_\theta^{-2} f_\theta' h e_0^\theta + f_\theta^{-1}\epsilon_0^\theta[-(\mu_4 + 1) f_\theta^{-1} f_\theta'' h^2/2 - f_\theta^{-1}(\epsilon_0^\theta + \epsilon_1^\theta) + f_\theta^{-2}(f_\theta')^2 h^2]$.

Lemma 11.5.2 (Summations). *Let* $\eta_n^\theta(x) = n^{-1}\sum_{i=1}^n K_h^\theta(X_{ix})X_{ix}\varepsilon_i$. *Under conditions (C1)–(C5), we have*

$$\begin{aligned}\mathcal{A}_n^\theta &\stackrel{def}{=} n^{-1}\sum_{j=1}^n \rho_n\{s_0^\theta(x_j)\}g'(\theta_0^\top X_j)\eta_n^\theta(X_j)/s_0^\theta(X_j) \\ &= \mathcal{E}_n^\theta + r_{n,0}^\theta(\theta - \theta_0) + \mathcal{Q}_n^\theta + \mathcal{O}(n^{2\varsigma}\gamma_n^3),\end{aligned}$$

$$\mathcal{B}_n^\theta \stackrel{def}{=} (nh)^{-1}\sum_{j=1}^n \rho_n\{s_0^\theta(X_j)\}e_k^\theta(X_j)\eta_n^\theta(X_j)/s_0^\theta(X_j) = \frac{\tilde{c}_{k,n}}{nh} + R_n^\theta + \mathcal{O}(n^{2\varsigma}\gamma_n^3),$$

$$\mathcal{C}_n^\theta \stackrel{def}{=} n^{-1}\sum_{j=1}^n \rho_n(s_0^\theta(X_j))\epsilon_k^\theta(X_j)\eta_n^\theta(X_j)/s_0^\theta(X_j) = M_n^\theta + \mathcal{O}(n^{2\varsigma}\gamma_n^3),$$

where $\mathcal{E}_n^\theta = \sum_{i=1}^n \rho_n\{f_\theta(X_j)\}g'(\theta^\top X_i)\nu_\theta(\theta^\top X_i)\varepsilon_i$, $r_{n,0}^\theta = o(1)$,

$$\mathcal{E}_n^\theta = \mathcal{O}\{(n/\log n)^{-1/2}\}, \quad \mathcal{Q}_n^\theta = \mathcal{O}\{(n/\log n)^{-1/2}\gamma_n\}, \quad R_n^\theta = \mathcal{O}\{n^{-1/2}\delta_n\},$$
$$M_n^\theta = \mathcal{O}\{n^{-1/2}\delta_n\},$$

with $E\{\mathcal{E}_n^\theta Q_n^\theta\} = o(h^8 + (nh)^{-2}), E\{\mathcal{E}_n^\theta R_n^\theta\} = o(h^8 + (nh)^{-2}), E\{\mathcal{E}_n^\theta M_n^\theta\} = o(h^8 + (nh)^{-2})$, *and* $\tilde{c}_{k,n} = \int v^{k+1} K^2(v) dv E[\rho_n(f_\theta(X_j)) f_\theta^{-1}(X_j)(\nu_\theta(X_j) f_\theta(X_j))'(X_j)]$ *if* k *is odd, 0 otherwise.*

Lemma 11.5.3 (Denominator). *Let* $\mathcal{D}_n^\theta = n^{-2}\sum_{i,j=1}^n \rho_n(s_0^\theta(X_j)) d_\theta^2(X_j) K_h^\theta(X_{ij}) X_{ij} X_{ij}^\top / s_0^\theta(X_j)$ *in the algorithm. Suppose* $(\theta, B) : p \times p$ *is an orthogonal matrix. Then under (C1)-(C5), we have almost surely*

$$(\mathcal{D}_n^\theta)^{-1} = \theta\theta^\top d_{11}^\theta h^{-2} - \theta d_{12}^\theta B^\top h^{-1} - B(d_{12}^\theta)^\top \theta^\top h^{-1} + B d_{22}^\theta B^\top,$$

where

$$d_{11}^\theta = (G_n^\theta)^{-1} + o(1), \quad d_{12}^\theta = H_n^\theta h + \mathcal{O}(\gamma_n), \quad d_{22}^\theta = \frac{1}{2}(B^\top W_n^\theta B)^{-1} + \mathcal{O}(\gamma_n),$$

with $G_n^\theta = n^{-1}\sum_{j=1}^n \rho_n(f_\theta(X_j)) f_\theta^{-1}(X_j)(g'(\theta_0 X_j))^2$ *and* $H_n^\theta = \frac{1}{2} n^{-1}\sum_{j=1}^n \rho_n(f_\theta(X_j)) f_\theta^{-1}(X_j)\{(f_\theta \nu_\theta)'(X_j)\}^\top (G_n^\theta)^{-1}(g'(\theta_0^\top X_j))^2 B(B^\top W_n^\theta B)^{-1}$ *and* $W_n^\theta = n^{-1}\sum_{j=1}^n \rho_n\{f_\theta(X_j)\}(g'(\theta^\top X_i))^2 \nu_\theta(X_j)\nu_\theta^\top(X_j)$.

11.5.1 Proof of Lemma 11.5.3

Let (θ, B) be an orthogonal matrix. It is easy to see that

$$n^{-1}\sum_{i=1}^n K_h^\theta(X_{ix})\theta^\top X_{ix} X_{ix}^\top \theta = s_2^\theta(x) h^2,$$

$$n^{-1}\sum_{i=1}^n K_h^\theta(X_{ix})\theta^\top X_{ix} X_{ix}^\top B = \{t_1^\theta(x) - s_1^\theta(x) x\}^\top B h,$$

$$n^{-1}\sum_{i=1}^n K_h^\theta(X_{ix}) B^\top X_{ix} X_{ix}^\top B = B^\top\{w_0^\theta(x) - t_0^\theta(x) x^\top - x(t_0^\theta(x))^\top + x x^\top s_0^\theta(x)\} B.$$

Thus

$$(\mathcal{D}_n^\theta)^{-1} = (\theta, B)\begin{pmatrix} D_{11}^\theta h^2 & (D_{12}^\theta)^\top B h \\ B^\top D_{12}^\theta h & B^\top D_{22}^\theta B \end{pmatrix}^{-1} (\theta, B)^\top,$$

where

$$D_{11}^\theta = n^{-1}\sum_{j=1}^n \rho_n(s_0^\theta(X_j))\{d^\theta(X_j)\}^2 s_2^\theta(X_j)/s_0^\theta(X_j),$$

$$D_{12}^\theta = n^{-1}\sum_{j=1}^n \rho_n(s_0^\theta(X_j))\{d_\theta(X_j)\}^2\{t_1^\theta(X_j) - s_1^\theta(X_j) X_j\}^\top / s_0^\theta(X_j),$$

$$D_{22}^{\theta} = n^{-1}\sum_{j=1}^{n}\rho_n(s_0^{\theta}(X_j))(d_{\theta}(X_j))^2\{w_0^{\theta}(X_j) - t_0^{\theta}(X_j)X_j^{\top} - X_j t_0^{\theta}(X_j)$$
$$+X_j X_j^{\top} s_0^{\theta}(X_j)\}/s_0^{\theta}(X_j).$$

By the matrix inversion formula in blocks (Schott, 1997), we have the equation in Lemma 11.5.3 with $d_{11}=\{D_{11}^{\theta}-(D_{12}^{\theta})^{\top}BB^{\top}(D_{22}^{\theta})^{-1}BB^{\top}D_{12}^{\theta}\}^{-1}$, $d_{12}=d_{11}^{\theta}(D_{12}^{\theta})^{\top}B(B^{\top}D_{22}^{\theta}B)^{-1}$, $d_{22}^{\theta} = \{B^{\top}D_{22}^{\theta}B\}^{-1} + d_{11}\{B^{\top}D_{22}^{\theta}B\}^{-1}B^{\top}D_{12}^{\theta}(D_{12}^{\theta})^{\top}B\{B^{\top}D_{22}^{\theta}B\}^{-1}$. By Lemma 11.8.1, we have

$$D_{11}^{\theta} = G_n^{-1} + o(1), \quad D_{12}^{\theta} = H_n h + \mathcal{O}(\gamma_n), \quad D_{22}^{\theta} = 2W_n + \mathcal{O}(\gamma_n).$$

Thus, Lemma 11.5.3 follows. □

Lemma 11.5.4 (Numerator). *Let* $\mathcal{N}_n^{\theta} = n^{-2}\sum_{i,j=1}^{n}\rho_n(s_0^{\theta}(X_j))K_h^{\theta}(X_{ij})X_{ij}\{Y_i - a_{\theta}(X_j) - d_{\theta}(X_j)\theta_0^{\top}X_{ij}\}/s_0^{\theta}(X_j)$. *Under assumptions (C1)–(C5), we have almost surely*

$$\mathcal{N}_n^{\theta} = \mathcal{E}_n^{\theta} + \frac{\tilde{c}_{1,n}}{nh} + \tilde{c}_{2,n}h^4 + \mathcal{R}_n^{\theta} + \mathcal{B}_n^{\theta}(\theta - \theta_0) + \mathcal{O}\{n^{2\varsigma}(\gamma_n^3 + \delta_{\theta}^3)\},$$

where $\mathcal{R}_n^{\theta} = \mathcal{O}\{n^{-1}(\log n/h)^{1/2} + (\log n/n)^{-1/2}h^2\}$, $\theta^{\top}\mathcal{R}_n^{\theta} = \mathcal{O}\{hn^{-1}(\log n/h)^{1/2} + (\log n/n)^{-1/2h^3}$ *and* $E\{\mathcal{R}_n^{\theta}\mathcal{E}_0^{\theta}\} = \mathcal{O}\{(nh)^{-2} + h^8\}$, $\mathcal{B}_n^{\theta} = W_n^{\theta} + o(1)$ *with* W_n^{θ} *defined in Lemma 11.5.3,* $\tilde{c}_{1,n}$ *and* $\mathcal{E}_0^{\theta}$ *are defined in Lemma 11.5.2 and*

$$\tilde{c}_{2,n} = \frac{1}{4}(\mu_4 - 1)\sum_{j=1}^{n}\rho_n\{f_{\theta}(X_j)\}g'(\theta_0^{\top}X_j)g''(\theta_0^{\top}X_j)\nu_{\theta}''(X_j).$$

11.6 Proof of Theorem 11.3.1

By assumption (C2), we have

$$\sum_{n=1}^{\infty}P\left(\bigcup_{i=1}^{n}\{X_i \notin \Lambda_n\}\right) \le \sum_{n=1}^{\infty}nP(X_i \notin \Lambda_n) \le \sum_{n=1}^{\infty}nP(|X_i| > n^c)$$
$$< \sum_{n=1}^{\infty}nn^{-6c}E|X|^6 < \infty$$

for any $c > 1/3$. It follows from the Borel-Cantelli lemma that

$$P\left(\bigcap_{n=1}^{\infty}\bigcup_{i=1}^{n}\{X_i \notin \Lambda_n\}\right) = 0. \tag{11.9}$$

Let $\tilde{\Lambda}_n = \{x : f_\theta(\theta^\top x) > 2n^{-\epsilon}\}$. Similarly, we have

$$P\left(\bigcap_{n=1}^{\infty}\bigcup_{i=1}^{n}\{X_i \notin \tilde{\Lambda}_n\}\right) = 0. \tag{11.10}$$

Thus, we can exchange summations over $\{X_j : j = 1,\cdots,n\}$, $\{X_j : X_j \in \Lambda_n, j = 1,\cdots,n\}$ and $\{X_j : X_j \in \tilde{\Lambda}_n, j = 1,\cdots,n\}$ in the sense of almost sure consistency. On the other hand, we have by (C2)

$$n^{-1}\sum_{|X_j|<n^c}(1+|X_j|^6) = O(1).$$

By the notation in Lemmas 11.5.3 and 11.5.4, after one iteration of Steps 1–3, the new θ is

$$\tilde{\theta} = \theta_0 + (\mathcal{D}_n^\theta)^{-1}\mathcal{N}_n^\theta. \tag{11.11}$$

Note that $\theta^\top\mathcal{E}_n^\theta = 0, \theta^\top c_{1,n}^\theta = 0, \theta^\top c_{2,n}^\theta = 0, \theta^\top W_n^\theta = 0, W_n^\theta(W_n^\theta)^+ = I - \theta\theta^\top$ and $\delta_\theta/h^2 \to 0$. We have

$$\begin{aligned}\tilde{\theta} &= \theta_0 + \theta[\theta^\top d_{11}^\theta h^{-2}\{\mathcal{R}_n^\theta + \mathcal{B}_n^\theta(\theta-\theta_0) + \mathcal{O}\{n^{2\varsigma}(\gamma_n^3+\delta_\theta^3)\}\} - d_{12}^\theta B^\top h^{-1}\mathcal{N}_n^\theta] \\ &\quad - B(d_{12}^\theta)^\top\theta^\top h^{-1}[\mathcal{R}_n^\theta + \mathcal{B}_n^\theta(\theta-\theta_0) + \mathcal{O}\{n^{2\varsigma}(\gamma_n^3+\delta_\theta^3)\}] + Bd_{22}^\theta B^\top\mathcal{N}_n^\theta \\ &= (1+a_n)\theta_0 + \left\{\frac{1}{2}(I-\theta_0\theta_0^\top) + b_n\right\}(\theta-\theta_0) + \frac{1}{2}\{W_n^\theta\}^+\mathcal{E}_n^\theta + \mathcal{O}(h^4),\end{aligned}$$

where $a_n = o(1)$ and $b_n = o(1)$.

By (11.25) below, we have $s_0^\theta(x) = f_\theta(\theta^\top x) + \mathcal{O}(\gamma_n)$. Thus by the smoothness of $\rho_n(\cdot)$ and (11.10), we have

$$\rho_n(s_0^\theta(x)) = \rho_n(f_\theta(\theta^\top x)) + \mathcal{O}(n^\varsigma\gamma_n) = 1 + \mathcal{O}(n^\varsigma\gamma_n). \tag{11.12}$$

Since $\rho_n(.)$ is bounded, we have $E\{\rho_n(\hat{f}_\theta(\theta^\top x)) - 1\}^2 = o(1)$. By (C3) and Lemma 11.8.1, we have

$$\mathcal{E}_n^\theta = n^{-1}\sum_{i=1}^{n} g'(\theta_0^\top X_i)\nu_{\theta_0}(X_i)\varepsilon_i + o(n^{-1/2}).$$

Note that $W_n = W_0 + o(\delta_\theta)$. It is easy to check that $|\tilde{\theta}| = 1 + a_n + b_n + \mathcal{O}(h^4) = 1 + o(1)$. Thus

$$\tilde{\theta}/|\tilde{\theta}| = \theta_0 + \left\{\frac{1}{2}(I - \theta_0\theta_0^{\top}) + o(1)\right\}(\theta - \theta_0) + \frac{1}{2}n^{-1}W_0^{+}\sum_{i=1}^{n} g'(\theta_0^{\top}X_i)\nu_{\theta_0}(X_i)\varepsilon_i$$
$$+ o(h^3 + n^{-1/2}).$$

Let $\theta^{(k)}$ be the value of θ after k iterations . Recall that $h_{k+1} = \max\{h_k/c_h, h\}$. Therefore,

$$|\theta_{k+1} - \theta_0|/h_{k+1}^2 \to 0,$$

for all $k > 1$. We have

$$\theta^{(k+1)} = \theta_0 + \left\{\frac{1}{2}(I - \theta_0\theta_0^{\top}) + o(1)\right\}(\theta^{(k)} - \theta_0)$$
$$+ \frac{1}{2}n^{-1}W_0^{+}\sum_{i=1}^{n} g'(\theta_0^{\top}X_i)\nu_{\theta_0}(X_i)\varepsilon_i + o(h_k^3 + n^{-1/2}).$$

Recursing the above equation, we have

$$\theta^{(k+1)} = \theta_0 + \left\{\frac{1}{2^k}(I - \theta_0\theta_0^{\top}) + o(1)\sum_{\iota=1}^{k}\frac{1}{2^\iota}\right\}(\theta^{(1)} - \theta_0)$$
$$+ \left\{\sum_{\iota=1}^{k}\frac{1}{2^\iota}\right\} n^{-1}W_0^{+}\sum_{i=1}^{n} g'(\theta_0^{\top}X_i)\nu_{\theta_0}(X_i)\varepsilon_i + o\left(\sum_{\iota=1}^{k}\frac{1}{2^\iota}h_{k-\iota}^3 + n^{-1/2}\right).$$

Thus as the number of iterations $k \to \infty$, Theorem 11.3.1 follows immediately from the above equation and the central limit theorem. □

11.7 Proof of Theorem 11.3.3

Based on Theorem 11.3.2, we can assume $\delta_\theta = (\log n/n)^{1/2}$. Note that $\theta^{\top}\{\mathcal{E}_n^\theta + c_{1,n}(nh)^{-1} + c_{2,n}h^4\} = 0$. We consider the product of each term in $(\mathcal{D}_n^\theta)^{-1}$ with $\mathcal{N}_n^\theta$. We have

$$\theta\theta^{\top}d_{11}^\theta h^{-2}\mathcal{N}_n^\theta = \theta\theta^{\top}d_{11}^\theta h^{-2}[\mathcal{R}_n^\theta + B_n^\theta(\theta - \theta_0) + \mathcal{O}\{n^{2\varsigma}(\gamma_n^3 + \delta_\theta^3)\}]$$
$$= a_n^\theta\theta_0 + a_n^\theta(\theta - \theta_0),$$
$$\theta d_{12}^\theta B^{\top}h^{-1}\mathcal{N}_n^\theta = b_n^\theta\theta_0 + b_n^\theta(\theta - \theta_0),$$

and hence,

$$(\mathcal{D}_n^\theta)^{-1}\mathcal{N}_n = \theta\{S_n^{\top}(\theta - \theta_0) + H_n\mathcal{E}_n^\theta + \mathcal{O}(n^{2\varsigma}\gamma_n^4)\}$$
$$= \theta_0\{S_n^{\top}(\theta - \theta_0) + H_n\mathcal{E}_0^\theta + \mathcal{O}(n^{2\varsigma}\gamma_n^4)\} + c_n(\theta - \theta_0),$$

where $S_n = \mathcal{O}(1)$ and $c_n = \mathcal{O}(\gamma_n/h)$. It is easy to see that $c_n = o(1)$ provided that $|\theta - \theta_0|/h^2 \to 0$. By Lemma 11.5.3 and 11.5.4, we have

$$\tilde{\theta} = \theta_0\{1 + S_n^\top(\theta - \theta_0) + H_n\mathcal{E}_0^\theta + \mathcal{O}(n^{2\varsigma}\gamma_n^4)\} + \frac{1}{2}W_n^\theta\left\{\mathcal{E}_0^\theta + \frac{c'_{1,n}}{nh} + c'_{2,n}h^4 + \mathcal{R}_n^\theta + \mathcal{Q}_n^\theta\right\}$$
$$+\left\{\frac{1}{2}(I - \theta\theta^\top) + c_n\right\}(\theta - \theta_0) + \mathcal{O}\{n^{2\varsigma}(\gamma_n^3 + h\log n/n)\}.$$

It is easy to see that $|\tilde{\theta}| = 1 + S_n^\top(\theta - \theta_0) + H_n\mathcal{E}_0^\theta + \mathcal{O}(n^{2\varsigma}\gamma_n^4)$. Thus

$$\tilde{\theta}/|\tilde{\theta}| = \theta_0 + \frac{1}{2}W_n^\theta\left\{\mathcal{E}_0^\theta + \frac{c'_{1,n}}{nh} + c'_{2,n}h^4 + \mathcal{R}_n^\theta + \mathcal{E}_0^\theta H_n^\top \mathcal{E}_0^\theta\right\}$$
$$+\left\{\frac{1}{2}(I - \theta\theta^\top) + c'_n\right\}(\theta - \theta_0) + \mathcal{O}\{n^{2\varsigma}(\gamma_n^3 + h\log n/n)\},$$

where $c'_n = o(1)$. Similar to the proof of Theorem 11.3.1, we complete the proof with $c_{1,n} = W_n^{-1}c'_{1,n}$ and $c_{2,n} = W_n^{-1}c'_{2,n}$. □

11.8 Proofs of the Lemmas

In this section, we first give some results about the uniform consistency. Based on these results, the Lemmas are proved.

Lemma 11.8.1. *Suppose $G_{n,i}(\chi)$ is a martingale with respect to $\mathcal{F}_i = \sigma\{G_{n,\ell}(\chi), \ell \le i\}$ with $\chi \in \mathcal{X}$ and $\mathcal{X}$ is a compact region in a multidimensional space such that (I) $|G_{n,i}(\chi)| < \xi_i$, where ξ_i are IID and $\sup E\xi_1^{2r} < \infty$ for some $r > 2$; (II) $EG_{n,k}^2(\chi) < a_n s(\chi)$ with $\inf s(\chi)$ positive, and (III) $|G_{n,i}(\chi) - G_{n,i}(\tilde{\chi})| < n^{\alpha_1}|\chi - \tilde{\chi}|M_i$, where $M_i, i = 1, 2, \ldots$ are IID with $EM_1^2 < \infty$. If $a_n = cn^{-\delta}$ with $0 \le \delta < 1 - 2/r$, then for any $\alpha'_1 > 0$ we have*

$$\sup_{|\chi| \le n^{\alpha'_1}} \left| n^{-1}s^{-1/2}(\chi)\sum_{i=1}^n G_{n,i}(\chi)\right| = O\{(n^{-1}a_n\log n)^{1/2}\}$$

almost surely. Suppose for any fixed n and k, $G_{n,i,k}(\theta)$ is a martingale with respect to $\mathcal{F}_{i,k} = \sigma\{G_{n,\ell,k}(\theta), \ell \le i\}$ such that (I) $|G_{n,i,k}(\theta)| \le \xi_i$, (II) $EG_{n,i,k}^2(\theta) < a_n$ and (III) $|G_{n,i,k}(\theta) - G_{n,i,k}(\tilde{\theta})| < n^{\alpha_2}|\theta - \tilde{\theta}|M_i$, where ξ_i, a_n and M_i are defined above. If $E|\varepsilon_k|^{2r} < \infty$ and $E\{\varepsilon_k | G_{n,i,j}(\theta), i < j, j = 1, \ldots, k-1\} = 0$, then

$$\sup_{\theta\in\Theta}\left| n^{-2}\sum_{k=2}^n\left\{\sum_{i=1}^{k-1} G_{n,i,k}(\theta)\right\}\varepsilon_k\right| = O\{(a_n\log n)^{1/2}/n\}$$

almost surely.

11.8.1 Proof of Lemma 11.8.1

We give the details for the second part of the Lemma. The first part is easier and can be proved similarly. Let $\Delta_n(\theta)$ be the expression between the absolute symbols in the equation. By (III) and the strong law of large numbers, it is easy to see that there are $n_1 = n^{\alpha_3}$ balls centered at $\theta_\iota : B_\iota = \{\theta : |\theta - \theta_\iota| < n^{-\alpha_4}\}$ with $\alpha_4 > \alpha_2 + 2$, such that $\bigcup_{\iota=1}^{n_1} B_\iota \supset \Theta$. By the strong law of large numbers, we have

$$\max_{1\le\iota\le n_1} \sup_{\theta\in B_\iota} |\Delta_n(\theta) - \Delta_n(\theta_\iota)| \le n^{\alpha_2} \max_{1\le\iota\le n_1} \sup_{\theta\in B_\iota} |\theta - \theta_\iota| n^{-2} \sum_{k=1}^{n} |\varepsilon_k| \times \sum_{i=1}^{n} M_i = O\{(a_n \log n)^{1/2}/n\}$$

almost surely. Let $\Delta_{n,k}(\theta_\iota) = \sum_{i=1}^{k-1} G_{n,i,k}(\theta_\iota)$. Next, we show that there is a constant c_1 such that

$$p_n \overset{def}{=} P\Big(\bigcap_{\ell=1}^{\infty}\bigcup_{n=\ell}^{\infty}\{\max_{1<k\le n}\max_{1<\iota\le n_1} |\Delta_{n,k}(\theta_\iota)| > c_1(na_n \log n)^{1/2}\}\Big) = 0. \quad (11.13)$$

Let $T_n = \{na_n \log(n)\}^{1/2}$, $G^I_{n,i,k}(\theta_\iota) = G_{n,i,k}(\theta_\iota) I(|G_{n,i,k}(\theta_\iota)| \le T_n)$ and $G^O_{n,i,k}(\theta_\iota) = G_{n,i,k}(\theta_\iota) - G^I_{n,i,k}(\theta_\iota)$. Write

$$\Delta_{n,k}(\theta_\iota) = \sum_{i=1}^{k-1}\{G^I_{n,i,k}(\theta_\iota)) - EG^I_{n,i,k}(\theta_\iota)\} + \sum_{i=1}^{k-1}\{G^O_{n,i,k}(\theta_\iota) - EG^O_{n,i,k}(\theta_\iota)\}. \quad (11.14)$$

Note that $E|G^O_{n,i,k}(\theta_\iota)| \le T_n^{-r+1} E|\xi_1|^r = E|\xi_1|^r\{na_n \log(n)\}^{-(r-1)/2}$. If $a_n = cn^{-\delta}$ with $0 \le \delta < 1 - 2/r$ and $k \le n$, we have

$$|\sum_{i=1}^{k-1} EG^O_{n,i,k}(\theta_\iota)| \le E|\xi_1|^r (k-1)\{na_n \log(n)\}^{-(r-1)/2} \le CE|\xi_1|^r\{na_n \log(n)\}^{1/2}. \quad (11.15)$$

Note that

$$\sum_{i=1}^{n} |G^O_{n,i,k}((\theta_\iota)| \le \sum_{i=1}^{n} |\xi_i| I(|\xi_i| > T_n) \le T_n^{-r+1} \sum_{i=1}^{n} |\xi_i|^r I(|\xi_i| > T_n).$$

For fixed T, by the strong law of large numbers, we have

$$n^{-1}\sum_{i=1}^{n}|\xi_i|^r I(|\xi_i|>T)\to E\{|\xi_1|^r I(|\xi_1|>T)\}$$

almost surely. The right hand side above is dominated by $E\{|\xi_1|^r\}$ and tends to zero as $T\to\infty$. Note that T_n increases to ∞ with n. For large n such that $T_n>T$, we have

$$n^{-1}\sum_{i=1}^{n}|\xi_i|^r I(|\xi_i|>T_n)\le n^{-1}\sum_{i=1}^{n}|\xi_i|^r I(|\xi_i|>T)\to 0$$

almost surely as $T\to\infty$. It follows that

$$\sum_{i=1}^{n}|G^{O}_{n,i,k}(\theta_\iota)|=o(nT_n^{-r+1})=o\{(na_n\log n)^{1/2}\} \tag{11.16}$$

almost surely. Thus by (11.15) and (11.16), if $c_1'>CE|\xi_1|^r$ we have

$$\begin{aligned}p_n'&\overset{def}{=}P\left(\bigcap_{\ell=1}^{\infty}\bigcup_{n=\ell}^{\infty}\left\{\max_{1<k\le n}\max_{1<\iota\le n_1}|\sum_{i=1}^{k-1}\{G^{O}_{n,i,k}((\theta_\iota)-EG^{O}_{n,i,k}(\theta_\iota)\}|>c_1'(na_n\log n)^{1/2}\right\}\right)\\&\le P\left(\bigcap_{\ell=1}^{\infty}\bigcup_{n=\ell}^{\infty}\left\{\sum_{i=1}^{n}|\xi_i|(|\xi_i|\ge T_n)>c_1'(na_n\log n)^{1/2}\right\}\right)\\&\quad+P\left(\bigcap_{\ell=1}^{\infty}\bigcup_{n=\ell}^{\infty}\left\{\max_{1<k\le n}\max_{1<\iota\le n_1}|\sum_{i=1}^{k-1}EG^{O}_{n,i,k}(\theta_\iota)|>c_1'(na_n\log n)^{1/2}\right\}\right)=0\end{aligned} \tag{11.17}$$

By condition (II), if $k\le n$ we have

$$\max_{1\le\iota\le n_1}\mathrm{Var}\sum_{i=1}^{k-1}\{G^{I}_{n,i,k}(\theta_\iota)-EG^{I}_{n,i,k}(\theta_\iota)\}\le c_2na_n\overset{def}{=}N_1, \tag{11.18}$$

where c_2 is a constant. By the condition on a_n and the definition of $G^{I}_{n,i,k}(\theta_\iota)$, we have constants c_3 and c_4 such that

$$\begin{aligned}&\max_{1\le\iota\le n^{\alpha}}|\{G^{I}_{n,i,k}(\theta_\iota)-EG^{I}_{n,i,k}(\theta_\iota)\}|\le c_3T_n\\&=c_3\{na_n/\log n\}^{1/2}\{a_n^{-r}\log^{r+1}n/n^{r-2}\}^{1/(2(r-1))}\\&\le c_4\{na_n/\log n\}^{1/2}\overset{def}{=}N_2.\end{aligned} \tag{11.19}$$

Let $N_3 = c_5\{na_n \log n\}^{1/2}$ with $c_5^2 > 2(\alpha_3+3)(c_2+c_4c_5)$. By Bernstein's inequality (cf. de la Peña 1999), we have from (11.18) and (11.19) that for any $k \le n$,

$$P\left(|\sum_{i=1}^{k-1}\{G_{n,i,k}^I(\theta_\iota) - EG_{n,i,k}^I(\theta_\iota)\}| > N_3\right) \le 2\exp\left(\frac{-N_3^2}{2(N_1 + N_2N_3)}\right)$$
$$\le 2\exp\{-c_5^2 \log n/(2c_2 + 2c_4c_5)\}$$
$$\le c_6 n^{-\alpha_3-3}.$$

Let $c_1 > \max\{c_5, c_1'\}$. We have

$$\sum_{n=1}^{\infty} P\left\{\max_{1<k\le n}\max_{1<\iota\le n_1}|\sum_{i=1}^{k-1}[G_{n,i,k}^I(\theta_\iota) - EG_{n,i,k}^I(\theta_\iota)]| > c_1(na_n\log n)^{1/2}\right\}$$
$$\le \sum_{n=1}^{\infty}\sum_{k=2}^{n}\sum_{\iota=1}^{n_1} P\left\{|\sum_{i=1}^{k-1}[G_{n,i,k}^I(\theta_\iota) - EG_{n,i,k}^I(\theta_\iota)]| > c_1(na_n\log n)^{1/2}\right\}$$
$$\le \sum_{n=1}^{\infty} c_6 n^{-\alpha_3-3} n^{1+\alpha_3} < \infty. \tag{11.20}$$

By (11.14), (11.17) and (11.20) and the Borel-Cantelli lemma, we have

$$p_n \le P\left\{\bigcap_{\ell=1}^{\infty}\bigcup_{n=\ell}^{\infty}\max_{1<k\le n}\max_{1<\iota\le n_1}|\sum_{i=1}^{k-1}[G_{n,i,k}^I(\theta_\iota) - EG_{n,i,k}^I(\theta_\iota)]| > c_1(na_n\log n)^{1/2}\right\}$$
$$+ p_n' = 0.$$

Therefore (11.13) follows.

Let $\Delta_{n,k}^I(\theta_\iota) = \Delta_{n,k}(\theta_\iota)I\{|\Delta_{n,k}(\theta_\iota)| \le c_1(na_n\log n)^{1/2}\}$ and $U_\ell(\theta_\iota) = \sum_{k=2}^{\ell}\Delta_{n,k}^I(\theta_\iota)\varepsilon_k$. Write

$$\Delta_n(\theta_\iota) = U_n(\theta_\iota) + \sum_{k=2}^{n}\Delta_{n,k}^O(\theta_\iota)\varepsilon_k,$$

where $\Delta_{n,k}^O(\theta_\iota) = \Delta_{n,k}(\theta_\iota) - \Delta_{n,k}^I(\theta_\iota)$. It is easy to see from (11.13) that for the second part on the right hand side above,

$$\max_{1<\iota\le n_1}|\sum_{k=2}^{n}\Delta_{n,k}^O(\theta_\iota)\varepsilon_k| = O\{n(a_n\log n)^{1/2}\} \tag{11.21}$$

almost surely, since for any constant $c > 0$,

$$\sum_{n=1}^{\infty} P\left\{\max_{1<\iota\leq n_1}|\sum_{k=2}^{n}\Delta_{n,k}^{O}(\theta_\iota)\varepsilon_k|>cn(a_n\log n)^{1/2}\right\}\leq\sum_{n=1}^{\infty}P\left(\max_{1<\iota\leq n_1}\max_{1<k\leq n}|\Delta_{n,k}^{O}(\theta_\iota)|>0\right)$$

$$\leq \sum_{n=1}^{\infty} P\{\max_{1<\iota\leq n_1}\max_{1<k\leq n}|\Delta_{n,k}(\theta_\iota)| > c_1(na_n\log n)^{1/2}\}$$

$$< \infty.$$

Now consider the first term. Let $T_n'^{1/2}/\log n$,

$$U_\ell^I(\theta_\iota) = \sum_{k=2}^{\ell}\Delta_{n,k}^I(\theta_\iota)tbI\varepsilon_k I(|\varepsilon_k| \leq T_n') - E[\varepsilon_k I(|\varepsilon_k| \leq T_n')]\}$$

and $U_\ell^O(\theta_\iota) = U_\ell(\theta_\iota) - U_\ell^I(\theta_\iota)$. Similar to the proof of (11.15) and (11.16), we have almost surely

$$|\sum_{k=2}^{\ell}\Delta_{n,k}^O(\theta_\iota)E\{\varepsilon_k I(|\varepsilon_k| > T_n')\}| = O\{n(a_n\log n)^{1/2}\}, \qquad (11.22)$$

$$|\sum_{k=2}^{\ell}\Delta_{n,k}^O(\theta_\iota)\varepsilon_k I(|\varepsilon_k| > T_n')| = O\{n(a_n\log n)^{1/2}\}. \qquad (11.23)$$

Note that

$$|\Delta_{n,k}^I(\theta_\iota)\{\varepsilon_k I(|\varepsilon_k| \leq T_n') - E[\varepsilon_k I(|\varepsilon_k| \leq T_n')]\}| < 2c_1(na_n\log n)^{1/2}T_n'$$
$$= 2c_1 n(a_n/\log n)^{1/2} \stackrel{def}{=} N_4$$

and by (II), $\mathrm{Var}\{U_\ell^I(\theta_\iota)\} = c_2'^2 a_n \stackrel{def}{=} N_5$, where c_2' is a constant. Let $N_6 = c_3' n (a_n\log n)^{1/2}$ with $c_3'^2 > 2(\alpha_3+3)(2c_1c_3' + c_2')$. By Bernstein's inequality, we have

$$P(|U_n^I(\theta_\iota)| \geq N_6) \leq 2\exp\left\{-\frac{N_6^2}{2(N_6N_4+N_5)}\right\} \leq 2n^{-\alpha_3-3}.$$

Therefore

$$\sum_{n=1}^{n} P\left\{\max_{1\leq\iota\leq n_1}|U_n^I(\theta_\iota)| \geq N_6\right\} < \sum_{n=1}^{n} n_1 P\{|U_n^I(\theta_\iota)| \geq N_6\} < \infty.$$

By the Borel-Cantelli lemma, we have

$$\max_{1\leq\iota\leq n_1}|U_n^I(\theta_\iota)| = O(N_6) \qquad (11.24)$$

almost surely. Lemma 11.8.1 follows from (11.21), (11.22), (11.23) and (11.24). □

11.8.2 Proof of Lemma 11.5.1

Write $s_k^\theta(x) = \epsilon_k^\theta(x) + Es_k^\theta(x)$. By Taylor expansion, we have

$$s_k^\theta(x) = \sum_{\tau=0}^{3} \mu_{k+\tau} f_\theta^{(\tau)}(x) h^\tau + \epsilon_k^\theta(x) + \mathcal{O}(h^4). \tag{11.25}$$

Because $Var\{\epsilon_k^\theta(x)\} = O\{(nh)^{-1}\}$, it follows from Lemma 11.8.1 that $\epsilon_k^\theta(x) = \mathcal{O}(\delta_n)$. It is easy to check that

$$\begin{aligned}
D_{n,0}^\theta(x) &= f_\theta^2 + \frac{1}{2}(\mu_4 + 1) f_\theta f_\theta'' h^2 - (f_\theta')^2 h^2 + f_\theta(\epsilon_0^\theta + \epsilon_2^\theta) - 2 f_\theta' h \epsilon_1^\theta + \mathcal{O}(\gamma_n^2). \\
D_{n,2}^\theta(x) &= f_\theta^2 + \mu_4(f_\theta f_\theta'' - (f_\theta')^2)h^2 + 2 f_\theta \epsilon_2^\theta - f_\theta' h \epsilon_3^\theta - \mu_4 f_\theta' h \epsilon_1^\theta + \mathcal{O}(\gamma_n^2). \\
D_{n,3}^\theta(x) &= f_\theta \epsilon_3^\theta + \mathcal{O}(h\gamma_n), \quad D_{n,4}^\theta(x) = \mu_4 f_\theta^2 + \mathcal{O}(\gamma_n), \quad D_{n,5}^\theta(x) = \mathcal{O}(h). \\
T_{n,0}^\theta(X|x) &= f_\theta^2 \nu_\theta(x) + \mathcal{O}(\gamma_n), \quad S_{n,0}^\theta(X|x) = \mathcal{O}(h), \quad T_{n,k}^\theta(X|x) = \mathcal{O}(1), \\
&\quad S_{n,k}^\theta(X|x) = \mathcal{O}(1), \text{ for } k \ge 1, \\
T_{n,0}^\theta(|\theta^\top X_{ix}|^6|x) &= \mathcal{O}(h^6), \; S_{n,0}^\theta(|\theta^\top X_{ix}|^6|x) = \mathcal{O}(h^6), \; T_{n,0}^\theta(XX^\top|x) = \mathcal{O}(1), \\
&\quad S_{n,0}^\theta(XX^\top|x) = \mathcal{O}(h), \\
E_{n,2}(x) &= (\mu_4 - 1) f_\theta f_\theta' h + f_\theta(\epsilon_3^\theta - \epsilon_1^\theta) + \mathcal{O}(h\gamma_n), \; E_{n,3}(x) = \mu_4 f_\theta^2 + \mathcal{O}(\gamma_n), \\
&\quad E_{n,4}(x) = \mathcal{O}(h).
\end{aligned}$$

Note that

$$a^\theta(x) = T_{n,0}^\theta(Y|x)/D_{n,0}^\theta(x), \quad d^\theta(x)h = S_{n,0}^\theta(Y|x)/D_{n,0}^\theta(x).$$

and

$$\begin{aligned}
A_n^\theta(x) &= \sum_{k=2}^{5} \frac{1}{k!} g^{(k)}(\theta_0^\top x) \frac{D_{n,k}^\theta(x)}{D_{n,0}^\theta(x)} h^{k-2}, \\
B_n^\theta(x) &= \sum_{k=0}^{4} \frac{1}{k!} g^{(k+1)}(\theta_0^\top x) \frac{T_{n,k}(X|x) - D_{n,k}(x)x}{D_{n,0}^\theta(x)} h^k, \\
C_n(x,\theta) &= \frac{1}{2} g''(\theta_0^\top x)\{T_{n,0}(XX^\top|x) - T_{n,0}(X|x)x^\top - xT_{n,0}(X^\top|x) \\
&\quad + xx^\top D_{n,0}^\theta(x)\}\{D_{n,0}^\theta(x)\}^{-1}, \\
\tilde{A}_n^\theta(x) &= \sum_{k=2}^{4} \frac{1}{k!} g^{(k)}(\theta_0^\top x) \frac{E_{n,k}^\theta(x)}{D_{n,0}^\theta(x)} h^{k-2},
\end{aligned}$$

$$\tilde{B}_n^\theta(x) = \sum_{k=1}^{4} \frac{k}{k!} g^{(k)}(\theta_0^\top x) \frac{S_{n,k}(X|x) - E_{n,k}(x)x}{D_{n,0}^\theta(x)} h^k,$$

$$\tilde{C}_n(x,\theta) = \frac{1}{2} g''(\theta_0^\top x)\{S_{n,0}(XX^\top|x) - S_{n,0}(X|x)x^\top - xS_{n,0}(X^\top|x)$$
$$+ xx^\top E_{n,0}^\theta(x)\}\{D_{n,0}^\theta(x)\}^{-1}.$$

Lemma 11.5.1 follows from simple calculations based on the above equations. □

11.8.3 Proof of Lemma 11.5.2

It follows from Lemma 11.8.1 that $\eta_n^\theta(x) = \mathcal{O}(\delta_n)(1+|x|)$ and $s_0^\theta = f_\theta + \tilde{\epsilon}_0^\theta$ where $\tilde{\epsilon}_0^\theta = \epsilon_k^\theta + (Es_k^\theta - f_\theta) = \mathcal{O}(\gamma_n)$. Because $|\rho_n''(.)| < n^{2\varsigma}$, we have

$$\rho_n(s_0^\theta(X_j)) = \rho_n(f_\theta(X_j)) + \rho_n'(f_\theta(X_j))\tilde{\epsilon}_0^\theta(X_j) + \mathcal{O}(n^{2\varsigma}\gamma_n^2). \quad (11.26)$$

Thus

$$\mathcal{A}_n^\theta = \tilde{\mathcal{E}}_n^\theta + \mathcal{Q}_{n,1}^\theta + \mathcal{O}(n^{2\varsigma}\gamma_n^3),$$

where $\tilde{\mathcal{E}}_n^\theta = n^{-2}\sum_{i=1}^n\sum_{j=1}^n \rho_n(f_\theta(X_j)) f_\theta^{-1}(X_j) g'(\theta^\top X_j) K_h^\theta(X_{ij}) X_{ij}\varepsilon_i$, and $\mathcal{Q}_{n,1}^\theta = n^{-1}\sum_{i=1}^n G_{n,i}^\theta$ with

$$G_{n,i}^\theta = n^{-1}\sum_{j=1}^n \Big[\frac{1}{2} f_\theta''(X_j)\{\rho_n'(f_\theta(X_j)) - \rho_n(f_\theta(X_j)) f_\theta^{-1}(X_j)\}h^2$$
$$+ \{1 - \rho_n(f_\theta(X_j)) f_\theta^{-1}(X_j)\}\tilde{\epsilon}_0^\theta(X_j)\Big]$$
$$\times f_\theta^{-1}(X_j) g'(\theta^\top X_j) K_h^\theta(X_{ij}) X_{ij}\varepsilon_i.$$

Simple calculations lead to $E\tilde{\mathcal{E}}^\theta = 0$, $E(\tilde{\mathcal{E}}_n^\theta)^2 = O(n^{-1})$, $E(G_{n,i}^\theta) = 0$ and $E(G_{n,i}^\theta)^2 = O\{h^4 + (nh)^{-1}\}$. By the first part of Lemma 11.8.1, we have

$$\tilde{\mathcal{E}}_n^\theta = \mathcal{O}\{(\log n/n)^{1/2}\}, \quad \mathcal{Q}_{n,1}^\theta = \mathcal{O}\{h^2(\log n/n)^{1/2} + n^{-1}(\log n/h)^{1/2}\}.$$

By Taylor expansion, $g'(\theta_0^\top x) = g'(\theta^\top x) + g''(v^*)(\theta_0 - \theta)^\top x$, where v^* is a value between $\theta^\top x$ and $\theta_0^\top x$. Write

$$\tilde{\mathcal{E}}_n^\theta = \mathcal{E}_n^\theta + \mathcal{Q}_{n,2}^\theta + r_{n,0}(\theta - \theta_0),$$

where $Q_{n,2}^{\theta}=n^{-2}\sum_{i=1}^{n}\sum_{j=1}^{n}\{\rho_n(f_\theta(X_j))f_\theta^{-1}(X_j)g'(\theta^\top X_j)K_h^\theta(X_{ij})X_{ij}-\rho_n(f_\theta(X_i))g'(\theta^\top X_i)\nu_\theta(X_i)\}\varepsilon_i$ and $r_{n,0}=\mathcal{O}(\gamma_n/h)$. By Lemma 11.8.1 and the fact that $Var(Q_{n,2}^{\theta})=O\{h^4+(nh)^{-1}\}$, we have

$$Q_{n,2}^{\theta}=\mathcal{O}\{(n/\log n)^{-1/2}\gamma_n\}.$$

Let $Q_n^{\theta}=Q_{n,1}^{\theta}+Q_{n,2}^{\theta}$. It is easy to check that $E\{Q_n^\theta\mathcal{E}_n^\theta\}=o(h^8+(nh)^{-2})$. Therefore, the first part of Lemma 11.5.2 follows.

Similarly, we have from (11.26) that

$$\mathcal{B}_n^\theta=(nh)^{-1}\sum_{j=1}^{n}\{\rho_n(f_\theta(X_j))+\rho_n'(f_\theta(X_j))\tilde{\epsilon}_0^\theta(X_j)\}e_k^\theta(X_j)\eta_n^\theta(X_j)/f_\theta(X_j)$$
$$+\,\mathcal{O}(n^{2\varsigma}\gamma_n^4/h).$$

Let $\tilde{R}_n^\theta$ be the first term on the right hand side above. Then

$$\begin{aligned}\tilde{R}_n^\theta &= n^{-3}\sum_{j=1}^{n}\{\rho_n(f_\theta(X_j))+\rho_n'(f_\theta(X_j))\tilde{\epsilon}_0^\theta(X_j)\}\\ &\quad\times\sum_{i=1}^{n}K_h^2(\theta^\top X_{ij})(\theta^\top X_{ij}/h)^kX_{ij}\varepsilon_i^2/f_\theta(X_j)\\ &\quad+n^{-3}\sum_{j=1}^{n}\{\rho_n(f_\theta(X_j))+\rho_n'(f_\theta(X_j))\tilde{\epsilon}_0^\theta(X_j)\}\\ &\quad\times\sum_{i\neq\ell}K_h(\theta^\top X_{ij})(\theta^\top X_{ij}/h)^kK_h(\theta^\top X_{\ell j})X_{\ell j}\varepsilon_i\varepsilon_\ell/f_\theta(X_j)\\ &\overset{def}{=}\tilde{R}_{n,1}^\theta+\tilde{R}_{n,2}^\theta+\tilde{R}_{n,3}^\theta+\tilde{R}_{n,4}^\theta.\end{aligned}$$

If ε is independent of X, then

$$\begin{aligned}E_\theta(x)&\overset{def}{=}E\{K_h^2(\theta^\top X_{ix})(\theta^\top X_{ij}/h)^kX_{ix}\varepsilon_i^2\}\\ &=h^{-1}\sum_{\ell=0}^{2}\frac{1}{\ell!}\tilde{\mu}_{k+\ell}\{f_\theta(x)\nu_\theta(x)\}^{(\ell)}h^\ell\sigma^2+O(h^2),\end{aligned}$$

where $\tilde{\mu}_k=\int K^2(v)v^kdv$. By Lemma 11.8.1, we have

$$n^{-1}\sum_{i=1}^{n}K_h^2(\theta^\top X_{ix})(\theta^\top X_{ix}/h)^kX_{ix}\varepsilon_i^2-E_\theta(x)=\mathcal{O}(h^{-1}\delta_n).$$

Thus

$$R^{\theta}_{n,0} \stackrel{def}{=} (n^2h)^{-1}\sum_{j=1}^{n}\rho_n(f_\theta(X_j))\left[n^{-1}\sum_{i=1}^{n}K_h^2(\theta^\top X_{ij})(\theta^\top X_{ij}/h)^k X_{ij}\varepsilon_i^2 - E_\theta(X_j)\right]$$

$$= \mathcal{O}\{(nh^2)^{-1}\delta_n\}. \tag{11.27}$$

It is easy to check that $E\{\mathcal{E}^{\theta}_n R^{\theta}_{n,0}\} = 0$. Write

$$(n^2h)^{-1}\sum_{j=1}^{n}\rho_n(f_\theta(X_j))E_\theta(X_j) = (nh)^{-1}E\{\rho_n(f_\theta(X_j))E_\theta(X_j)\} + R^{\theta}_{n,1},$$

where $E\{R^{\theta}_{n,1}\mathcal{E}^{\theta}_n\} = 0$ and

$$R^{\theta}_{n,1} = (n^2h)^{-1}\sum_{j=1}^{n}[\rho_n(f_\theta(X_j))E_\theta(X_j) - E\{\rho_n(f_\theta(X_j))E_\theta(X_j)\}]$$

$$= \mathcal{O}\{(nh^2)^{-1}(n/\log n)^{-1/2}\}. \tag{11.28}$$

Note that $E\{\rho_n(f_\theta(X))v_\theta(X)\} = 0$. We have

$$(nh)^{-1}E\{\rho_n(f_\theta(X_j))E_\theta(X_j)\} = \frac{\tilde{c}_{k,n}}{nh} + R^{\theta}_{n,2}, \tag{11.29}$$

where $R^{\theta}_{n,2} = O(n^{-1})$ and $E\{R^{\theta}_{n,2}\mathcal{E}^{\theta}_0\} = 0$. By (11.27)–(11.29) and the fact that $(n/\log n)^{-1/2} = o(\gamma_n)$, we have

$$\tilde{R}^{\theta}_{n,1} = \frac{\tilde{c}_k}{nh} + R^{\theta}_{n,1} + R^{\theta}_{n,2}. \tag{11.30}$$

Similarly

$$\tilde{R}^{\theta}_{n,2} = \mathcal{O}\{(nh)^{-1}\gamma_n\}. \tag{11.31}$$

Let $G^{\theta}_{n,i,\ell}=n^{-1}\sum_{j=1}^{n}\rho_n(f_\theta(X_j))K_h(\theta^\top X_{ij})(\theta^\top X_{ij}/h)^k K_h(\theta^\top X_{\ell j})X_{\ell j}/f_\theta(X_j)$. Write $\tilde{R}^{\theta}_{n,3}$ as

$$\tilde{R}^{\theta}_{n,3} = n^{-2}\sum_{i\neq\ell}^{n}\frac{1}{2}(G^{\theta}_{n,i,\ell} + G^{\theta}_{n,\ell,i})\varepsilon_i\varepsilon_\ell = n^{-2}\sum_{\ell=1}^{n}\left\{\sum_{i<\ell}\frac{1}{2}(G^{\theta}_{n,i,\ell} + G^{\theta}_{n,\ell,i})\varepsilon_i\right\}\varepsilon_\ell.$$

By the second part of Lemma 11.8.1, we have

$$\tilde{R}^{\theta}_{n,3} = \mathcal{O}\{n^{-1/2}\delta_n\}. \tag{11.32}$$

Similarly, we have

$$\tilde{R}^{\theta}_{n,4} = \mathcal{O}\{n^{-1/2}\delta_n\}. \tag{11.33}$$

Thus the second part of Lemma 11.5.2 follows from (11.30) and (11.31).

The third part of Lemma 11.5.2 can be proved similarly as the proof of the second part. □

11.8.4 Proof of Lemma 11.5.4

By (11.8), Lemma 11.5.1 and $\theta_0 = \theta + (\theta_0 - \theta)$, simple calculations lead to

$$Y_i - a_\theta(x) - d_\theta(x)\theta_0^\top X_{ix} = \varepsilon_i + \{\tilde{A}^\theta(x, X_i) - A_n^\theta(x)h^2\} + \{\tilde{B}^\theta(x, X_i) - B_n^\theta(x)\}^\top(\theta_0 - \theta) - V_n^\theta(x) + \mathcal{O}\{h^2\gamma_n^2 + \delta_\theta^3\},$$

where $\tilde{A}^\theta(x, X_i) = A^\theta(x, X_i) - d_\theta(x)\theta^\top X_{ix}$ and $\tilde{B}^\theta(x, X_i) = B^\theta(x, X_i) - d_\theta(x)X_{ix}$. It follows from the Taylor expansion that

$$C^\theta_{n,k}(x) \overset{def}{=} n^{-1}\sum_{i=1}^{n} K_h^\theta(X_{ix})(\theta^\top X_{ix}/h)^k X_{ix} = \sum_{\ell=0}^{5}\frac{1}{\ell!}\mu_{k+\ell}(f\mu_\theta)^{(\ell)}h^\ell + \tilde{\xi}_k^\theta + \mathcal{O}(h^6),$$

where $\tilde{\xi}_k^\theta = n^{-1}\sum_{i=1}^n\{K_h^\theta(X_{ix})(\theta^\top X_{ix}/h)^k X_{ix} - EK_h^\theta(X_{ix})(\theta^\top X_{ix}/h)^k X_{ix}\} = \xi_k^\theta - x\epsilon_k^\theta$. We have

$$\begin{aligned}
&n^{-1}\sum_{i=1}^{n} K_h^\theta(X_{ix})X_{ix}\tilde{A}^\theta(x, X_i)\\
&= \{g'(\theta_0^\top x) - d_\theta(x)\}C_{n,1}(x)h + \sum_{k=2}^{5}\frac{1}{k!}g^{(k)}(\theta_0^\top x)C^\theta_{n,k}(x)h^k\\
&= -\frac{1}{2}g''(\mu_4 - 1)f_\theta' f_\theta^{-1}(\nu_\theta f_\theta)'h^4 + \frac{1}{2}g''(\nu_\theta f_\theta)'(\epsilon_3^\theta - \epsilon_1^\theta)h^2\\
&\quad + \tilde{V}_n^\theta\{(\nu_\theta f_\theta)'h + \frac{1}{6}\mu_4(\nu_\theta f_\theta)''h^3 + \tilde{\xi}_1^\theta\}\\
&\quad + \frac{1}{2}g''h^2\{f_\theta\nu_\theta + \frac{1}{2}\mu_4(f_\theta\nu_\theta)''h^2\} + \frac{1}{24}g^{(4)}\mu_4 f_\theta\nu_\theta h^4 + \frac{1}{2}g''h^2\tilde{\xi}_2^\theta\\
&\quad + \frac{1}{6}g'''h^3\tilde{\xi}_3^\theta + \mathcal{O}(h^2\gamma_n^2).
\end{aligned}$$

Thus

$$n^{-1}\sum_{i=1}^{n} K_h^{\theta}(X_{ix})X_{ix}\{\tilde{A}^{\theta}(x,X_i)-A_n^{\theta}(x)h^2\}=\frac{1}{4}(\mu_4-1)g''f_{\theta}v_{\theta}''+B_{n,1}^{\theta}(\theta-\theta_0)$$

$$+H_{2,n}^{\theta}+\mathcal{O}(h^2\gamma_n^2), \tag{11.34}$$

where $B_{n,1}^{\theta}=\{(v_{\theta}f_{\theta})'h+\frac{1}{6}\mu_4(v_{\theta}f_{\theta})'''h^3+\tilde{\xi}_1^{\theta}\}B_n^{\theta}(x)^{\top}$ with $B_n^{\theta}(x)$ defined in Lemma 11.5.1, and

$$\begin{aligned}H_{2,n}^{\theta}=&\frac{1}{2}g''(\epsilon_3^{\theta}-\epsilon_1^{\theta})h^2(v_{\theta}f_{\theta})'+f_{\theta}^{-1}e_1^{\theta}(f_{\theta}v_{\theta})'h+\frac{1}{2}f_{\theta}^{-2}f_{\theta}''(fv_{\theta})'h^3e_1^{\theta}\\&+f_{\theta}^{-2}(fv)'h(\epsilon_0^{\theta}e_1^{\theta}-\epsilon_1^{\theta}e_0^{\theta})-f^{-2}f'(fv)'h^2e_0^{\theta}+f^{-1}(fv)'h\epsilon_0^{\theta}\\&\left\{-\frac{1}{2}(\mu_4+1)f^{-1}f_{\theta}''h^2-f^{-1}(\epsilon_0^{\theta}+\epsilon_1^{\theta})+f^{-2}(f')^2h^2\right\}\\&+\frac{1}{6}\mu_4f^{-1}e_1^{\theta}(fv)''h^3+f^{-1}e_1^{\theta}\tilde{\xi}_1^{\theta}-f^{-2}f'he_0^{\theta}\tilde{\xi}_1^{\theta}+\frac{1}{2}g''h^2\tilde{\xi}_2^{\theta}\\&+\frac{1}{6}g'''h^3\tilde{\xi}_3^{\theta}-\frac{1}{2}g''h^2\tilde{\xi}_0^{\theta}-\frac{1}{2}g''(\theta_0^{\top}x)\{f_{\theta}^{-1}(\epsilon_2^{\theta}-\epsilon_0^{\theta})\\&+(2-\mu_4)f_{\theta}^{-2}f_{\theta}'h\epsilon_1^{\theta}-f_{\theta}^{-2}f_{\theta}'h\epsilon_3^{\theta}\}v_{\theta}f_{\theta}h^2-\frac{1}{6}g'''\epsilon_3^{\theta}v_{\theta}h^3.\end{aligned}$$

By the expansions of $d_{\theta}(x)$ in Lemma 11.5.1, $\rho_n(s_0^{\theta}(x))$ in (11.26), and (11.34), we have

$$n^{-2}\sum_{j=1}^{n}\rho_n(s_0^{\theta}(X_j))d_{\theta}(X_j)\sum_{i=1}^{n}K_h^{\theta}(X_{ix})X_{ij}\{\tilde{A}^{\theta}(X_j,X_i)-A_n^{\theta}(X_j)h^2\}/s_0^{\theta}(X_j)$$

$$=\tilde{c}_{2,n}h^4+(B_{n,2}^{\theta})^{\top}(\theta-\theta_0)+\tilde{R}_{n,1}^{\theta}+\mathcal{O}\{n^{2\varsigma}(h^2\gamma_n^2+\delta_{\theta}^2h+\delta_{\theta}^3)\},$$

where $B_{n,2}^{\theta}=n^{-1}\sum_{j=1}^{n}\rho_n(s_0^{\theta}(X_j))d_{\theta}(X_j)B_{1,n}^{\theta}(X_j)/s_0^{\theta}(X_j)$ and $\tilde{R}_{n,1}^{\theta}=n^{-1}\sum_{j=1}^{n}\rho_n(s_n^{\theta}(X_j))d_{\theta}(X_j)H_{2,n}^{\theta}(X_j)\,/s_0^{\theta}(X_j)$. Again by the expansion of d_{θ} and the fact that $\tilde{\xi}_1^{\theta}=\mathcal{O}(\delta_n)$, we have $B_{n,2}^{\theta}=\mathcal{O}(h+\delta_n)$. It is easy to check that $H_{2,n}^{\theta}=\mathcal{O}(h\delta_n+\delta_n^2)$. We have

$$\tilde{R}_{n,1}=n^{-1}\sum_{j=1}^{n}[\rho_n(f_{\theta}(X_j))+\rho_n'(f_{\theta}(X_j))\{f_{\theta}(X_j)+\frac{1}{2}f_{\theta}''(X_j)h^2$$

$$+\epsilon_0^{\theta}(X_j)\}][\{g'(\theta_0^{\top}X_j)+\frac{1}{6}g'''(\theta_0^{\top}X_j)h^2$$

$$
\begin{aligned}
&+ \tilde{V}_n^\theta(X_j)/h\}H_{2,n}^\theta(X_j)f_\theta^{-1}(X_j)\{1 - \frac{1}{2}f_\theta^{-1}(X_j)f_\theta''(X_j)h^2 \\
&- f_\theta^{-1}(X_j)\epsilon_0^\theta(X_j)\} + \mathcal{O}(n^{2\varsigma}\gamma_n^3) \\
&\overset{def}{=} R_{n,1} + \mathcal{O}(n^{2\varsigma}\gamma_n^3).
\end{aligned}
$$

Next, we need to consider the terms in $R_{n,1}$ one by one. Write

$$
\begin{aligned}
R_{n,1,1}^\theta &\overset{def}{=} n^{-1}\sum_{j=1}^n \rho_n(f_\theta(X_j))f_\theta^{-1}(X_j)(f_\theta(X_j)\nu_\theta(X_j))'e_1^\theta h \\
&= hn^{-2}\sum_{i=1}^n\Big\{\sum_{j=1}^n K_h^\theta(X_{ij})\rho_n(f_\theta(X_j))f_\theta^{-1}(X_j)(f_\theta(X_j)\nu_\theta(X_j))'\Big\}\varepsilon_i.
\end{aligned}
$$

Note that $E\{\rho_n(f_\theta(X))f_\theta^{-1}(X)(f_\theta(X)\nu_\theta(X))'|\theta^\top X\} = 0$. We have by Lemma 11.8.1

$$
R_{n,1,1}^\theta = \mathcal{O}\{hn^{-1}(h^{-1}\log n)^{-1/2}\}
$$

and

$$
\begin{aligned}
E\{\mathcal{E}_0^\theta R_{n,1,1}^\theta\} &= hn^{-3}E\sum_{i=1}^n\Big\{\sum_{j=1}^n K_h^\theta(X_{ij})\rho_n(f_\theta(X_j))f_\theta^{-1}(X_j)(f_\theta(X_j)\nu_\theta(X_j))'\Big\} \\
&\quad \times \rho_n(f_\theta(X_i))g'(\theta^\top X_i)\nu_\theta(X_i)\varepsilon_i^2 \\
&= hn^{-3}E\left\{\sum_{j=1}^n K_h^\theta(0)\rho_n(f_\theta(X_j))f_\theta^{-1}(X_j)(f_\theta(X_j)\nu_\theta(X_j))'\right. \\
&\quad \left. \times \rho_n(f_\theta(X_j))g'(\theta^\top X_j)\nu_\theta(X_j)\varepsilon_j^2\right\} \\
&= O(n^{-2}).
\end{aligned}
$$

Applying a similar approach to all the terms in $R_{n,1}^\theta$, we have

$$
R_{n,1}^\theta = \mathcal{O}\{n^{-1}(\log n/h)^{1/2} + (\log n/n)^{1/2}h^2\} \text{ and } E\{\mathcal{E}_n^\theta R_{n,1}^\theta\} = o\{(nh)^{-2} + h^8\}. \tag{11.35}
$$

By Lemmas 11.5.1 and 11.8.1, we have

$$B_{n,3}^{\theta} \stackrel{def}{=} n^{-2}\sum_{j=1}^{n}\rho(s_0^{\theta}(X_j))d_{\theta}(X_j)\sum_{i=1}^{n}K_h^{\theta}(X_{ij})X_{ij}\{\tilde{B}^{\theta}(X_j,X_i)$$

$$- B_n^{\theta}(X_j)\}^{\top}/s_0^{\theta}(X_j)$$

$$= W_n^{\theta} + \mathcal{O}\{(\gamma_n+\delta_{\theta})/h\}.$$

By Lemma 11.5.2, we have

$$n^{-2}\sum_{j=1}^{n}\rho(s_0^{\theta}(X_j))d_{\theta}(X_j)\sum_{i=1}^{n}K_h^{\theta}(X_{ij})X_{ij}\varepsilon_i/s_0^{\theta}(X_j)$$

$$= \mathcal{E}_0^{\theta} + \frac{\tilde{c}_{1,n}}{nh} + B_{n,4}^{\theta}(\theta_0-\theta) + R_{n,2}^{\theta} + \mathcal{O}(n^{2\varsigma}\gamma_n^3),$$

where $\tilde{c}_{1,n}$ is defined in the lemma, and

$$B_{4,n}^{\theta} = n^{-1}\sum_{j=1}^{n}\{\rho_n(f_{\theta}(X_j)) + \rho_n'(f_{\theta}(X_j))\epsilon_0^{\theta}(X_j)\}\eta_n^{\theta}(X_j)(\tilde{B}_n^{\theta}(X_j))^{\top}/h$$

and

$$R_{n,2}^{\theta} = n^{-1}\sum_{j=1}^{n}\Big[\frac{1}{6}\rho_n(f_{\theta}(X_j))g'''(\theta_0^{\top}X_j)h^2 + \rho_n'(f_{\theta}(X_j))\epsilon_0^{\theta}(X_j)\{g'(\theta_0^{\top}X_j)$$

$$+ \tilde{V}_n^{\theta}(X_j)/h\}\Big]\eta_n^{\theta}(X_j).$$

Noting that $\eta_n^{\theta} = \mathcal{O}(\delta_n)$, we have $B_{4,n}^{\theta} = \mathcal{O}(\delta_n/h)$. Similarly, we have

$$n^{-2}\sum_{j=1}^{n}\rho_n(s_0^{\theta}(X_j))d_{\theta}(X_j)V_n^{\theta}(X_j)\sum_{i=1}^{n}K_h^{\theta}(X_{ij})X_{ij}/s_0^{\theta}(X_j) = R_{n,3}^{\theta} + \mathcal{O}(n^{2\varsigma}\gamma_n^3),$$

where

$$R_{n,3}^{\theta} = n^{-1}\sum_{j=1}^{n}\rho_n(s_0^{\theta}(X_j))d_{\theta}(X_j)V_n^{\theta}(X_j)[\nu_{\theta}(X_j) + \frac{1}{2}f_{\theta}^{-1}\{(f_{\theta}\nu_{\theta})''$$

$$- f_{\theta}^{-1}f_{\theta}''\nu_{\theta}(X_j)\}h^2 + \xi_0^{\theta}(X_j) - \epsilon_0^{\theta}(X_j)].$$

By the same arguments leading to (11.35), we have

$$R_{n,2}^{\theta} = \mathcal{O}\{n^{-1}(\log n/h)^{1/2} + (\log n/n)^{1/2}h^2\} \text{ and } E\{\mathcal{E}_n^{\theta}R_{n,2}^{\theta}\} = o\{(nh)^{-2}+h^8\}, \tag{11.36}$$

$$R_{n,3}^{\theta} = \mathcal{O}\{n^{-1}(\log n/h)^{1/2} + (\log n/n)^{1/2}h^2\} \text{ and } E\{\mathcal{E}_n^{\theta} R_{n,3}^{\theta}\} = o\{(nh)^{-2} + h^8\}. \tag{11.37}$$

Lemma 11.5.4 follows from the above equations with $\mathcal{R}_n^{\theta} = R_{n,1}^{\theta} + R_{n,2}^{\theta} + R_{n,3}^{\theta}$ and $\mathcal{B}_n^{\theta} = B_{n,2}^{\theta} + B_{n,3}^{\theta} + B_{n,4}^{\theta} = W_n^{\theta} + \mathcal{O}\{n^{2\varsigma}(\gamma_n + \delta_{\theta})/h\}$. □

Acknowledgements The first author is most grateful to Professor V. Spokoiny for helpful discussions and NUS RMI for support. The second author thanks the Deutsche Forschungsgemeinschaft SFB 649 "Ökonomisches Risiko" for financial support. The third author thanks the ESRC for financial support.

References

Bickel, P., Klaassen, A.J., Ritov, Y., & Wellner, J.A. (1993). *Efficient and adaptive inference in semiparametric models*. Baltimore: Johns Hopkins University Press.

Carroll, R.J., Fan. J., Gijbels, I., & Wand, M.P. (1997). Generalized partially linear single-index models. *Journal of American Statistical Association*, *92*, 477–489.

de la Peña, V.H. (1999) A general class of exponential inequalities for martingales and ratios. *The Annals of Probability*, *27*, 537–564.

Delecroix, M., Härdle, W., & Hristache, M. (2003). Efficient estimation in conditional single-index regression. *Journal of Multivariate Analysis*, *86*, 213–226.

Delecroix, M., Hristache, M., & Patilea, V. (2006). On semiparametric M-estimation in single-index regression. *Journal of Statistical Planning and Inference*, *136*, 730–769.

Fan, J., & Gijbels, I. (1996). *Local polynomial modeling and its applications*. London: Chapman & Hall.

Fan, J., & Yao, Q. (2003). *Nonlinear time series: nonparametric and parametric methods*. New York: Springer.

Friedman, J.H. (1984). SMART User's Guide. Laboratory for Computational Statistics, Stanford University Technical Report No. 1.

Hall, P. (1989). On projection pursuit regression. *Annals of Statistics*, *17*, 573–588.

Härdle, W., Hall, P., & Ichimura, H. (1993). Optimal smoothing in single-index models. *Annals of Statistics*, *21*, 157–178.

Härdle, W., & Stoker, T.M. (1989) Investigating smooth multiple regression by method of average derivatives. *Journal of American Statistical Association*, *84* 986–995.

Härdle, W., & Tsybakov, A.B. (1993). How sensitive are average derivatives? *Journal of Econometrics*, *58*, 31–48.

Horowitz, J.L. & Härdle, W. (1996) Direct semiparametric estimation of single-index models with discrete covariates. *Journal of American Statistical Association*, *91*, 1632–1640.

Hristache, M., Juditsky, A., & Spokoiny, V. (2001) Direct estimation of the single-index coefficients in single-index models. *Annals of Statistics 29*, 1–32.

Hristache, M., Juditsky, A., & Spokoiny, V. (2002). Direct estimation of the index coefficient in a single-index model. *Annals of Statistics*, *29*, 593–623.

Ichimura, H. (1993). Semiparametric least squares (SLS) and weighted SLS estimation of single-index models. *Journal of Econometrics*, *58*, 71–120.

Li, K.C. (1991). Sliced inverse regression for dimension reduction (with discussion). *Journal of American Statistical Association*, *86*, 316–342.

Li, K.C. (1992). On principal Hessian directions for data visualization and dimension reduction: another application of Stein's lemma. *Journal of American Statistical Association*, *87*, 1025–1039.

Linton, O. (1995) Second order approximation in the partially linear regression model. *Econometrica, 63*, 1079–1112.

Nishiyama, Y., & Robinson, P.M. (2000). Edgeworth expansions for semiparametric average derivatives. *Econometrica, 68*, 931–980.

Nishiyama, Y., & Robinson, P.M. (2005). The bootstrap and the Edgeworth correction for semiparametric average derivatives. *Econometrica, 73*, 903–948.

Powell, J.L., Stock, J.H., & Stoker, T.M. (1989). Semiparametric estimation of index coefficients. *Econometrica, 57*, 1403–1430.

Powell, J.L., & Stoker, T.M. (1996). Optimal bandwidth choice for density weighted averages. *Journal of Econometrics, 755*, 291–316.

Robinson, P. (1988). Root-N consistent semi-parametric regression. *Econometrica, 156*, 931–954.

Ruppert, D., Sheather, J., & Wand, P.M. (1995). An effective bandwidth selector for local least squares regression. *Journal of American Statistical Association, 90*, 1257–1270.

Schott, J.R. (1997) *Matrix analysis for statistics*. New York: Wiley.

Weisberg, S., & Welsh, A.H. (1994). Estimating the missing link functions, *Annals of Statistics, 22*, 1674–1700.

Xia, Y., Tong, H., Li, W.K., & Zhu, L. (2002). An adaptive estimation of dimension reduction space (with discussion). *Journal of the Royal Statistical Society Series B, 64*, 363–410.

Xia, Y. (2006). Asymptotic distributions for two estimators of the single-index model. *Econometric Theory, 22*, 1112–1137.

Xia, Y., & Li, W.K. (1999) On single-index coefficient regression models. *Journal of American Statistical Association, 94*, 1275–1285.

Yin, X., & Cook, R.D. (2005). Direction estimation in single-index regressions. *Biometrika, 92*, 371–384.

Index

I. van Keilegom and P.W. Wilson (eds.), *Exploring Research Frontiers in Contemporary Statistics and Econometrics*, DOI 10.1007/978-3-7908-2349-3,